Mechanics of Paper Products

Edited by Kaarlo Niskanen

Mechanics of Paper Products

Edited by Kaarlo Niskanen

DE GRUYTER

Editor

Professor
Kaarlo Niskanen
FSCN, Mid Sweden University
SE-851 70 Sundsvall
Sweden

This book has 203 figures and 12 tables.

Front cover image

With kind permission by Joanna Hornatowska and Catherine Östlund, Innventia AB, Stockholm.

ISBN 978-3-11-025461-7
e-ISBN 978-3-11-025463-1

Library of Congress Cataloging-in-Publication Data

> Mechanics of paper products / by Kaarlo Niskanen . . . [et al.]. — 1st ed.
> p. cm.
> Includes bibliographical references and index.
> ISBN 978-3-11-025461-7
> 1. Paper—Mechanical properties. 2. Paper products.
> I. Niskanen, Kaarlo.
> TS1121.M45 2012
> 676'.28–dc23   2011020332

Bibliografic information published by the Deutsche Nationalbibliothek

The Deutsche Nationalbibliothek lists this publication in the Deutsche Nationalbibliografie; detailed bibliographic data are available in the Internet at http://dnb.d-nb.de.

© 2012 by Walter de Gruyter GmbH & Co. KG, Berlin/Boston

The citation of registered names, trade names, trade marks, etc. in this work does not imply, even in the absence of a specific statement, that such names are exempt from laws and regulations protecting trade marks etc. and therefore free for general use.

Typesetting: Apex CoVantage, LLC, Herndon, Virginia, USA
Printing and Binding: Hubert & Co., Göttingen
Printed in Germany.
www.degruyter.com

Preface

This book was initiated by the *Alf de Ruvo Memorial Foundation* and is dedicated to the memory of Alf de Ruvo (1938–2000).

After his Master of Science in Chemical Engineering and a Licentiate of Technology degree in Paper Technology, Alf de Ruvo spent a couple of years in the USA before becoming the Group Leader of the Packaging Materials Group at the Swedish Pulp and Paper Research Institute (STFI) in 1969. In 1975, he was appointed Director of Research at the Paper Technology department. Later, he took on greater challenges and in 1983 he became the Director of Research for Sunds-Defibrator (a machinery manufacturer). In 1987 he was employed in a similar position at SCA (a paper products company), where he was Vice President of Research and Technology and later on Executive Vice President. For his scientific achievements, in 1988 the Swedish Association of Pulp and Paper Engineers (SPCI) awarded him the Ekman Medal. He was also a Tappi Fellow.

The career of Alf de Ruvo may be described in a few lines, but his efforts for the Swedish forest products industry in general, and for SCA in particular, as a scientist, technical innovator, mentor and source of inspiration were unparalleled. Alf de Ruvo had a degree in Chemical Engineering, but his own research during the time at STFI covered paper physics, composite materials, product properties and converting of paperboard materials. His work was characterized by his cross-disciplinary approach to science. It is not an exaggeration to say that he brought science to the art of paper mechanics and brought paper mechanics at STFI to the absolute frontier of paper technology. Hence, there has been a move from a trial and error approach to a well-funded methodology based on fracture mechanics to an analysis of statistical variations as a systematic tool in the evolution of the science of paper mechanics.

Hence, following the tradition of Alf de Ruvo, this book is aimed at, paper mechanics from a solid and continuum mechanics point of view and not from a paper technology perspective. It is the hope of the Alf de Ruvo Memorial Foundation and the authors that the book will fill a knowledge gap, considering the essential role of solid and continuum mechanics in understanding papermaking, converting and the end-use of paper and board materials. All the authors have contributed on a voluntary basis, with a never failing enthusiasm and solid belief in the value of this book, and it is for this reason that it was possible to produce the final manuscript within one year.

Professor Tom Lindström

Royal Institute of Technology
and Innventia, on behalf of
the Alf de Ruvo Memorial
Foundation

List of Contributing Authors

Professor **Lars Berglund**, Wallenberg Wood Science Centre and Department of Fiber and Polymer Technology, KTH Royal Institute of Technology, SE-100 44 Stockholm, Sweden, email: blund@kth.se

Professor **Leif Carlsson**, College of Ocean & Mechanical Engineering, Engineering Building 36, Florida Atlantic University, Boca Raton, FL 33431-0991, USA, email: carlsson@fau.edu

Professor **Douglas W. Coffin**, Department of Chemical and Paper Engineering, 64 Engineering Building, Miami University, Oxford, OH 45056, USA, email: coffindw@muohio.edu

Professor **Per-Johan Gustafsson**, Department of Construction Sciences, Lund University, Box 118, SE-221 00 Lund, Sweden, email: Per-Johan.Gustafsson@construction.lth.se

Dr. **Rickard Hägglund**, SCA R&D Centre AB, Box 716, SE-851 21 SUNDSVALL, Sweden, email: rickard.hagglund@sca.com

Professor **Artem Kulachenko**, KTH Royal Institute of Technology, Department of Solid Mechanics, SE-100 44 Stockholm, Sweden, email: artem@kth.se

Dr. **Petri Mäkelä**, Tetra Pack Packaging Solutions AB, Ruben Rausings gata, SE-221 86 Lund, Sweden, email: petri.makela@tetrapak.com

Professor **Kaarlo Niskanen**, FSCN, Mid Sweden University, SE-851 70 Sundsvall, Sweden, email: kaarlo.niskanen@miun.se

Professor **Mikael Nygårds**, KTH Royal Institute of Technology, Department of Solid Mechanics, SE-100 44 Stockholm, Sweden and Innventia AB, Drottning Kristinas väg 61, SE-114 86 Stockholm, Sweden, email: nygards@kth.se

Professor **Sören Östlund**, KTH Royal Institute of Technology, Department of Solid Mechanics, SE-100 44 Stockholm, Sweden, email: soren@kth.se

Professor **Tetsu Uesaka**, FSCN, Mid Sweden University, SE-851 70 Sundsvall, Sweden, email: tetsu.uesaka@miun.se

Contents

1	**The challenge**	1
	Kaarlo Niskanen	
2	**Paper as an engineering material**	5
	Per-Johan Gustafsson and Kaarlo Niskanen	
2.1	Introduction	5
2.2	Linear elasticity of paper	6
2.2.1	Elastic constants	6
2.2.2	Typical stiffness values for paper	9
2.3	Stress–strain behavior of paper	11
2.3.1	In-plane tensile loading	11
2.3.2	Visco-elastic effects	15
2.3.3	Other loading modes	15
2.4	Multi-axial strength	17
2.5	Mechanical properties in relation to the papermaking process	19
2.5.1	Preparation of papermaking fibers	19
2.5.2	Effect of the paper machine	22

Part I: Structural strength

3	**Packaging performance**	29
	Rickard Hägglund and Leif A. Carlsson	
3.1	Introduction	29
3.2	Paper-based packaging materials	30
3.2.1	Corrugated board	30
3.2.2	Box manufacturing process	33
3.2.3	Carton board	34
3.3	Loads imposed on boxes	35
3.4.	Strength of boxes	38
3.4.1	Short-term compressive loading	38
3.4.2	Empirical models for static box strength	42
3.4.3	Finite element models	44
3.4.4	Long-term loading	47
3.5	Summary	48

4	**Behavior of corners in carton board boxes**	53
	Mikael Nygårds	
4.1	Introduction	53
4.2	Folding of a multiply carton board	55
4.3	Creasing	59
4.4	Important material properties	63
4.5	Final remarks	63
5	**Fracture properties**	67
	Sören Östlund and Petri Mäkelä	
5.1	Introduction	67
5.2	Examples of practical applications of fracture mechanics	69
5.2.1	Mode I failure under in-plane tension	69
5.2.2	Out-of-plane delamination	72
5.3	Crack tip modeling in paper materials	73
5.3.1	Characteristic length scale and the basis of crack tip modeling	73
5.3.2	Linear elastic fracture mechanics LEFM	75
5.3.3	Nonlinear fracture mechanics using J-integral	76
5.3.4	Cohesive zone models	79
5.3.5	Continuum damage mechanics modeling of paper	81
5.3.6	Delamination of paper materials	83
5.4	Compressive failure	85
5.5	Summary	87

Part II: Dynamic stability

6	**Web dynamics in paper transport systems**	93
	Tetsu Uesaka	
6.1	Introduction	93
6.2	Dynamics of web transport	94
6.2.1	Basic formulation of web transport problems	94
6.2.2	The case of an axially moving web	95
6.2.3	Moving thread problem	100
6.2.4	Fluttering of a two-dimensional web	105
6.3	Concluding remarks	107
7	**Creep and relaxation**	111
	Douglas W. Coffin	
7.1	Introduction	111
7.2	Relaxation and creep as phenomena	112
7.3	Modeling of time-dependence	114

7.3.1	Linear behavior	114
7.3.2	Nonlinearity	115
7.3.3	Recoverability	116
7.3.4	Time scales	117
7.4	Creep and relaxation properties of paper	119
7.4.1	Creep	120
7.4.2	Stress relaxation	122
7.4.3	Tensile versus compressive creep	123
7.4.4	Effect of the papermaking process and furnish	124
7.5	Moisture effects	126
7.5.1	Softening with moisture	126
7.5.2	Accelerated creep	127
7.6	Prediction of box lifetime	129
7.6.1	Creep response of a box	129
7.6.2	Previous equations for box lifetime	131
7.6.3	Derivation of a new equation for box lifetime	132
7.6.4	Accounting for variability	133
7.7	Summary	134

8	**Statistical aspects of failure of paper products**	139
	Tetsu Uesaka	
8.1	Introduction	139
8.2	Practical examples	139
8.2.1	Web breaks in a printing press and on a paper machine	139
8.2.2	Stacking performance of boxes	142
8.3	Statistical approaches for failure in materials or systems	143
8.3.1	The chain model	144
8.3.2	The bundle model	145
8.3.3	Time-dependent, statistical failure model	147
8.4	Statistical failure of paper	149
8.4.1	Strength distributions	150
8.4.2	Factors controlling strength distributions	151
8.4.3	Strength scaling	152
8.4.4	Web break prediction	154
8.5	Research front of statistical failure of paper	156
8.6	Concluding remarks	157

Part III: Reactions to moisture and water

9	**Moisture-induced deformations**	163
	Artem Kulachenko	
9.1	Introduction	163
9.2	Moisture-induced deformations	163

9.2.1	Hygroexpansion of paper	163
9.2.2	Effect of moisture history	167
9.3	Fluting	171
9.3.1	Tension wrinkling	171
9.3.2	Effect of small scale strain variations	174
9.3.3	Fluting vs. cockling	177
9.4	Summary	178
10	**Mechanics in printing nip for paper and board**	**181**
	Tetsu Uesaka	
10.1	Introduction	181
10.2	Nip mechanics in offset printing of paper	182
10.3	Nip mechanics in flexo post printing of corrugated board	186
10.4	Micro-fluidics of ink in printing nip	188
10.5	Concluding remarks	191

Part IV: Material properties

11	**Micromechanics**	**197**
	Kaarlo Niskanen	
11.1	Introduction	197
11.2	Fiber network structure	198
11.2.1	Two-dimensional network	198
11.2.2	Densification mechanisms	200
11.2.3	Statistical geometry of real fiber networks	201
11.2.4	Key structural factors when engineering the mechanical properties of paper	205
11.3	Elastic modulus	206
11.3.1	The effect of paper density	206
11.3.2	The shear-lag mechanism	208
11.3.3	The activation mechanism	210
11.3.4	Elastic modulus of activated fiber network	211
11.3.5	Key factors when engineering the elastic modulus of paper	215
11.4	Stress–strain behavior, creep, and bond opening	216
11.5	Fracture process in the fiber network	220
11.5.1	Microscopic observations	220
11.5.2	Micromechanical description of the fracture process	221
11.6	Hygroexpansion	225
11.7	Final remarks	227

12	**Wood biocomposites – extending the property range of paper products**	231
	Lars Berglund	
12.1	Introduction	231
12.2	Material components: fibers and polymers	236
12.2.1	Plant fiber structure	236
12.2.2	Polymer matrices and binders	240
12.3	Micromechanics of fiber composites	242
12.3.1	Weight fraction and volume fraction	242
12.3.2	Elastic properties in unidirectional composites	243
12.3.3	Elastic properties in short fiber composites	244
12.3.4	Interfacial strength in short fiber composites	246
12.4	Composites data: wood fiber/thermoplastic	247
12.5	Composites data: wood fiber/thermoset	248
12.6	Nano-fibrillated cellulose materials	250
12.6.1	Cellulosic "nano-paper"	250
12.6.2	Nano-composites	252
12.7	Conclusions	252
Index		255

1 The challenge

Kaarlo Niskanen

This book discusses the mechanical properties of products made from wood fibers. Printing papers and office papers are the ubiquitous examples of such products; packaging made of paper or board and tissue paper are also commonly made of wood fibers. Fiberboards, which have a similar structure but different physical properties than paper, are used in furniture and construction. The renewable wood raw material used in all these products makes them attractive to modern society, and, therefore, many believe that wood fibers will increasingly be used also in other applications. We have written this textbook, in part, in anticipation of such developments.

Throughout the text we use current paper and board products as concrete examples because the properties of these products illustrate what may be required for future applications. The cases discussed in this book give a broad picture of the various challenges that are faced by anyone working with materials based on wood fibers. We expect that the reader will extrapolate from these examples how to solve analogous challenges in other situations.

Because we focus on the mechanical properties of paper and board products, this textbook discusses the practical applications of mechanical engineering. The reader is assumed either to have a good understanding of engineering mechanics or to be able to acquire such an understanding as needed. Readers should also be comfortable with mathematical concepts and be able to appreciate the power of numerical modeling methods. The reader is not expected to know much of paper or papermaking. The main features are described in chapter 2. Building a deeper understanding of materials based on wood fibers is in itself a goal of this book.

Before one can solve a practical performance problem or develop a better product, one has to understand what the problem really is. In an effort to underline the importance of problem definition, we have chosen a somewhat unorthodox and admittedly difficult approach. We have tried to write a book that centers around two practical cases (the "box" and the "web"), identifies the important process requirements and the related material or systems properties before exploring the material properties, and, ultimately, discusses how the material properties or systems behavior could be improved. The result is not perfect because, as it turns out, little research has been done on real-world problems and their relation to the material properties.

The approach means that the book passes over much of the material traditionally discussed in the context of the mechanical properties of paper and board. Many excellent books on papermaking, paper physics, and the mechanical properties of paper products explain what is already known. In this text we are more interested in unveiling what *should* be known.

That said, we are confident that this book will be useful not only when developing new products and materials based on wood fibers but also – and most readily – when solving performance problems of contemporary paper and board products.

After the introduction to paper materials in chapter 2, the text is divided into four parts that clarify the underlying logic. Part I considers the basic structural strength of paper products. Wood fibers are particularly beneficial in applications where stiffness at low weight is required; for example, in packaging applications. The first practical case introduced in chapter 3 is the "box"; the manufacturing and usage requirements of paper-based containers and boxes. The performance of a box depends on its structure and on the changes in the material that are the result of the process of converting board into a box. Chapter 4 then illustrates the latter aspect by considering the corners of paperboard boxes. The quality of corners is crucial, for example, in liquid packaging applications where defects can result in leakage or trigger box fracture. Chapter 5 then discusses the machinery that has been developed specifically to analyze fractures that are triggered by defects or other discontinuities in a structure.

Part I is concerned primarily with the "static" or "instantaneous" strength of products. Often, in reality, the dynamics of the process are crucial. This extension to structural strength is discussed in part II. One major advantage of paper materials is the high speed at which the "web," the second practical case, can be manufactured and converted – but only provided that the web can be held stable in the fast process. Chapter 6 explains what determines the short-time dynamic stability of "running" paper or board webs. The requirements posed by fast dynamic loading are different from the requirements posed on boxes. A stack of boxes should hold through the entire logistic chain, which may last for months. Chapter 7 then explores the rheological properties of paper materials, applicable in both the short-time web stability and the long-time box endurance.

From chapters 6 and 7 we conclude that the strength problems of paper products are also a systemic issue. Product performance cannot be comprehensively explained by just average material properties. Spatial and temporal variability in material properties and process conditions are often crucial. Therefore, chapter 8 defines the general methodology for obtaining statistically reliable information about box failures and web breaks.

After tackling the "static strength" and "dynamic or systemic strength" of paper products, part III considers the dimensional instability caused by changes in the moisture content of paper. The moisture sensitivity of paper is a direct result of the biological origin of wood fibers. It enables the preparation of paper without any added adhesives and makes it possible to recycle paper by simply soaking it in water. Thus, moisture sensitivity is a challenge in all uses of wood fibers; and one has to learn to minimize the problems. This is the focus of chapters 9 and 10.

Moisture changes are practically impossible to avoid in the normal usage of paper products. Where chapter 7 discusses the effects of moisture on the creep rate of boxes, chapter 9 explains how changes in moisture lead can lead to permanent out-of-plane deformations, especially if the paper surface is exposed to liquid water. Chapter 10 then explains the mechanics of a printing nip where water and other liquids are applied on paper or board.

Finally, having illustrated some of the requirements that usage poses on paper material, part IV discusses how the relevant material properties can be controlled. In chapter 11 we outline the micromechanical concepts that can be used to connect the material properties of paper to the papermaking fibers and the papermaking process. Chapter 12 continues from there by asking what functionality can be achieved if one goes beyond paper, using wood as an ingredient in biocomposites.

This book is by no means a comprehensive account of the performance of paper products or materials based on wood fibers. We have not intended to give such an account. What we hope to have accomplished is a text that illustrates how one can systematically tackle challenges in the development of better products. Asking the right questions is critical to finding relevant answers; we have certainly not uncovered all the questions.

2 Paper as an engineering material

Per-Johan Gustafsson and Kaarlo Niskanen

2.1 Introduction

This textbook concerns the mechanical properties of products that are made of paper or board. Throughout the book, therefore, it is important to know the properties of paper as an engineering material. In this chapter, we describe the general mechanical properties of paper and the papermaking process. The manufacturing process, including the wood raw material used, naturally governs these mechanical properties, but here we limit the discussion to what is most important to know when tackling problems of product performance. An overview of the underlying mechanisms and raw materials effects is presented in Chapters 11 and 12. Those interested in learning more are referred to the many textbooks on papermaking, such as the series *Papermaking Science and Technology*, published by the Finnish Paper Engineers' Association.

Paper is a thin, almost two-dimensional material. Everyday papers, such as office paper and newsprint, have a thickness of 0.1 mm. The mass per unit area of paper, called the *basis weight*, is usually between 40 and 100 g/m^2 depending on the type of paper. Specially prepared paper can have a thickness as low as 0.01 mm and a basis weight of a few grams per square meter. On the other end, paper material used for book covers or fixtures to display products in stores can be more than 1 mm thick. Thick paper grades are called *board* or *paperboard*. The trade terminology for paper and board grades refers primarily to the applications where the materials are used, not to their structure. In this book, *paper* is a general term used for all kinds of paper and board materials.

A paper machine creates a continuous web that is 5–10 m wide. A finished roll may contain 10 km of paper. The coordinate system used throughout this book is defined in Figure 2.1. The running direction of the web is customarily referred to as the machine direction (MD), and the lateral direction as the cross-machine direction (CD). The thickness direction is the Z-direction (ZD).

The main constituent of paper is wood or plant fibers that are specially prepared into a pulp, as is outlined in Section 2.5. The water suspension of the fibers and various additives form what is called *furnish*. When a thin layer of furnish is dried, the fibers bond to each other and paper is formed. Most published data on the mechanical properties of paper comes from measurements done on paper sheets made in a laboratory. The properties of these laboratory sheets, or handsheets, differ from machine-made commercial paper even if the same raw material is used.

Constitutive laws are used to analyze the mechanical properties of paper. When a constitutive law is combined with the equations for equilibrium, geometrical compatibility, and boundary conditions, one can calculate the magnitude, distribution,

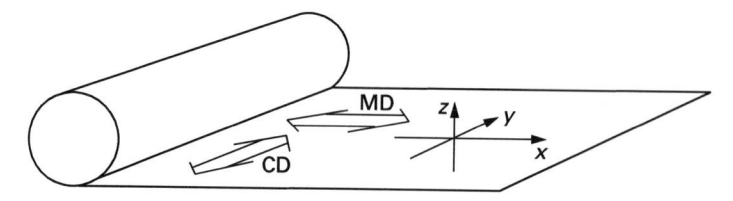

Fig. 2.1 The coordinate system for paper that we use in this book.

and variation of stress and deformations in a specific product structure. In general, constitutive laws or theories describe how stress and temperature at a point interact with the deformation or straining of the material. In the case of paper, the effect of temperature is relatively small, but the effect of moisture is very important.

Stress and *strain* are quantities that are defined for continuous materials. Usually paper can be regarded as continuous all the way down to the centimeter scale, after which the fibers network structure starts to dominate. Thus, we can use phenomenological models derived from macroscopic experiments to represent material behavior in products made of paper. However, special care is needed when determining phenomenological models for z-directional deformations and loads and for fracture processes in general. A typical paper sheet consists of only 10 fiber layers (here layer refers to the typical number of fibers in the thickness direction of paper but does not mean that fibers would actually form clear layers), so that experimental boundary effects can be large in ZD. In fracture processes, deformations take place at very small scales close to a crack, and the material behavior at that length scale must be determined. Chapters 11 and 12 examine the effects of microscopic structure and composition on the constitutive behavior of paper. More information on the mechanical properties of paper and their measurement can be found in and Mark et al. (2002) and Niskanen (2008).

2.2 Linear elasticity of paper

2.2.1 Elastic constants

The *elastic constants* give the stress to strain relation for paper when the performance is linear elastic. In the general 3D case, the state of stress is defined by the six independent stress components: the three normal stresses, σ_x, σ_y, and σ_z, and the three shear stresses, $\tau_{xy} = \tau_{yx}$, $\tau_{xz} = \tau_{zx}$, and $\tau_{yz} = \tau_{zy}$ (see Fig. 2.2). A positive value of shear stress τ_{ij} corresponds to stress acting on the surface with its normal in the positive i-direction and directed in the positive j-direction. Because of moment equilibrium, shear stresses are equal in pairs, $\tau_{ij} = \tau_{ji}$. The alternative notations, σ_{MD}, σ_1, σ_{11}, or σ_{xx} etc are sometimes used instead of σ_x etc when the coordinates coincide with MD etc.

The state of strain in the 3D case also is defined by six components, the normal strains ε_i, and the shear strains γ_{ij}. The strains referred to in this book are the conventional small strain theory engineering strains. This means that the normal strains are defined as elongation Δl divided by the initial length l, and the shear strains are given by a sum of two deformation angles expressed in radians, for example, $\gamma_{xz} = \varphi_{xz} + \varphi_{zx}$

2.2 Linear elasticity of paper

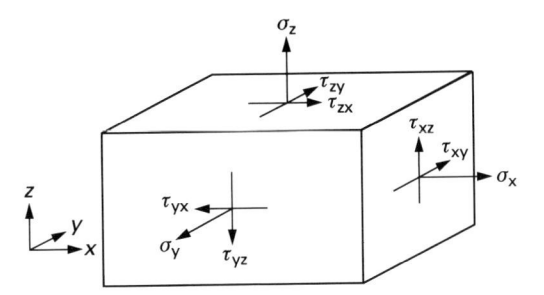

Fig. 2.2 The stress components in a 3D state of stress.

(Fig. 2.3). The strains can also be defined by means of the local displacements u_x, u_y, and u_z of the material, $\varepsilon_x = \partial u_x/\partial x$, and so on, and $\gamma_{xz} = \partial u_z/\partial x + \partial u_x/\partial z$, and so on.

Unlike many other materials, the elastic modulus E of paper is significantly anisotropic. This arises from the manufacturing process, giving $E_x > E_y$ (see Section 2.5). The anisotropy between the in-plane and thickness directions comes from the low thickness of paper. Because a typical fiber's length, 1–3 mm, is more than 10 times larger than paper thickness, fibers must be aligned in the plane of the paper. The z-directional straining of paper creates primarily transverse stresses in fibers, whereas in-plane straining creates longitudinal stresses. The longitudinal elastic modulus of fibers is larger than the transverse modulus (see Chapter 12).

A reasonable approximation is that the anisotropy of paper is orthotropic, that is, the stiffness properties are symmetric with respect to the x, y, and z axes, even though there may be a slight deviation in the symmetry axes because of skewness in fiber orientation (see Section 2.5). The components of the stress and strain vectors σ and ε components are coupled by the elastic compliance tensor $\varepsilon = S\sigma$, or

$$\begin{bmatrix} \varepsilon_x \\ \varepsilon_y \\ \varepsilon_z \\ \gamma_{xy} \\ \gamma_{xz} \\ \gamma_{yz} \end{bmatrix} = \begin{bmatrix} 1/E_x & -v_{yx}/E_y & -v_{zx}/E_z & 0 & 0 & 0 \\ -v_{xy}/E_x & 1/E_y & -v_{zy}/E_z & 0 & 0 & 0 \\ -v_{xz}/E_x & -v_{yz}/E_y & 1/E_z & 0 & 0 & 0 \\ 0 & 0 & 0 & 1/G_{xy} & 0 & 0 \\ 0 & 0 & 0 & 0 & 1/G_{xz} & 0 \\ 0 & 0 & 0 & 0 & 0 & 1/G_{yz} \end{bmatrix} \begin{bmatrix} \sigma_x \\ \sigma_y \\ \sigma_z \\ \tau_{xy} \\ \tau_{xz} \\ \tau_{yz} \end{bmatrix}, \quad (2.1)$$

or by the stiffness tensor $\sigma = C\varepsilon$,

$$\begin{bmatrix} \sigma_x \\ \sigma_y \\ \sigma_z \\ \tau_{xy} \\ \tau_{xz} \\ \tau_{yz} \end{bmatrix} = A \begin{bmatrix} E_x(1-v_{yz}v_{zy}) & E_x(v_{yx}+v_{zx}v_{yz}) & E_x(v_{zx}+v_{yx}v_{zy}) & 0 & 0 & 0 \\ E_y(v_{xy}+v_{zy}v_{xz}) & E_y(1-v_{xz}v_{zx}) & E_y(v_{zy}+v_{xy}v_{zx}) & 0 & 0 & 0 \\ E_z(v_{xz}+v_{xy}v_{yz}) & E_z(v_{yz}+v_{xz}v_{yx}) & E_z(1-v_{yx}v_{xy}) & 0 & 0 & 0 \\ 0 & 0 & 0 & G_{xy}/A & 0 & 0 \\ 0 & 0 & 0 & 0 & G_{xz}/A & 0 \\ 0 & 0 & 0 & 0 & 0 & G_{yz}/A \end{bmatrix} \begin{bmatrix} \varepsilon_x \\ \varepsilon_y \\ \varepsilon_z \\ \gamma_{xy} \\ \gamma_{xz} \\ \gamma_{yz} \end{bmatrix}, \quad (2.2)$$

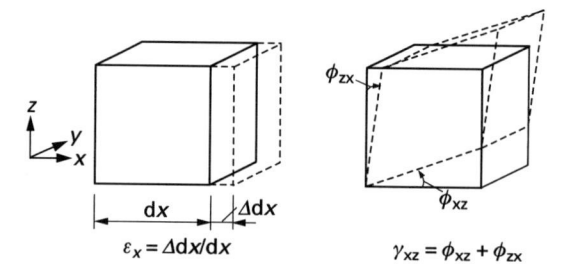

Fig. 2.3 The normal strain ε_x and the shear strain γ_{xz} in a 3D state of small strain.

where $A = 1/(1 - v_{xy}v_{yx} - v_{xz}v_{zx} - v_{yz}v_{zy} - v_{xy}v_{zx}v_{yz} - v_{yx}v_{xz}v_{zy})$. Linear elasticity implies that the compliance tensor in (2.1) is symmetric (Malvern, 1969) and accordingly

$$\begin{cases} v_{xy} = v_{yx} E_x/E_y \\ v_{xz} = v_{zx} E_x/E_z \\ v_{yz} = v_{zy} E_y/E_z \end{cases} \tag{2.3}$$

Thus, linear elasticity reduces the number of independent material parameters from 12 to 9. Equations 2.1–2.3 define the generalized Hooke's law for paper. Strictly speaking, the elastic moduli E, shear moduli G, and Poisson ratios v are applicable only in the ideal case of perfectly linear elastic (i.e., reversible and loading-rate independent) behavior. Nevertheless, it is customary to apply the notation in Equations 2.1–2.3 for measurement results, and, for example, to denote the slope of a measured stress–strain curve as elastic modulus E even when the result may depend on the experimental conditions. In the case of paper, measured stress–strain curves are linear elastic within measurement accuracy when the load level is well below the failure load.

In many practical situations, paper is exposed to pure plane stresses ($\sigma_z = \tau_{xz} = \tau_{yz} = 0$) In this special case,

$$\begin{bmatrix} \sigma_x \\ \sigma_y \\ \tau_{xy} \end{bmatrix} = \frac{1}{1 - v_{xy}v_{yx}} \begin{bmatrix} E_x & E_x v_{yx} & 0 \\ E_y v_{xy} & E_y & 0 \\ 0 & 0 & G_{xy}(1 - v_{xy}v_{yx}) \end{bmatrix} \begin{bmatrix} \varepsilon_x \\ \varepsilon_y \\ \gamma_{xy} \end{bmatrix}. \tag{2.4}$$

The influence of an eventual change of temperature Θ and moisture content (MC) χ can be taken into account by adding temperature- and moisture-induced strains to Equations 2.1 and 2.2:

$$\varepsilon = S\sigma + \alpha\Delta\Theta + \beta\Delta\chi \text{ and } \sigma = C(\varepsilon - \alpha\Delta\Theta - \beta\Delta\chi). \tag{2.5}$$

This formulation ignores the moisture and temperature dependence of the elastic constants. In orthotropic materials the thermal expansion and hygroexpansion coefficients in the principal directions are

$$\boldsymbol{\alpha} = [\alpha_x \ \alpha_y \ \alpha_z \ 0 \ 0 \ 0]^T \text{ and } \boldsymbol{\beta} = [\beta_x \ \beta_y \ \beta_z \ 0 \ 0 \ 0]^T. \tag{2.6}$$

The thermal strains $\alpha\Delta\Theta$ of paper are, in general, much smaller than the hygroscopic strains. For example, a 20°C temperature change may lead to $\alpha\Delta\Theta \approx 0.01\%$, while a change in the relative humidity (RH) of air from 10% to 60% leads to $\beta \Delta \chi \geq$ 0.1%. Thus, thermal expansion can usually be excluded. Relative humidity RH gives the concentration of water vapor in air relative to the saturation concentration that is possible at the given temperature.

2.2.2 Typical stiffness values for paper

Table 2.1 shows a collection of directly measured values of the elastic stiffness parameters for a few paper grades. Many values are missing because of measurement difficulties caused by the small thickness of paper. Various estimation schemes have been developed to escape direct measurement (Baum, 1987).

Table 2.1 demonstrates that the ZD stiffness of paper is generally low compared to the in-plane values. The negative value of the Poisson ratio v_{xz} for the paperboard shows that uniaxial tensile loading in MD increased thickness in this case, which is not uncommon. In compression, at least the elastic moduli, perhaps even the Poisson ratios, are usually equal to the corresponding tensile values.

The high density of the coated paper is caused by the coating. Without the coating the paper would have similar density as the other samples. In general, the density of paper is between 300 and 900 kg/m³. The in-plane elastic modulus usually increases with density (cf. Section 11.3) and ranges from 1000–9000 MPa when the effect of

Tab. 2.1 Measured values of elastic stiffness parameters in tensile loading for some machine-made papers.

	Paperboard (Persson, 1991)	Carton (Baum, 1987)	Linerboard (Baum, 1987)	Coated paper, middle of web (Stålne, 2006)	Coated paper, web edge (Stålne, 2006)
Density, kg/m³	640	780	691	1140	1140
MD modulus E_x, MPa	5420	7440	7460	7690	7660
CD modulus E_y, MPa	1900	3470	3010	3050	2570
ZD modulus E_z, MPa	17	40	29		140
Poisson ratio v_{xy}	0.38	0.15	0.12	0.33	0.27
Poisson ratio v_{xz}	−2.20	0.008	0.011		
Poisson ratio v_{yx}	0.14			0.07	0.10
Poisson ratio v_{yz}	0.54	0.021	0.021		
Poisson ratio v_{zx}	0.05				−0.04
Poisson ratio v_{zy}	0.05				0.03
Shear modulus G_{xy}, MPa	1230	2040	1800	1910	1820
Shear modulus G_{xz}, MPa	8.8	137	129		
Shear modulus G_{yz}, MPa	8.0	99	104		

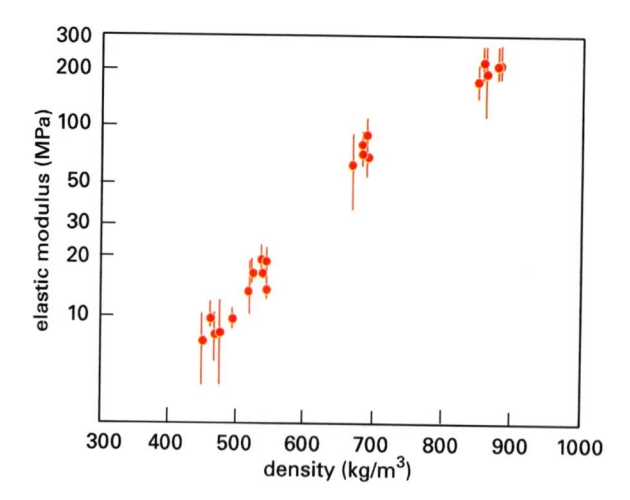

Fig. 2.4 ZD elastic modulus (logarithmic scale) of laboratory sheets against density for mechanical pulps (density ca. 500 kg/m^3) and chemical pulps (density > 700 kg/m^3), using data from Girlanda and Fellers (2007).

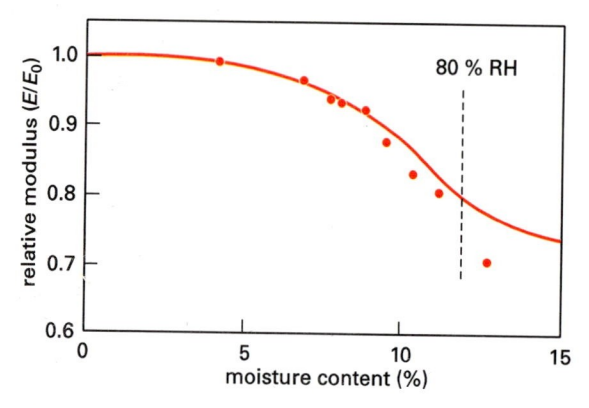

Fig. 2.5 Elastic modulus against moisture content for a set of laboratory sheets. The modulus values are given relative to the value in dry paper (dots). The curve shows a theoretical prediction. Reprinted from Salmén et al. (1984) with permission from Elsevier.

anisotropy is removed by averaging over MD and CD. Typical values of the MD/CD ratio of elastic moduli in machine-made papers are 2–4, but can be as high 5–6. From the variation of the z-directional elastic modulus against density in Figure 2.4, one can see that the data for different pulps (cf. Section 2.5) can fall on one single line. The same is not true for in-plane elastic moduli.

Water acts as a softener of paper. Thus, the elastic modulus of paper depends on the moisture content (Figs. 2.5 and 2.6). Ultimately at high moisture contents, the modulus of paper goes to zero as the bonding between fibers opens, and one returns to a state that prevailed when drying started in the papermaking process. We note in passing that it is this reversibility of the papermaking process that makes the recycling of paper possible.

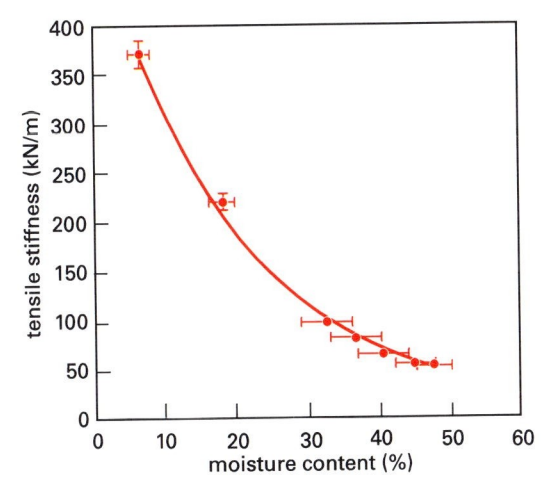

Fig. 2.6 Tensile stiffness against moisture content for a machine-made paper, measured with a cyclic small-strain excitation. Tensile stiffness is equal to elastic modulus multiplied by paper thickness, the latter being a slightly increasing function of moisture content. Drawn using data of Ketoja et al. (2007).

The softening effect makes paper increasingly visco-elastic and visco-plastic, which means that, especially at higher moisture contents, the slope of the measured stress–strain curves depends on the strain rate. The apparent modulus (slope of the stress–strain curve) increases if strain rate is increased. At moisture contents of 50% or higher, it is governed by interactions between fibers that are mediated by liquid water. Therefore, any stress created by constrained deformations would rapidly relax to zero. In addition to elastic modulus, the softening effect of moisture is evident in the stress–strain behavior of paper, discussed next.

2.3 Stress–strain behavior of paper

2.3.1 In-plane tensile loading

In principle, a stress increment may cause an instant or delayed and reversible (i.e., elastic) or irreversible (i.e., inelastic or plastic) strain increment (Fig. 2.7). The presence of a delayed response implies that the stress–strain behavior is time-dependent or rate-dependent. Furthermore, the relationship between stress and strain can be linear or nonlinear. The stress–strain curve of paper exhibits all these behaviors. The time-dependence seen in the creep and stress relaxation of paper is discussed in Chapter 7. This section gives a general overview of the different aspects of the three-dimensional stress–strain behavior of paper.

A recursive tensile in-plane stress–strain measurement of paper usually gives a result of the type shown in Figure 2.8. One can see that the elastic modulus changes very little even though part of the strain is irreversible or plastic. This is typical of

12 | 2 Paper as an engineering material

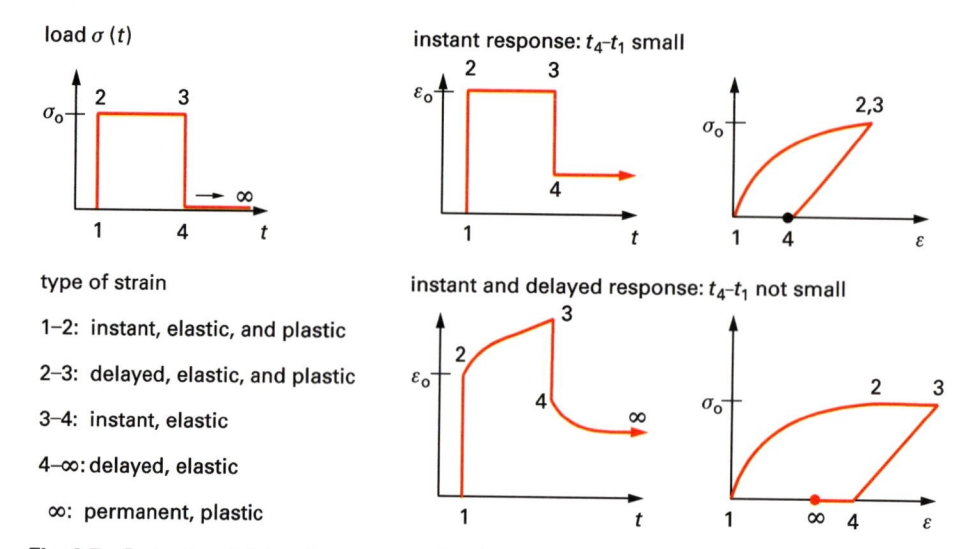

Fig. 2.7 Instant and delayed response to load.

load $\sigma(t)$

type of strain

1–2: instant, elastic, and plastic
2–3: delayed, elastic, and plastic
3–4: instant, elastic
4–∞: delayed, elastic
∞: permanent, plastic

instant response: t_4-t_1 small

instant and delayed response: t_4-t_1 not small

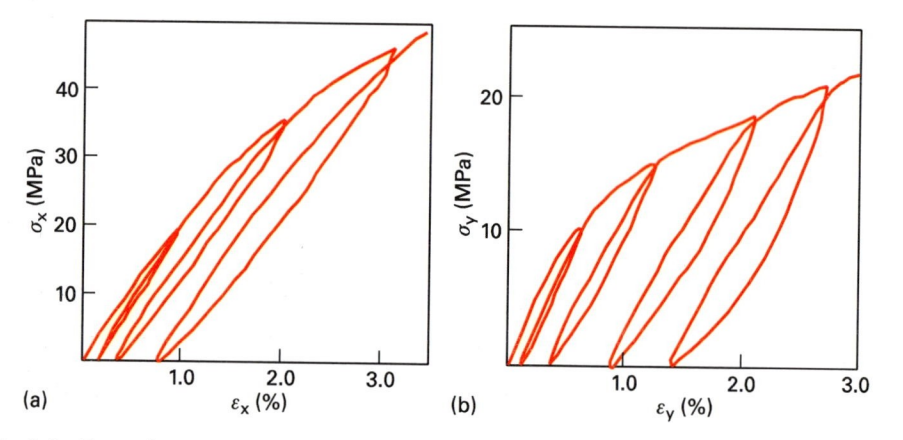

Fig. 2.8 Recursive stress–strain curves of a paperboard in MD (a) and CD (b), from Persson (1991). Reproduced with permission from the author.

almost all paper grades: the elastic modulus decreases by a maximum of 10% before the breaking point is reached. Brittle paper grades, such as baking paper or glassine, exhibit a larger loss in the elastic modulus, while ductile paper grades, such as sack paper, show a modest increase. Corresponding stress–strain curves are illustrated in Figure 2.9.

The fact that the elastic modulus of paper changes only a little before the peak stress suggests that the microscopic fibers' network structure undergoes permanent plastic deformations that do not weaken elastic stiffness of the fibers (cf. Chapter 11.3). However, after the peak stress the elastic modulus decreases. This is apparent in the post-peak unloading–reloading cycles shown in Figure 2.10. The post-peak behavior in general can be recorded only when short specimens are used (Hillerborg

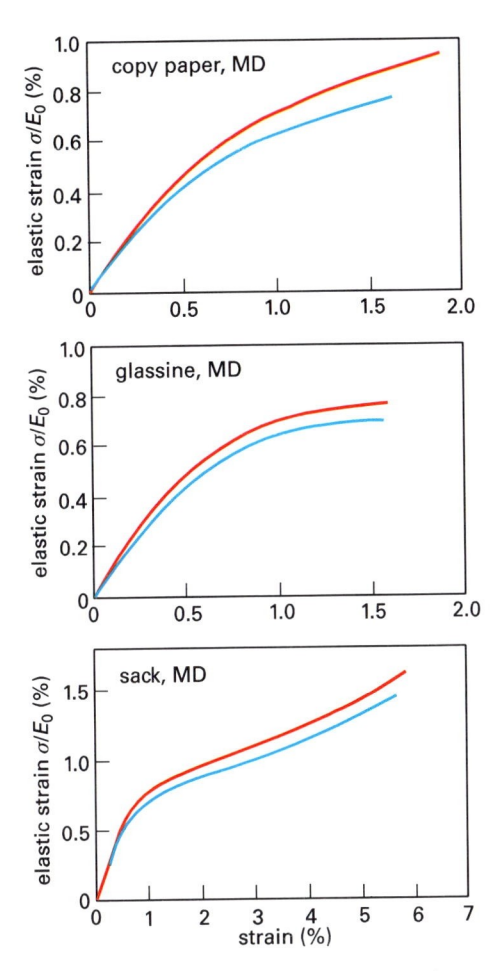

Fig. 2.9 Examples of MD stress–strain curves of some machine-made paper grades. Stress values are divided by the elastic modulus E_0 measured initially at zero strain, giving an estimate of the elastic strain. Data courtesy of Lauri Salminen.

et al., 1976; Tryding and Gustafsson, 2000); long specimens show sudden failure at the peak stress, which is discussed in Chapter 5.3.4.

If one measures how the post-peak cohesive stress decays as the fracture process zone widens, one gets results illustrated in the right-hand side of Figure 2.11. The fracture zone widens to more than 1 mm before all stiffness is lost. The large ductility presumably comes from the length of fibers. Stiffness does not go to zero before the fibers crossing the fracture line have been pulled out of either half of the specimen. In Figure 2.11, the newsprint is made of mechanical pulp, which has shorter fibers than the chemical pulp used in the strong kraftpaper. In brittle paper grades, such as glassine and baking paper, inter-fiber bonding is unusually strong, and therefore, fibers break during the fracture process, and the fracture zone widening is small (cf. Section 11.5).

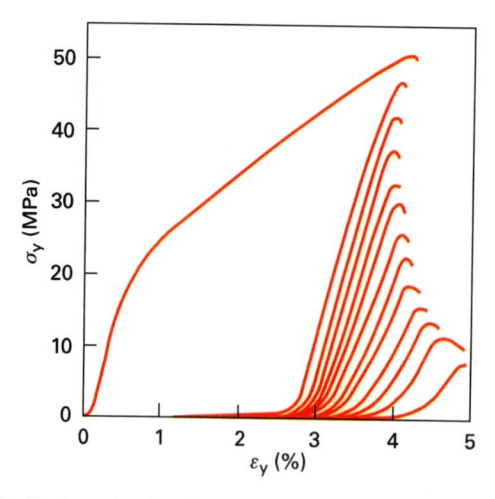

Fig. 2.10 Post-peak reloading stress–strain behavior of paperboard in CD using 5 mm long and 15 mm wide specimens (Tryding, 1996). Reproduced with permission from the author.

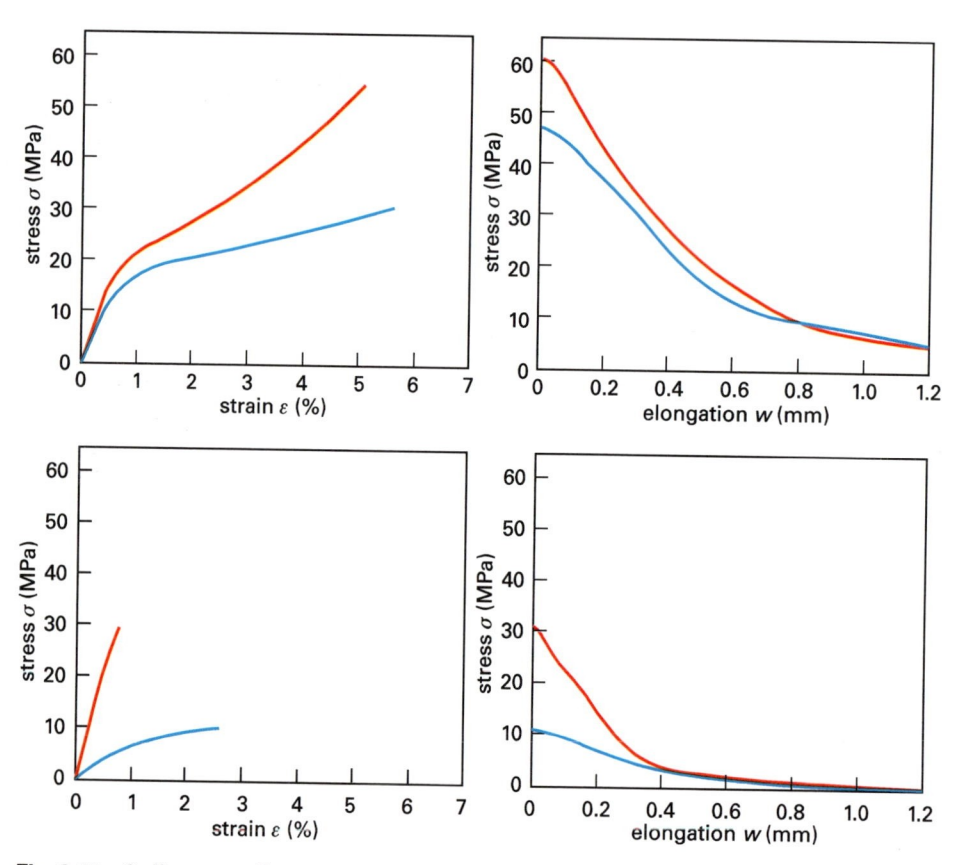

Fig. 2.11 Ordinary tensile stress–strain curves (left) and post-peak cohesive stress vs fracture zone widening (right) in MD (red) and CD (blue) in a 70g/m^2 kraftpaper (top) and 45g/m^2 newsprint (bottom). The post-peak measurement used 5 mm long and 15 mm wide specimens. Data of Tryding (1996). Reproduced with permission from the author.

2.3.2 Visco-elastic effects

The effect of strain rate on the in-plane tensile stress–strain curve is illustrated in Figure 2.12. In this particular example, the breaking strain does not decrease when the strain rate is increased. However, often the increase in strain rate leads to a reduction in the breaking strain of paper, which is expected if the material becomes more brittle at high strain rates.

The softening effect of moisture is shown in Figure 2.13. The elastic modulus and breaking stress are lower and the breaking strain is higher at the higher relative humidity, corresponding to the higher moisture content in the paper.

In the in-plane tensile stress–strain curves displayed in Figures 2.9 and 2.13, the breaking strain of paper ranges from 1%–5%, which is quite typical. The values increase with increasing moisture content, and they may decrease with increasing strain rate. The breaking strain of paper falls below 1% only in very special cases. One can also see that the apparently linear part of the curves ends somewhere in the neighborhood of 0.5%. The breaking stress is usually strongly correlated with the elastic modulus so that the ratio of the two is close to 1%, and the in-plane tensile breaking stress values range from 10–100 MPa.

2.3.3 Other loading modes

Because paper is a thin planar material, the measurement of in-plane compression is complicated. Buckling of the specimen must be prevented with some fixture that creates in-plane forces. Even if buckling is prevented, paper fails under compressive

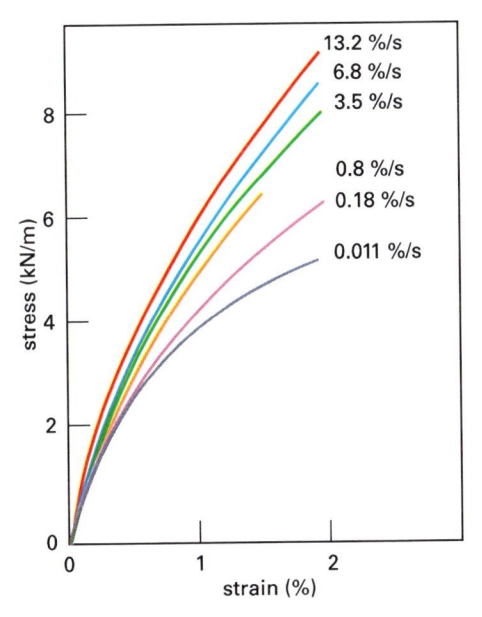

Fig. 2.12 Stress–strain curves at different strain rates for a wrapping paper, after Andersson and Sjöberg (1953). Stress values are multiplied by the thickness of paper. Reproduced with permission from Svenska Pappers- och Cellulosaingeniörsföreningen (SPCI).

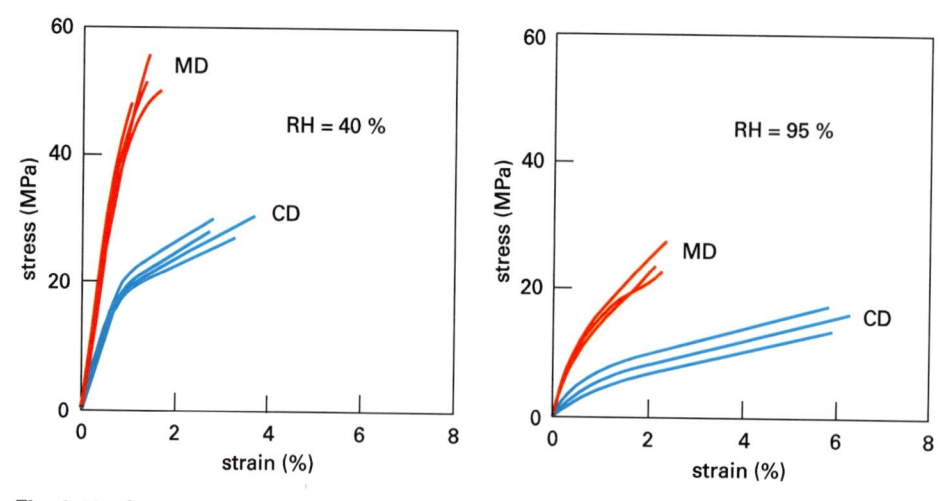

Fig. 2.13 Stress–strain curves in MD and CD of a paperboard at relative humidity of 40% and 95%. The corresponding moisture contents are 6.6% and 20%. Drawn using data of Yeh, Considine and Suhling (1991).

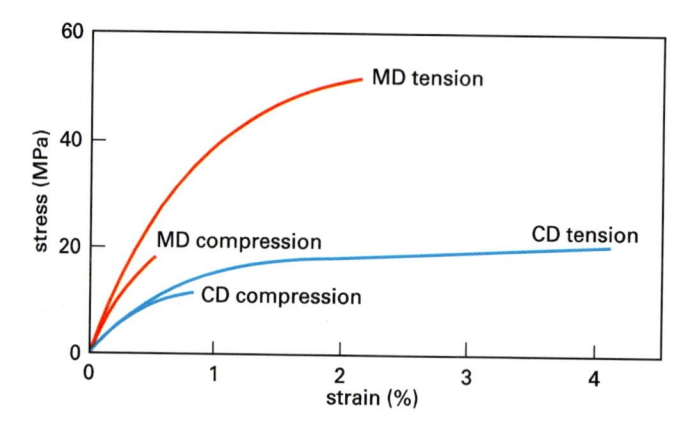

Fig. 2.14 Comparison of compressive and tensile behavior of a paperboard in MD and CD, after Fellers (1980). Reproduced with permission from the author.

stress much sooner than under tensile stress (Fig. 2.14). In the z-directional testing the situation is the opposite, and compressive behavior is easy to measure. The ZD compressive strain is determined by the pore volume fraction and surface roughness, which are both pressed away by the applied stress. As the pore volume closes, the apparent stiffness of the material increases rapidly toward infinity, giving an exponential compressive stress–strain curve in ZD.

Tensile and shear testing in ZD requires an adhesive layer on the surfaces to transfer stress to the specimen. Because paper is thin, the adhesive contributes to the measured displacement and influences the deformations of paper. Tensile strain values in the z-direction of paper are large, and stress values are small compared to the in-plane directions (Fig. 2.15). The maximum stress is observed relatively early in

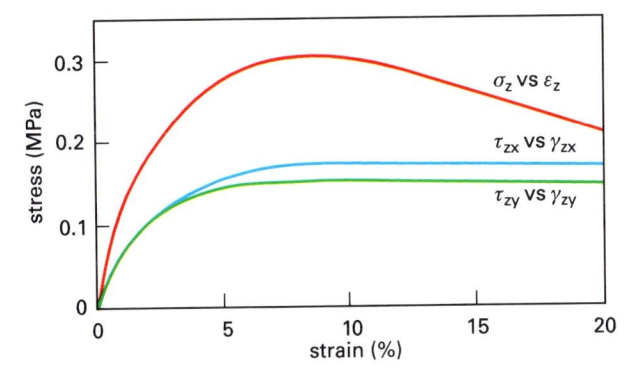

Fig. 2.15 Out-of-plane stress vs. strain for tensile and shear loading of a paperboard. Data of Persson (1991). Reproduced with permission from the author.

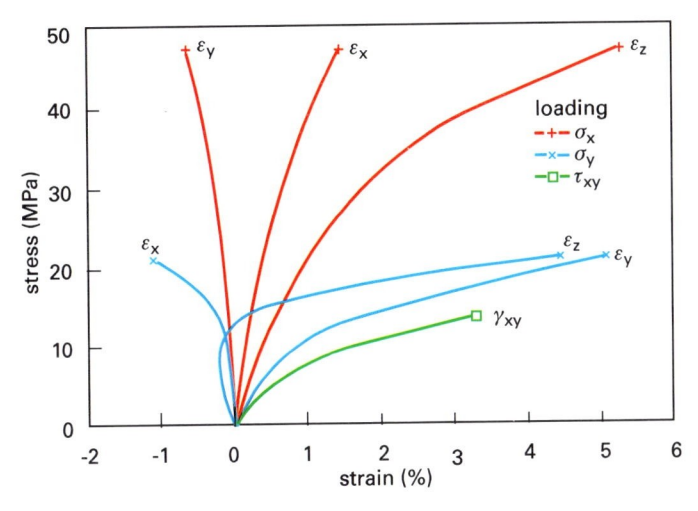

Fig. 2.16 Strains in the three principal directions of a paperboard, created by tensile stress in MD or CD and the shear strain created by a MD-CD shear stress, after Persson (1991). The shear behavior was calculated from tensile loading at 45° angle from MD, assuming orthotropy. Reproduced with permission from the author.

the stress–strain curve, before a long post-peak tail. Typical values for the maximum tensile stress in ZD are 0.1–1 MPa, two orders of magnitude smaller than in MD. Three-dimensional strains measured for tensile loading in MD and CD and shear loading in the MD-CD plane are illustrated in Figure 2.16.

2.4 Multi-axial strength

In packages and other structural uses, multi-axial stresses on the material often arise from contents and stacking. In order to analyze the material performance in such situations, it is necessary to model the multi-axial constitutive behavior and ultimate strength of the material. As has been shown previously, the constitutive laws of paper can be highly nonlinear, especially in ZD. For a representative description of the multi-axial nonlinear loading–unloading behavior, the elastic and plastic strain components must be separated (Xia et al., 1994).

In spite of the complicated constitutive behavior of the paper materials, the ultimate strength values under multi-axial stresses can be estimated in a relatively simple manner from the Tsai-Wu criterion (Tsai and Wu, 1971). One calculates an effective scalar stress index f^{TW} according to

$$f^{TW} = \begin{bmatrix} \sigma_x \\ \sigma_y \\ \sigma_z \\ \tau_{xy} \\ \tau_{xz} \\ \tau_{yz} \end{bmatrix}^T \begin{bmatrix} a_x \\ a_y \\ a_z \\ 0 \\ 0 \\ 0 \end{bmatrix} + \begin{bmatrix} \sigma_x \\ \sigma_y \\ \sigma_z \\ \tau_{xy} \\ \tau_{xz} \\ \tau_{yz} \end{bmatrix}^T \begin{bmatrix} b_{xx} & b_{xy} & b_{xz} & 0 & 0 & 0 \\ b_{yx} & b_{yy} & b_{yz} & 0 & 0 & 0 \\ b_{zx} & b_{zy} & b_{zz} & 0 & 0 & 0 \\ 0 & 0 & 0 & b_{xyxy} & 0 & 0 \\ 0 & 0 & 0 & 0 & b_{xzxz} & 0 \\ 0 & 0 & 0 & 0 & 0 & b_{yzyz} \end{bmatrix} \begin{bmatrix} \sigma_x \\ \sigma_y \\ \sigma_z \\ \tau_{xy} \\ \tau_{xz} \\ \tau_{yz} \end{bmatrix}, \quad (2.7)$$

and assumes that failure occurs when $f^{TW} = 1$. Here a_i and b_{ij} are material parameters, and b is symmetric, $b_{ij} = b_{ji}$. The zeros in Equation 2.7 arise because the sign of shear stress cannot affect the effective stress for an orthotropic material, and therefore, shear stresses are decoupled from other stresses. The 12 independent material parameters in Equation 2.7 must then be determined from experiments under different loading modes.

In the case where only xy-plane stresses are applied, the Tsai-Wu criterion simplifies into

$$f^{TW} = a_x \sigma_x + a_y \sigma_y + b_{xx} \sigma_x^2 + 2 b_{xy} \sigma_x \sigma_y + b_{yy} \sigma_y^2 + b_{xyxy} \tau_{xy}^2 = 1. \quad (2.8)$$

In the biaxial $\sigma_x - \sigma_y$ stress plane this equation corresponds to an ellipse. Comparison with measured data for $\tau_{xy} = 0$ is shown in Figure 2.17. If the shear stress τ_{xy} is increased, the size of the ellipse predicted by Equation 2.8 decreases. For the linerboard material in Figure 2.17, the failure stress in pure shear mode can be estimated from network mechanics (Gustafsson et al., 2001) to occur at $\tau_{xy} \approx 2.4$ MPa.

If only out-of-plane stresses are applied, the Tsai-Wu criterion (Eq. 2.7) predicts failure when

$$f^{TW} = a_z \sigma_z + b_{zz} \sigma_z^2 + b_{xzxz} \tau_{xz}^2 + b_{yzyz} \tau_{yz}^2 = 1. \quad (2.9)$$

This is compared with measured values in Figure 2.18. Clearly, the Tsai-Wu criterion does not work for compressive normal stresses σ_z large than 1 MPa (compressive stresses are negative in Fig. 2.18). For larger compressive normal stresses, the Coulomb friction law seems to apply, or

$$\sqrt{\tau_{xz}^2 + \tau_{yz}^2} = c - \mu \sigma_z, \quad (2.10)$$

where μ is the coefficient of friction, and c the maximum cohesive stress.

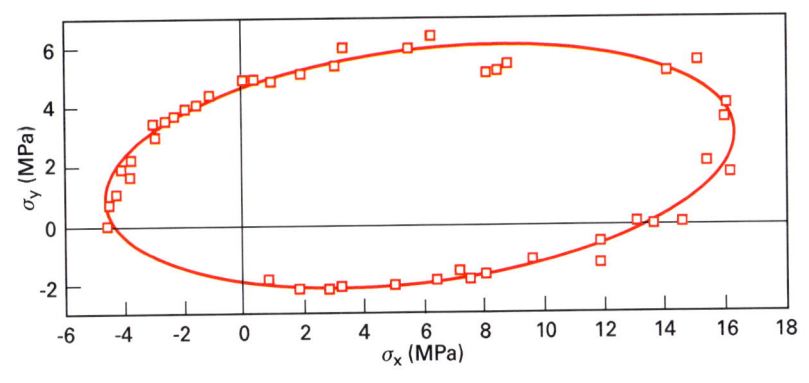

Fig. 2.17 Biaxial strength data (squares) for a linerboard compared with the Tsai-Wu criterion (Eq. 2.8, line), after Fellers et al. (1983). The material parameters are $a_x = -0.158$ MPa^{-1}, $a_y = -0.314$ MPa^{-1}, $b_{xx} = 0.0172$ MPa^{-2}, $b_{yy} = 0.1120$ MPa^{-2} and $b_{xy} = -0.011$ MPa^{-2}. Reproduced with permission from The Pulp and Paper Fundamental Research Society (www.ppfrs.org).

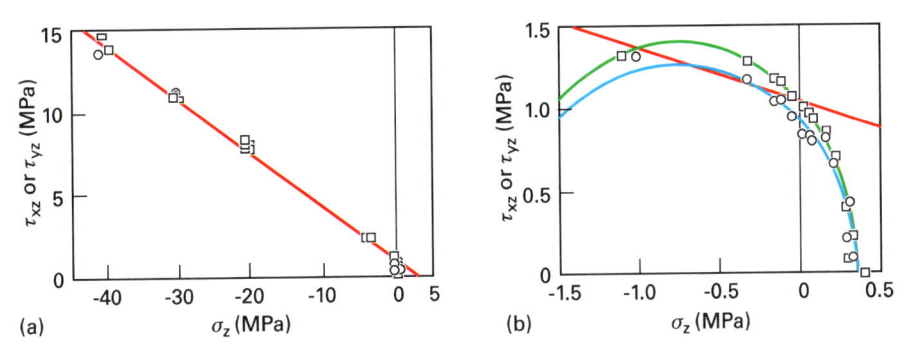

Fig. 2.18 Failure stress in out-of-plane loading of a paperboard for $\tau_{xz} = 0$ (circles) and $\tau_{yz} = 0$ (squares), compared with the Tsai-Wu criterion (Eq. 2.9) and Coulomb criterion (Eq. 2.10) at large and small normal stresses, (a) and (b), respectively (Stenberg, 2002). The parameter values are $a_z = 2.25$ MPa^{-1}, $b_{zz} = 1.48$ MPa^{-2}, $b_{xzxz} = 0.94$ MPa^{-2}, $b_{yzyz} = 1.16$ MPa^{-2}; and $\mu = 0.32$, $c = 1.05$ MPa. Reproduced with permission from the author.

2.5 Mechanical properties in relation to the papermaking process

2.5.1 Preparation of papermaking fibers

When paper is manufactured from wood, one usually starts by cutting the wood into chips (Fig. 2.19). The chips are then disintegrated in fibers either mechanically or chemically to prepare what is called pulp; depending on the process, the end product is *mechanical pulp fibers* or *chemical pulp fibers*. In the native wood, typical fiber dimensions are 1–3 mm in length and 20–40 μm in cross-sectional width. Different pulp fibers, other plant fibers, and man-made fibers is compared in Chapter 12.2.1.

Fig. 2.19 Pine wood chips prepared for papermaking. The grainy structure comes from the annual variation in wood growth. Chip length in the grain direction is 25 mm. Figure courtesy of Lisbeth Hellström, FSCN.

In a typical mechanical pulp manufacturing process, the wood chips are sheared between rotating steel plates. The pattern of the plate surface is designed to optimize the pulp quality. Steam is applied to soften the lignin that holds fibers together in the wood material, and also the cellulose of the fiber cell wall, so as to reduce fiber damage in the disintegration process. Pressure and chemicals may also be used for this purpose. Depending on the specific manufacturing process, different types of mechanical pulp are obtained, such as TMP (thermomechanical pulp), CTMP (chemi-thermomechanical pulp), and their variants. One can also start with solid wood and grind fibers of the wood surface, which leads to pulps such as GW (groundwood) and PGW (pressure groundwood). A thorough coverage of mechanical pulping can be found in Lönnberg (2009).

In the chemical pulping process, wood chips are cooked with chemicals to dissolve the lignin that holds fibers together in wood. Also, water-soluble hemicelluloses get extracted in the process. Depending on the chemicals used, different types of chemical pulps are obtained, such as the kraft pulp (also called sulfate pulp) and sulfite pulp (Gullichsen and Fogelholm, 1999). The cooking process leaves the structure of fibers rather intact (Fig. 2.20) except for the removal of lignin and hemicelluloses (see Chapter 12.2.1 for more information). A typical *yield* of the chemical pulping process is slightly more than 50%, meaning that about half of the original dry mass of wood is retained in the fibers (compared with mechanical pulping where the yield is usually more than 90%). In chemical pulping, the chemicals dissolved from the wood material have been traditionally used for energy. However, recently increasing development efforts have been directed to various biorefinery concepts that convert the extracted chemicals to renewable fuels of polymeric raw materials.

After the fibers are separated, both chemical and mechanical pulp can be bleached with chemicals to increase the whiteness of the final product. Then the pulp is treated further in a mechanical process called *refining* or *beating*, which increases the flexibility and conformability of the fibers and opens up the fibrillar structure of the fiber surface. This is necessary to achieve good bonding between the fibers so that the

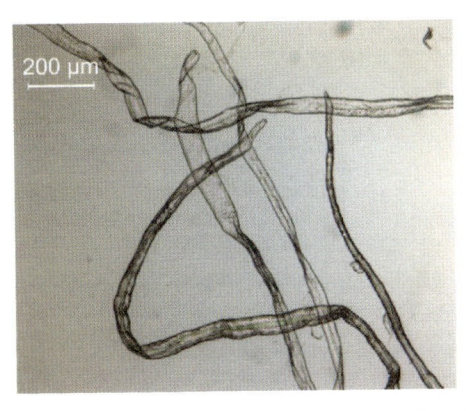

Fig. 2.20 Mildly refined chemical pulp fibers made of pine wood; courtesy of Boel Nilsson, SCA R&D.

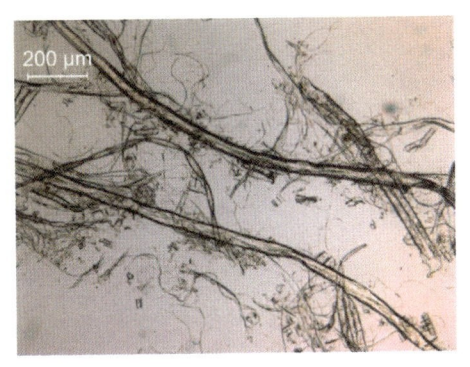

Fig. 2.21 Spruce wood mechanical TMP pulp fibers, fiber fragments, and fines; courtesy of Boel Nilsson, SCA R&D.

resulting paper has sufficient strength. Especially in the mechanical pulping process a large fraction of the fibers is damaged and broken into fragments (Fig. 2.21). Sections of the fiber wall structure break off as flat lamella and in narrow ribbon-like fibrils; these small fiber fragments are called *fines*.

The fragmentation of fibers in mechanical pulping is intentional. Smaller particles give a more uniform paper structure, reduce transparency, and improve the sharpness of print on the paper. Chemical pulping causes some fiber damage also, but not as much as in mechanical pulping. Thus, chemical pulp is generally stronger than mechanical pulp. Also, because lignin and some of the hemicelluloses are removed, chemical pulp does not turn yellow as easily as mechanical pulp when exposed to light or heat.

After the pulp manufacture, one adds other components, such as mineral fillers and chemicals, to obtain a water-based furnish ready for papermaking. Fillers increase the opacity and whiteness of paper. Chemicals are added to help retain the fines particles of pulp along with the fibers when water is removed on the paper machine ("retention aids"), to improve bonding between fibers ("bonding aids"), and to control ink penetration into the paper ("sizing"). Tuning of the pulp properties and furnish composition is the main method used to tailor paper properties.

2.5.2 Effect of the paper machine

Figure 2.22 demonstrates the structure of a paper machine, here cut in two parts to fit the figure on the page. The total length of a paper machine is typically a few hundred meters. The machine begins at top left with a forming section where the furnish at about 1% solids content is spread from a "headbox" (red in Fig. 2.23) on a moving wire. The low solids content is necessary so that fibers can be spread uniformly on the wire. Water is then removed by suction units (yellow) through a top and bottom wire and by wet pressing, where cylinders (dark green) press the wet paper web between two wires or felts. The forming and pressing sections together are called the "wet end" of the paper machine (see Paulapuro, 2008, for more information). When leaving the wet end, the solids content of the web is around 50%. Then the paper web is moved to the dryer section and pressed against hot cylinders so that water evaporates (red). A large number of dryer cylinders are needed because of the high speed of the process, which can reach 2000 m/min. Bonding between fibers forms spontaneously when water disappears from the web.

In the "dry" end of the paper machine (bottom half of Fig. 2.22), water suspensions of sizing and pigments can be spread on the web surfaces to improve paper appearance and performance in printing (the blue rolls on the left). Further drying is

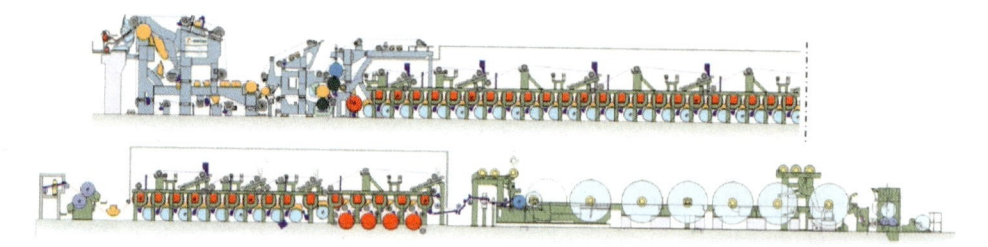

Fig. 2.22 Containerboard machine equipped with a gap former and two-layer headbox; courtesy of Metso Paper Inc.

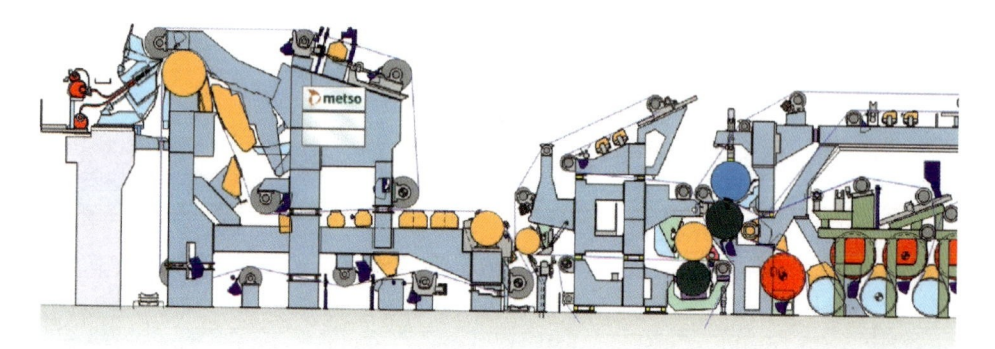

Fig. 2.23 Details of the forming section of the machine in Figure 2.22.

2.5 Mechanical properties in relation to the papermaking process

then needed before winding to paper rolls. There is no coating or calendaring in the papermaking line of Figure 2.22 because it is designed for containerboards that do not need high surface quality. In printing papers and packaging boards, one or several mineral coating layers can be used to maximize the product quality.

The initial forming section of the paper machine determines the network structure of fibers in the paper (Fig. 2.24). The fiber distribution is disordered but not completely random because the fibers have a tendency to form bundles or *flocs*. Furnish is diluted to less than 1%, and turbulence is induced in the headbox and on the wire to reduce the flocculation of fibers. Nevertheless, the mass distribution of paper always shows some variability, starting from a few centimeters down to fractions of a millimeter. At small length scales similar to the paper thickness (0.1 mm and higher), one has to consider the microscopic three-dimensional structure, which is also disordered (Chapter 11.2).

The nonuniform in-plane mass distribution of paper is called *formation*. It can best be seen with bare eyes in thin paper grades such as newsprint or some office papers. Aside from being a visual imperfection, formation can cause out-of-plane deformations to paper if the moisture content changes (Chapter 9), and in rare cases, it can reduce the strength properties of paper, such as creep resistance. The nonuniform paper formation also shows up in the orientation distribution of the fibers (Fig. 2.24) and the mass density, thickness, and porosity of paper from the microscopic to macroscopic. The nonuniformity of fiber orientation can accentuate eventual problems with out-of-plane deformations of paper because fiber orientation has a strong effect on the hygroscopic strains.

Fiber orientation in paper arises from the forming process. A small speed difference is generally needed between the wire and the furnish jet that comes from the headbox to enhance smooth spreading of the furnish and thereby reduce the nonuniformity in mass distribution. At the same time, the speed difference creates a shear field through the thickness of the furnish layer, which in turn rotates fibers more parallel to the machine direction. The anisotropy of fiber orientation is one of two factors that cause anisotropy in the mechanical properties and hygroexpansion

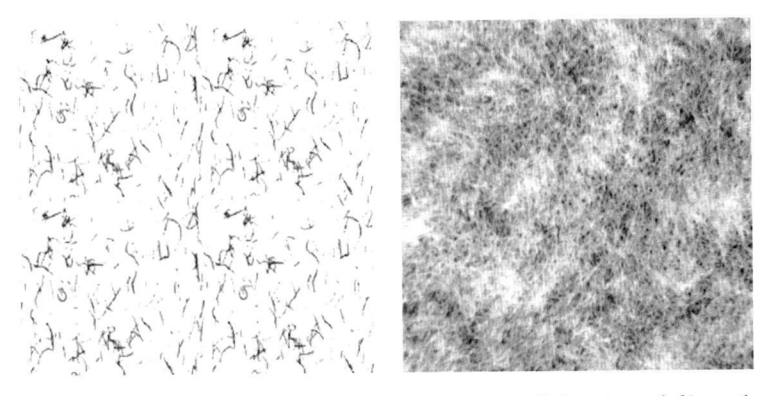

Fig. 2.24 Surface image of a paper sheet containing a small fraction of fibers dyed black (left), and a layer split from a sheet, showing fibers and fiber bundles (right, courtesy of Pekka Pakarinen).

of paper (Niskanen, 1993), the other being the drying effect discussed below. On a broad paper machine, slight CD deviations can occur in the flow direction of furnish so that the local symmetry axis of the fiber orientation distribution may be a few degrees off the machine direction. The nonzero fiber orientation angle can cause diagonal curl in products where paper or board is used in sheet form, and this can lead to, for example, paper jams in a copy machine.

The direction of initial water removal creates a z-directional profile of fines and filler concentration in paper. The smaller particles are flushed with the water, generally toward the wires. Because of the flushing, layers close to the wire surfaces can be depleted of fines and fillers. A ZD variation arises even in the fiber orientation distribution because the shear field in the suspension layer changes as the water removal progresses.

The wet pressing stage of a paper machine determines the thickness and density of paper. Intensive wet pressing is favorable for water removal and reduces the energy needed in the dryer section, but it also leads to a densification of the paper. Low paper thickness gives low bending stiffness, which is often a problem, for example, in the handling of paper sheets or in the strength of paperboard boxes.

The forming and wet pressing stages of the paper machine determine the structure of paper from centimeters down to the microscopic fiber network structure. More detailed information can be found in Chapter 1 of Niskanen (2008).

On the dryer section of the machine, the paper web shrinks because water is removed from the fibers. Earlier in the process, water is removed only from the pore space between fibers. At this stage of drying, a tension must be applied on the web to prevent fluttering and to improve contact with the drying cylinders. The drying tension on the paper machine prevents paper shrinkage in the MD, which occurs almost exclusively in the CD. Drying tension versus shrinkage is the second factor that influences the anisotropy in the mechanical properties and hygroexpansion of paper.

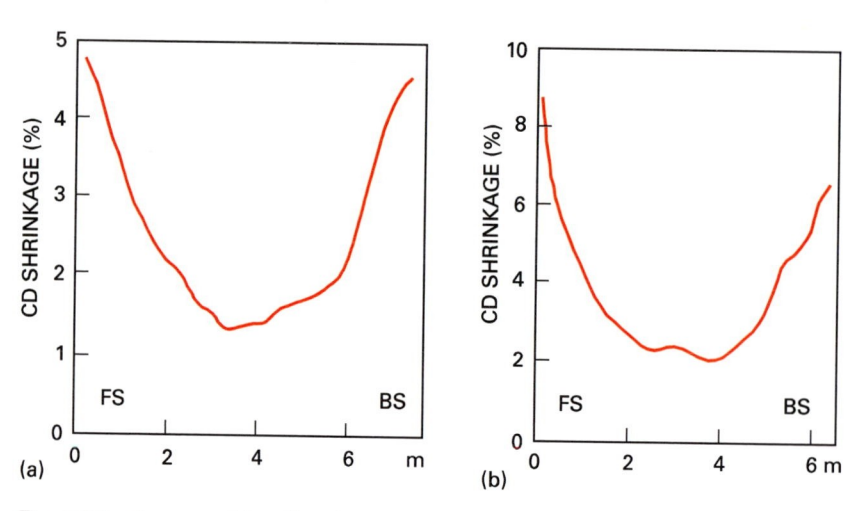

Fig. 2.25 Cross-machine direction shrinkage that occurred from the forming section to dry paper on a commercial newsprint paper machine (a) and office paper machine (b). Reproduced from Niskanen (1993) with permission from Paperi ja Puu Oy.

The CD drying shrinkage has a parabolic profile across the web (Fig. 2.25). The edges of the web shrink without any constraint, while in the center of the web shrinkage is partially prevented by adhesion or friction against the drying cylinders. This results in parabolic profiles of material properties and web tension. For example, the CD elastic modulus has the lowest values and the CD hygroexpansion has the largest values at the web edges. The mechanisms behind this variation are explained in Chapter 11.

References

Andersson, O., and Sjöberg, L. (1953). Tensile studies of paper at different rates of elongation. Svensk Pappertid. 56(16), 615–624.
Baum, G. A. (1987). The elastic properties of paper, a review. In: Design Criteria for Paper Performance, P. Kolseth, C. Fellers, L. Salmén, and M. Rigdahl, eds. 1984 seminar on progress in paper physics (Stockholm: Svenska Träforskningsinstitutet (STFI)-Meddelande A969), pp. 1–27.
Baum, G.A., Brennan, D.C., and Habeger, C.C. (1981). Orthotropic elastic constants of paper. Tappi. 64(8), 97–101.
Fellers, C. (1980). The significance of structure on the compression behavior of paper. Doctoral dissertation. Royal Institute of Technology, Stockholm, Sweden.
Fellers, C., Westerlind, B., and De Ruvo, A. (1983). An investigation of the biaxial failure envelope of paper. Presented at the British Paper and Board Industry Federation, "The Role of Fundamental Research in Papermaking Symposium," vol. 2, September 1981, Cambridge, UK (London: Mechanical Engineering Publications), pp. 527–559.
Girlanda, O., and Fellers, C. (2007). Evaluation of the tensile stress-strain properties in the thickness direction of paper materials. Nordic Pulp Pap. Res. J. 22(1), 49–56.
Gullichsen, J., and Fogelholm, C-J. (1999). Chemical Pulping (Helsinki, Finland: Fapet Oy).
Gustafsson, P. J., Nyman, U., and Heyden, S. (2001). A network mechanics failure criterion. Report TVSM-7128, Div of Structural Mechanics, Lund University.
Hillerborg, A., Modeer, M., and Petersson. P.-E. (1976). Analysis of crack formation and crack growth in concrete by means of fracture mechanics and finite elements. Cement and Concrete Research 6, 773–782.
Ketoja, J. A., Tanaka, A., Asikainen, J., and Lehti, S. T. (2007). Creep of wet paper. Presented at 61st Appita Annual Conference and Exhibition, Gold Coast, Australia, May 6–9. International Paper Physics Conference (Carlton, Australia: Appita).
Lönnberg, B., ed. (2009). Mechanical Pulping, 2nd Edition (Helsinki, Finland: Paperi ja Puu Oy).
Malvern, L. E. (1969). Introduction to the Mechanics of a Continuous Medium (Englewood Cliffs, NJ, USA: Prentice-Hall).
Mark, R. E., Habeger, C. H., Borch, J., and Lyne, M. B. (2002) Handbook of Physical Testing of Paper, Volume 1 (New York: Marcel Decker).
Niskanen, K. J. (1993). Anisotropy of laser paper. Paperi ja Puu 75, 321–328.
Niskanen, K., ed. (2008). Paper Physics, 2nd Edition (Helsinki, Finland: Paperi ja Puu Oy).
Paulapuro, H., ed. (2008). Papermaking, Part I Stock Preparation and Wet End, 2nd Edition (Helsinki, Finland: Paperi ja Puu).
Persson, K. (1991). Material model for paper; experimental and theoretical aspects. Diploma Thesis. Division of Solid Mechanics, Lund Institute of Technology, Lund, Sweden.

Salmén, L., Carlsson, L., de Ruvo, A., et al. (1984). Treatise on the elastic and hygroexpansional properties of paper by a composite laminate approach. Fibre Sci. Technol. 20(4), 283–296.

Stenberg, N. (2002). On the out-of-plane mechanical behavior of paper materials. Doctoral Thesis. Department of Solid Mechanics, Royal Institute of Technology, Stockholm, Sweden.

Tryding, J. (1996). In-plane fracture of paper. Doctoral Thesis, Division of Structural Mechanics, Lund University.

Tryding, J., and Gustafsson, P. J. (2000). Characterization of tensile fracture properties of paper. Tappi J. 83(2), 84–89.

Tsai, S., and Wu, E. (1971). A general theory of strength for anisotropic materials. J. of Composite Materials 5, 58–80.

Xia, Q., Boyce, M., and Parks, D. (1994). A constitutive model for the anisotropic elastic-plastic deformation of paper and paperboard. Int. J. of Solids and Structures 39, 58–80.

Yeh, K. C., Considine, J. M., and Suhling, J. C. (1991). The influence of moisture content on the nonlinear constitutive behavior of cellulosic materials. In: Proceedings of the International Paper Physics Conference (Atlanta, GA, USA: Tappi Press), Book 2, p. 695.

Part I
Structural strength

3 Packaging performance

Rickard Hägglund and Leif A. Carlsson

3.1 Introduction

Packaging is an important application of paper materials. The main purpose of packaging is to facilitate shipment of goods from the producer to the consumer. Packaging has many other important functions, such as protecting the packaged goods from hazards such as contamination in the distribution environment, facilitating transportation and storing of products, and carrying printed information and graphics. The design and development of packaging are often linked to the product to be shipped. The packaging design process starts by establishing mechanical loads, marketing aspects, shelf life, quality assurance, distribution environment, and legal, regulatory, graphic design, end-use, and environmental factors. The design criteria, packaging performance, completion time targets, resources, and cost constraints need to be established and agreed upon.

This chapter focuses on the performance requirements that relate to the strength of the package. An important example of such requirements is the strength of the container in compression (stacking strength) during high humidity exposure. The packaging must withstand the load during the specified time of storage. Packaging for cooler/freezer products must be able to support loads at low temperatures in a moist environment without failing.

Packaging may be categorized in one of three groups depending on its role in the distribution chain.

– Primary packaging, or consumer packaging, is the material that first envelops the product. This is usually the smallest unit of distribution or use, and it is the package in direct contact with the contents. Examples are carton board packaging of small items, a glass jar, or a plastic bottle. Carton board boxes are often referred to as folding boxes or simply cartons.
– Secondary packaging, or transport packaging, is outside the primary packaging and is often used to group primary packages together. One of the most common forms of secondary packaging is the corrugated board box.
– Tertiary packaging refers to materials used in bulk handling, warehouse storage, and transport and normally includes materials such as pallets and stretch film.

This chapter discusses mainly corrugated board packaging and its performance, but parallels are made with carton board packaging. The secondary packaging carries most of the load during transport, although in some systems the primary packaging supports significant load and requires strength.

The first commercial carton board box was produced in England in 1817. Folding cartons were introduced in the 1860s. They were shipped flat to save space, ready to be erected by customers when required. Mechanical die cutting and creasing was developed soon after. In 1915 the first milk carton was patented, and in 1929 machinery was developed for commercial production of cartons. In 1935 the milk carton was implemented in a dairy plant. The first corrugating machine, that is, the machine that makes corrugated board, appeared in the United States in the early 1900s. Until the year 1919, the majority of products were shipped in wooden crates and often transported by train. The corrugated box was relatively new, and the railroads had no experience in handling and transporting them. Railroad companies did not assume the liability for damage while shipping items in corrugated boxes until Rule 41 was established in 1919. In the original version of Rule 41, the most important requirement for the box was that it contain and protect the shipped goods. As a result, the box materials and structures were designed primarily for burst and puncture resistance, which are measures of the tensile strength and ability to resist penetration. Later, World War II contributed to establishing corrugated board packaging when it was called upon to deliver goods to all corners of the earth. After World War II the market for corrugated board expanded rapidly, and the range of sizes and capabilities of packaging grew to fit the large number of new products developed.

Today, corrugated paperboard is the most popular material for transport packages for a wide variety of products, varying from fresh fruit and vegetables, consumer electronics, and industrial machinery to semi-bulk transports of various commodities in large bins. The global corrugated board market in 2009 was 77 million tons with a value of almost $82 billion, and the market is forecast to grow significantly in the coming years (PIRA, 2011). More than 60% of corrugated board is used to package nonfood products. The largest single end-use sector is the electrical goods market. It is equally suitable for any mode of transport, for example, shipping by sea, railroad, truck, or air. This versatility is largely due to the possibility of combining different material and thereby accommodating particular requirements in the distribution system in a tailor-made way. Corrugated board can be recycled, and currently more than 80% is.

3.2 Paper-based packaging materials

Packaging paper materials are made from cellulose fibers produced either from virgin wood or recycled fibers, or both. The paper surface is sometimes coated in order to enhance printability. The paper is often combined with polymers or metal-foils to form laminates for packaging of products such as juices and dairy products. This section describes the structure and manufacture of corrugated board and the process of converting corrugated board into boxes. Carton board materials are also discussed. The manufacturing process of carton board differs from that of corrugated board, but the converting process of the carton board is similar.

3.2.1 Corrugated board

Corrugated board is a sandwich construction with a web core and face sheets made from paper. *Container board* is the common name for the paper materials used to manufacture corrugated board, and includes linerboard, used for the facings, and fluting, which is the paper used in the core. The face sheets and core are typically glued together with a starch-based adhesive. The main function of the core is to separate the face sheets in order to achieve a structure with high bending stiffness. The core must also provide shear transfer between the face sheets to minimize sliding deformation during bending. Further, the corrugated board should be stiff and strong in the out-of-plane direction in order to keep the face sheets apart and parallel at the correct distance during in-plane and transverse normal loading. A weak core may fail due to its inability to support the face sheets against local buckling or wrinkling.

Corrugated board is made of a number of layers depending on the packaging requirements. The single-faced corrugated board consists of two layers: one linerboard and the corrugated fluting layer (Fig. 3.1). The single-wall corrugated board is a true sandwich consisting of three layers: two linerboards and the corrugated fluting core. The double- and triple-wall corrugated boards consist of two and four additional layers, respectively. About 80% of corrugated board is produced as single-wall board.

Figure 3.2 provides the geometrical parameters of the corrugated sandwich panel. In this case the board is symmetric, but asymmetric board configurations are also used. The core has a sine wave shape. The main dimensions of a corrugated board are its thickness d_{corr}, as well as the wavelength λ_{core} and height h_{core} of the flutings. The cross-machine direction CD of the face sheets and core is oriented parallel to the flutes.

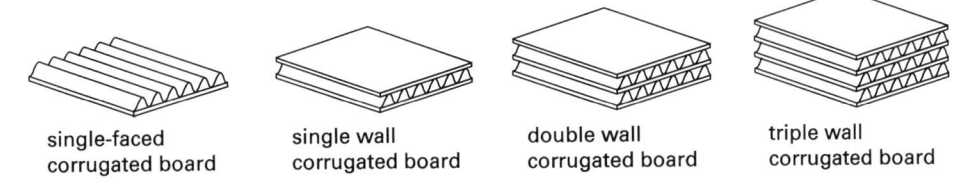

Fig. 3.1 Single wall, double wall, and triple wall corrugated board.

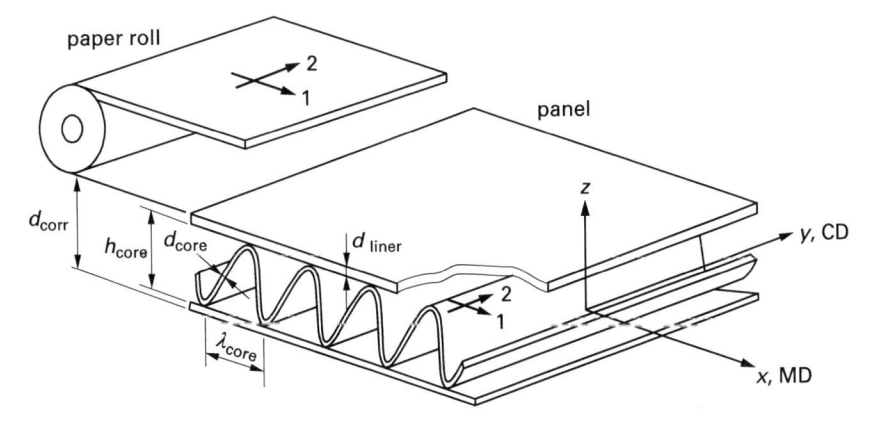

Fig. 3.2 Geometric parameters of a corrugated board. The wavelength and height of the fluting waves in the core are λ_{core} and h_{core}; the thicknesses of the fluting and linerboard are $d_{fluting}$ and d_{liner}, and the total thickness of the corrugated board is $d_{corr} = 2\, d_{liner} + h_{core}$.

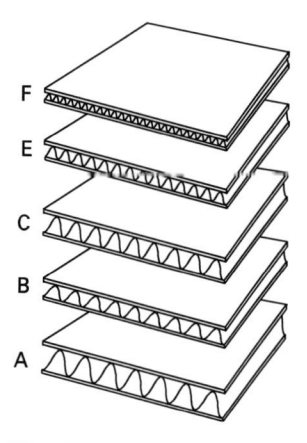

profile	wave length λ_{core} (mm)	flute height h_{core} (mm)
F	2.3 - 2.5	0.7 - 0.8
E	3.2 - 3.6	1.0 - 1.8
C	7.1 - 8.3	3.2 - 3.9
B	6.1 - 6.9	2.2 - 3.0
A	8.3 - 10.0	4.0 - 4.8

Fig. 3.3 Wavelength and height intervals of common flute profiles.

Figure 3.3 lists the most common commercial single-wall corrugated boards. The letter designations have been assigned in the order the different geometries were introduced to the market, not in the order of size. The wave length and flute height are not defined exactly, and may vary between production sites. Boards with smaller flutes have more flutes per unit length. Generally, large flute height improves box compression strength because it gives high bending stiffness of the board. Small flutes are used mainly for high-quality print and when the requirement on box strength is lower. E- and F-flutes are often used as a replacement for carton board. C-flute is typical in conventional transport packages where compression strength is required. B-flute is used, for example, for canned goods and other products not requiring high compression strength. A-flute is used as cushion pads or in heavy double- or triple-wall boards.

Many package styles and structural designs are possible and often specified using the international standard of the Federation of Corrugated Board Manufacturers (FEFCO, 2011). One of the most common boxes is the Regular Slotted Container (RSC) denoted FEFCO 0201 (Fig. 3.4). It is manufactured from a single sheet of corrugated board that is scored and slotted to permit folding. Flaps attached to the side and end panels become the top and bottom panels of the box. A flap is defined by the area from the edge of the sheet to the flap folding lines. Opposite pairs of flaps are long enough so that they can close the bottom and top of the box.

Corrugated board is manufactured in a continuous operation where the fluting is first plasticized by hot steam, then fed between a pair of sine-wave shaped gears to form the wavy web that is finally glued to the face sheets (Kuusipalo, 2008). The direction of manufacture in Figure 3.5 is from left to right: (1) The fluting unwinds from the reel stand and is heated and moistened with steam. (2) The fluting then passes between a pair of hot gears to get its sine-wave shape. (3) Glue is applied onto the flute tips that are glued to the preheated liner (single facer) to form a single-faced web. (4) This is then heated again, and glue is applied on the opposite flute tips. (5) The second liner (double backer) is heated and bonded to the single face web

Fig. 3.4 A Regular Slotted Container (RSC).

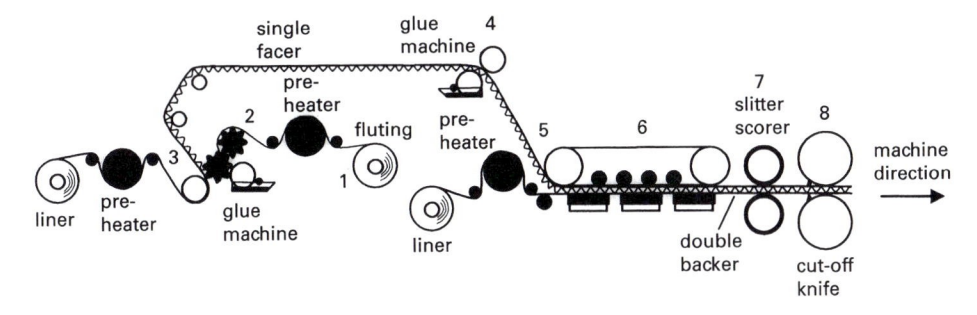

Fig. 3.5 The manufacturing process of corrugated board.

to form a corrugated board sandwich panel. (6) The board is dried under pressure between hot plates. For double-wall and triple-wall boards, the process is repeated, normally in line with the original single back operation. (7) Finally, the corrugated board is cut in the MD and CD according to the package design.

3.2.2 Box manufacturing process

The converting of corrugated board refers to all the process stages that transform the corrugated board into the final corrugated container (Kuusipalo, 2008). Typically this involves printing, slotting, creasing, die-cutting, folding, and gluing. Printing can be post-printing, done directly to the corrugated board, or pre-printing, in which case the liner is printed before it is assembled into the corrugated board. Most of the RSCs are produced with one in-line operation, that converts a flat sheet into its final form, ready to be shipped to the customer.

Die-cutting is used when the package demands very precise cutting or has a complicated shape. The die-cut tool may also be equipped with creasing rules (dies) that

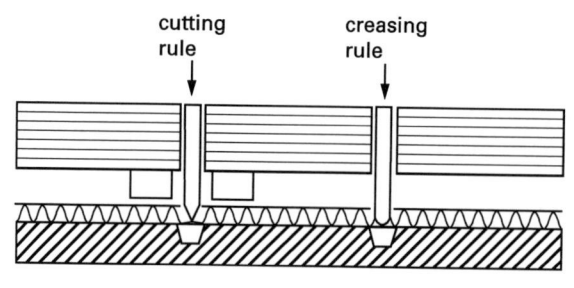

Fig. 3.6 Flat-bed die-cutting and creasing.

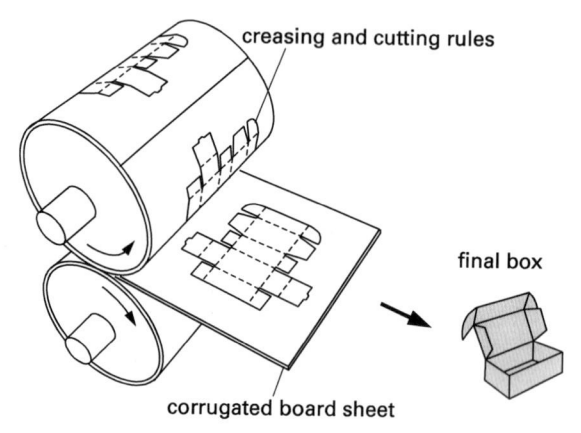

Fig. 3.7 A rotary cutting and creasing tool.

make scores to the board to define folding lines in the box. The principles of flat-bed cutting and creasing are shown in Figure 3.6. In rotary die-cutting (Fig. 3.7), the cutting and creasing rules are mounted on the surface of a wooden cylinder, enabling continuous operation.

In the manufacture of the corrugated board, the material is subject to various lateral loads in, for example, feed roller and printing nips. Depending on board properties such as thickness and strength, the die-cutting process may damage the corrugated board, (Cavlin and Edholm, 1988). As the male die (Fig. 3.6) penetrates the board, tensile, compressive, and shear stresses are created in the board. Shear deformation in particular may fail the bond between layers in the board. This delamination alters the strength properties of the creases (discussed in more detail in Chapter 4). If the crease is too deep, the tension may break the upper liner. On the other hand, if the crease is too shallow, then the bottom linerboard may break during the box folding phase (Isaksson and Hägglund, 2005; Hägglund and Isaksson, 2008; Thakkar et al., 2008).

3.2.3 Carton board

Carton board is the common name for paper used in packaging cartons. The material consists of three or more furnish layers manufactured simultaneously on a multilayer paperboard machine (see, e.g., Fig. 2.22). Cardboard may be coated with polymers to achieve a material that can be used in ovens, microwaves, and other demanding conditions, or it may be laminated with metal films to enhance appearance and protect the content. According to the composition and thickness of the layers, the main types of carton boards are (Confederation of European Paper Industries [CEPI], 2011) as follows:

- Solid bleached board (SBB) is typically made from pure bleached chemical pulp with two or three layers of coating on the top and one layer on the bottom. It is a medium density board with good printing properties for graphical and packaging end uses, and it is perfectly white throughout the thickness. It is used in such markets as cosmetics, graphics, pharmaceuticals, tobacco, and luxury packaging. It can also be combined with other materials for liquid packaging applications.
- Solid unbleached board (SUB) is typically made from pure unbleached chemical pulp with two or three layers of coating on the top surface. In some cases a white bottom surface is applied. SUB is primarily used in packaging of beverages such as bottles and cans because it is very strong and can be made resistant to water. It is also used in a wide variety of general packaging areas where strength is important.
- Folding boxboard (FBB) is typically made from a number of layers of mechanical pulp sandwiched between two layers of chemical pulp with up to three layers of coating on the top or printing surface and one layer of coating on the bottom. This is a low density material with high bending stiffness with a yellowish color on the inside. It is used in markets such as pharmaceuticals, frozen and chilled foods, and confectionary.
- White lined chipboard (WLC) consists predominantly of recycled pulp. WLC typically has three layers of coating on the top printing surface and one layer on the reverse. Because of its recycled content, the back side is grey. WLC is used in a range of applications such as packages for frozen and chilled foods, breakfast cereals, shoes, tissues, and toys.

The main types of cartons are tube and tray. In a tube the machine direction of the board runs horizontally around the package in order to maximize resistance against bulging of the side panels. The European Carton Makers Association (ECMA) Code of Folding Carton Design Styles is a reference standard for folding carton package designs (ECMA, 2011). The process of converting carton board to packaging is analogous to the converting of corrugated board discussed previously, except that the carton board material is continuous. The carton board material behavior is discussed in detail in Chapter 4.

3.3 Loads imposed on boxes

Boxes are subject to a number of different loading conditions in the filling, stacking, transportation, and storage operations. The loads are especially important for corrugated boxes that are used as secondary packages to support loads and protect

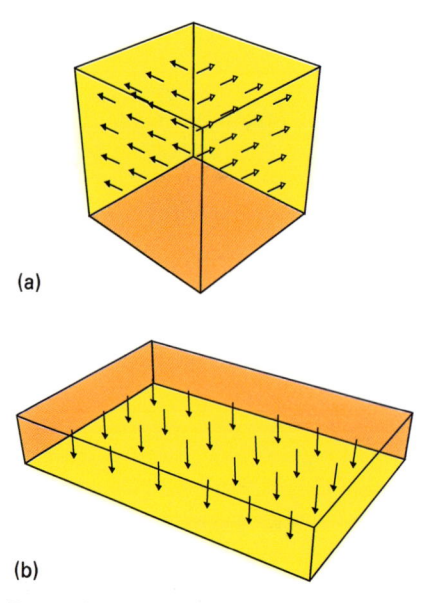

Fig. 3.8 Internal pressure on box walls (a) and tray bottom (b).

primary packages, which in turn typically are not supporting loads. The loads discussed in this section, however, should apply also for primary packages.

The package content imposes internal loads that affect the service-life of the box. When a box is filled with a fluid or a granular material, pressure exerts on the box walls (Fig. 3.8a). The pressure caused by a fluid increases linearly with depth. The pressure caused by a granular material grows nonlinearly, reaching asymptotically a maximum value. The pressure causes the box panels to bulge. The amount of bulging is governed by the bending stiffness and dimensions of the box panels. Bulging has a negative impact on the load carrying capability of the box. Excessive bulging may also cause undesirable contacts with surrounding objects and reduce the perceived quality of the box. In trays, the weight of the content acts on the bottom (Fig. 3.8b), causing the bottom panel to sag. In tight stacks, this may cause problems if the bottom of a tray gets in contact with the contents of the tray below.

Boxes are often stacked on a pallet (Fig. 3.9). The boxes at the bottom are thus subject to a top-to-bottom compressive loading by the overlying boxes. If the content is rigid and of similar dimensions to the box, the content may carry part of the load, increasing the load carrying capability. The distribution of loads on the boxes depends on the pattern in which the boxes are stacked on the transport pallet. Kellicutt (1963) found that the column stacking pattern in Figure 3.9a gives maximum strength utilization because forces are transferred mainly along the perimeters of the boxes. Interlock stacking (Fig. 3.9b) provides better stability, but the load carrying capability is significantly less because rigid corners rest on flexible panel centers.

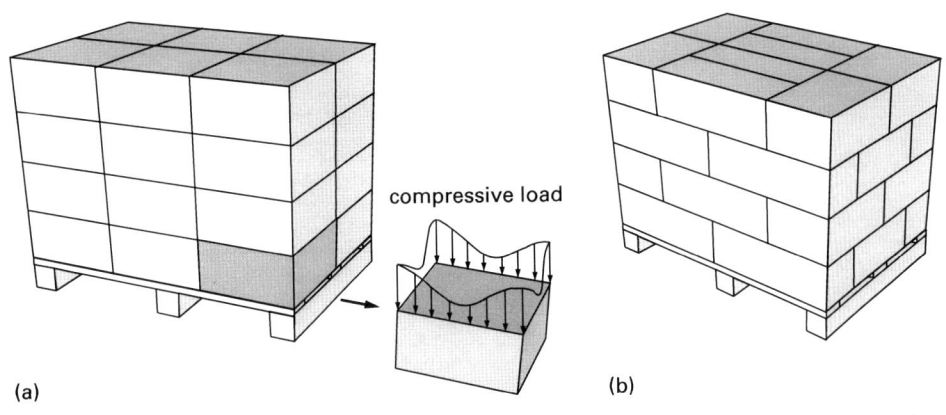

Fig. 3.9 Column stacking (a) and interlock stacking (b), with the top-to-bottom compressive load on a box shown in (a).

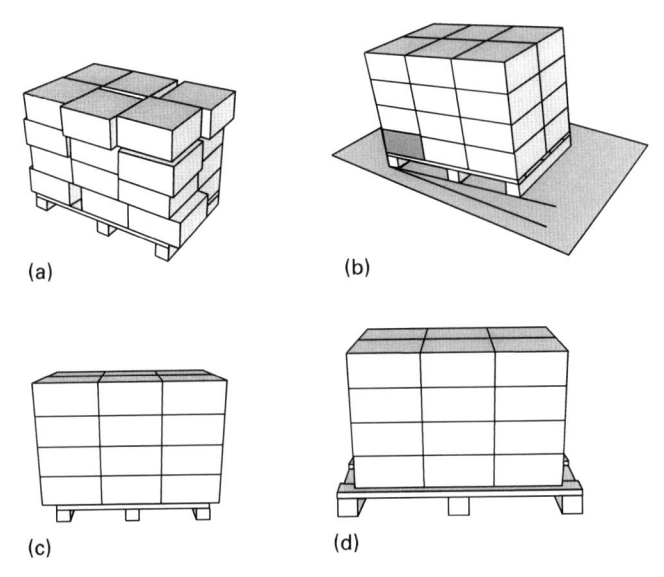

Fig. 3.10 Offset in column stacking (a), inclined stack (b), pallet "overhang" (c), and pallet "underhang" (d).

Deviations from perfect stacking reduce the load carrying capability. Misalignment of the stack by a few centimeters (Fig. 3.10a) or a tilt of a few degrees (Fig. 3.10b) can reduce the load carrying capacity significantly. The same is true with overhangs, when the pallet is smaller than the stack area (Fig. 3.10c). Underhangs (Fig. 3.10d) do not influence strength, but they occupy more space.

3.4 Strength of boxes

Knowledge about the compression loading of corrugated board panels and structures has increased considerably in the last decades (Maltenfort, 1956; McKee and Gander, 1957; Westerlind and Carlsson, 1992). This section describes the strength of corrugated board and carton board boxes under compressive loading.

3.4.1 Short-term compressive loading

The short-term compression strength is the single most important property specification of a corrugated board box. It is a direct measure of a box performance in a stack. The standardized Box Compression Test (BCT) measures the failure load. BCT is a pure top-to-bottom compression load test between flat parallel steel plates that is carried out on an empty sealed corrugated board box using a constant deformation speed (Fig. 3.11). The compressive load and crosshead displacement are recorded continuously until collapse occurs. The maximum load attained is reported as the BCT value. Testing is conducted at a constant 23°C temperature and 50% RH.

Several parameters are used to characterize the mechanical performance of the corrugated board material. The most important properties are the Edge Crush Test (ECT) and the bending stiffness. The ECT value is a measure of the compressive strength of corrugated board in the direction of the flutes, while bending stiffness is a measure of resistance to bending. Similarly, the most important mechanical properties for the strength of carton board packaging are the bending stiffness of the carton board and its compressive strength. The compressive strength of carton board is measured at a short distance between the grips, in order to avoid buckling of the specimen. This test is referred to as Short Span Compression Test (SCT).

The manner in which a corrugated box fails in compression depends on the board constituents and the box dimensions. Buckling and compression failure are the two

Fig. 3.11 Box Compression Test (BCT).

principal failure modes of a corrugated board box. Usually the boxes fail through a post-buckling collapse that occurs after buckling of the panels (Fig. 3.12a). This failure mode is particularly common for tall boxes and boxes made of relatively thin board of low bending stiffness. Typically two of the four panels bow outwards and the other two inwards. Boxes that are short and very stiff in bending do not buckle and fail by forming horizontal wrinkles (Fig. 3.12b). The failure of carton board boxes is dominated by the corners as discussed in Chapter 4.

Figure 3.13 shows the load-deformation responses of the boxes in Figure 3.12. The initial stiffness is very low due to the creases. The shallow box is stiffer and stronger than the tall box. Once the faces of the tall box have buckled, the corners carry most of the applied load. Buckling occurs at approximately two-thirds of the maximum load for a tall box. If the fluting is strong and stiff enough to support the linerboards, then the failure typically initiates on the concave side linerboard (facing), close to the corners (Fig. 3.14). The shallow box that does not buckle has a fairly even load distribution along its perimeter at failure, which explains its higher strength.

It is necessary to distinguish between local and global buckling. Local buckling occurs when the linerboard buckles between the flute crests. Global buckling occurs when the entire panel buckles. At collapse the local buckles coalesce into a wrinkle that starts near a corner and runs diagonally toward the center of the panel (Fig. 3.14). Figure 3.15 illustrates the evolution of local strain distribution on the concave side of a panel. Figure 3.16 shows the corresponding load-displacement curve

Fig. 3.12 Compressive failure of a C-flute corrugated board box of height 400 mm (a) and 70 mm (b).

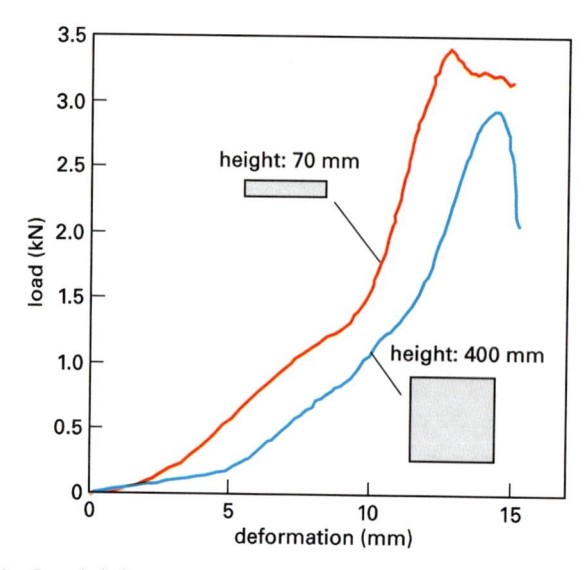

Fig. 3.13 Load-deformation response in compression of the C-flute corrugated board boxes shown in Figure 3.12.

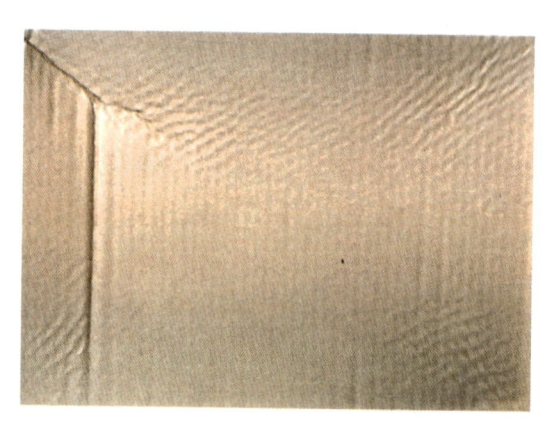

Fig. 3.14 Local buckling and wrinkling pattern on the concave side of a collapsing panel of a corrugated board box.

of the panel under in-plane compression. The strain field is initially fairly uniform, but then a diagonal strain pattern emerges, and finally a wrinkle is formed after which the board fails.

The compression test results shown in Figures 3.15 and 3.16 refer to panel tests because box compression tests are relatively costly and involve a number of uncontrolled variables that result from the creasing and folding operations. Therefore, it has become customary to conduct compression testing of panels to estimate the strength of corrugated board boxes. The test rig in Figure 3.17 simulates the

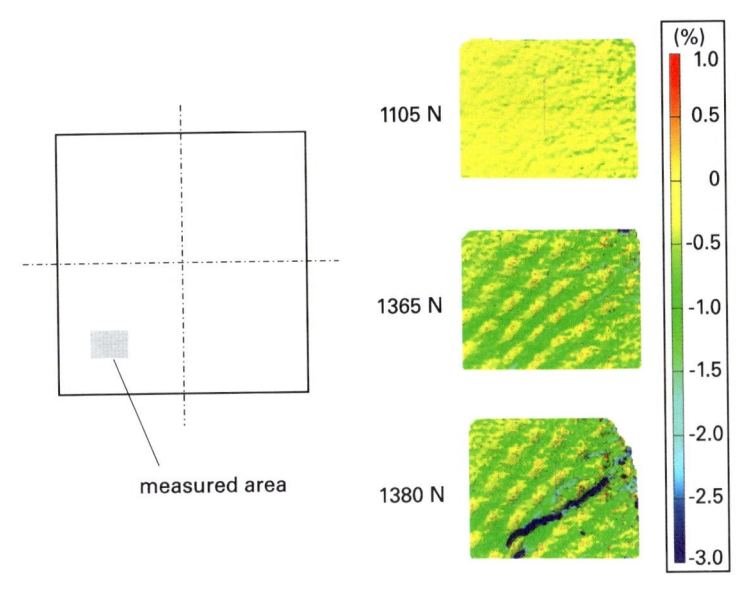

Fig. 3.15 Strain distributions in a 60 × 40 mm² area on the concave side of a 400 × 400 mm² C-flute panel under compressive loading, measured using digital image correlation. The three load levels are indicated on the load-displacement curve shown in Figure 3.16.

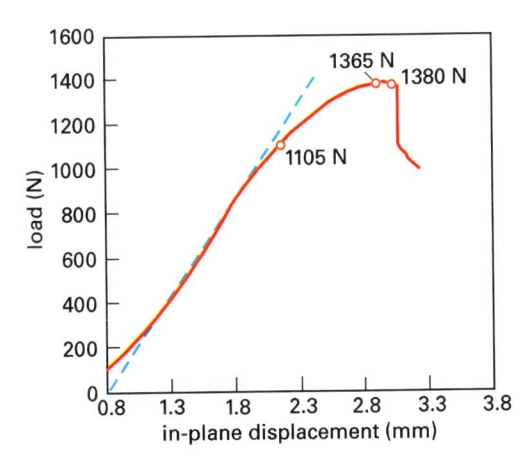

Fig. 3.16 Load versus in-plane displacement in an edge-wise loaded C-flute panel.

boundary conditions and loading of the panels in a corrugated board box under top-to-bottom loading. Out-of plane deformations are prevented along the perimeter, but rotations are allowed. In a panel compression test the in-plane deformations are generally much smaller than in a box because stronger constraints are imposed on the edges in the panel test.

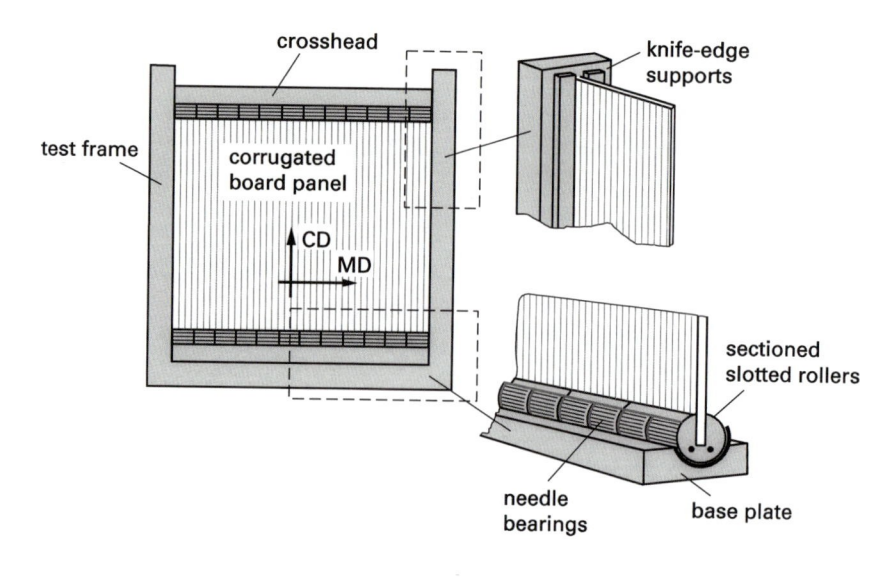

Fig. 3.17 A panel compression test rig (Nordstrand, 2004). Reproduced with permission from the author.

3.4.2 Models for static box strength

Models of box strength provide guidance for package design and optimization. Although corrugated board has been used for more than a century, the design process of packages is still mostly empirical or semi-empirical and based on box testing, which is time-consuming and expensive. Finite element simulations of box loading offer a modeling strategy that requires only the dimensions of the box and the mechanical properties of the face sheets and core.

Before World War II, engineers at the British Air Ministry had established an empirical relationship between the collapse load of a panel and its buckling load and material compression strength. This formula is often called the Cox formula (Cox, 1933). McKee et al. (1963) developed a design formula (3.1) for corrugated board boxes based on the Cox formula and published the classical article "Compression Strength Formula for Corrugated Boxes". This formula incorporates an empirical relation between the bending stiffness of the corrugated board, its geometry, and the material properties of the fluting and the linerboard, and the box perimeter.

$$\text{BCT} = a \cdot \text{ECT}^b (\sqrt{S_{\text{corr, MD}} S_{\text{corr, CD}}})^{1-b} Z^{2b-1}, \quad (3.1)$$

where a and b are empirical constants, Z is the perimeter of the box, and $S_{\text{corr, MD}}$ and $S_{\text{corr, CD}}$ are the bending stiffnesses of the corrugated board in MD and CD. The Edge Crush strength (ECT) is the compressive strength of the corrugated board in CD. Both the bending stiffness and edge crush strength can be estimated from the fluting and linerboard properties and the corrugated board dimensions (Patel, 1996). The model is simple to use and often gives satisfying accuracy for boxes when all four walls have similar dimensions. The McKee formula has several limitations. It underestimates BCT of shallow boxes and other boxes that do not buckle. It is inaccurate for boxes with panels of very different size, or when one linerboard is significantly heavier than the other. Extensive box testing is required for calibrating the parameters a and b for new board types, and the model cannot predict the optimum balance between linerboard and fluting.

For carton board boxes it has been shown that the compression strength correlates well with the strength of its panels. Based on this observation and the theory for post-buckling collapse of edge-loaded simply-supported plates, Grangård (1970) derived the following formula for box strength,

$$P_{\text{collapse}} = k \cdot \sqrt{\text{SCT}} \cdot \sqrt{S_{b, \text{MD}} S_{b, \text{CD}}}, \quad (3.2)$$

where P_{collapse} is the ultimate strength of the panel, SCT is the strength value obtained from the Short Span Compression Test, $S_{b, \text{MD}}$ and $S_{b, \text{CD}}$ are the bending stiffnesses in the MD and CD, and k is a constant that depends on the panel dimensions, and geometry. Thus, k is usually fitted to experimental data (Fig. 3.18). As the McKee

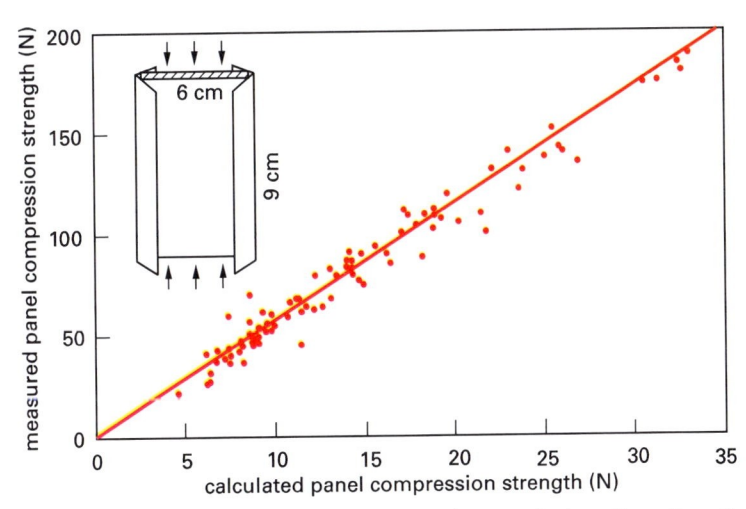

Fig. 3.18 Measured panel compression strength against predictions from Equation 3.2 with $k = 1$. The regression fit implies that $k = 5.7$. Reproduced from Fellers et al. (1983) with permission from Innventia AB.

formula for corrugated board, Equation 3.2 is limited to boxes with panels without any cut-outs.

3.4.3 Finite element models

Finite element models of corrugated board boxes may be conducted in a number of different ways (Pommier et al., 1991; Patel, 1996; Rahman, 1997; Gilchrist et al., 1999; Urbanik and Saliklis, 2003; Nordstrand, 2004; Nyman, 2004). In this section, we compare a highly detailed and refined model of corrugated board with a homogenized model that has fewer elements (Allansson and Svärd, 2001). The corrugated board considered is a 400 × 400 mm^2 C-flute that is simply supported and subject to edge-wise compressive loading (Fig. 3.19).

The refined model describes the corrugated structure of the board in detail. Each of the fluting and linerboard layers is modeled by 4-node quadrilateral rectangular isoparametric shell elements (Bathe, 1982). *Isoparametric* means that the element can take the form of a distorted rectangle. The elements are flat, which makes the corrugated fluting core piece-wise linear in a side view (Fig. 3.20). The facings and core are

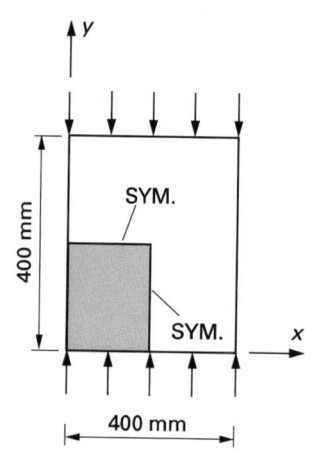

Fig. 3.19 Geometry of the corrugated board panel analyzed using both the refined model and the homogenized model of Allansson and Svärd (2001). Only one-quarter of the panel is considered due to symmetry. For each symmetry section, one translation and two rotations are allowed. Reproduced with permission from the first author.

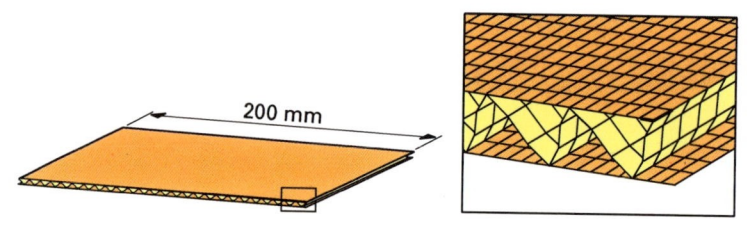

Fig. 3.20 Details of the refined model of Allansson and Svärd (2001). Reproduced with permission from the first author.

connected by common nodes at the fluting crests. Allansson and Svärd (2001) found that an even distribution of six elements per wave length in each layer was the coarsest mesh that can produce reliable results. This model has 73,008 elements. A nonlinear solution scheme for large deformations and rotations is necessary for a realistic analysis of stress state and deformations. Besides geometric parameters, all the elastic constants of the constituent paper layers are required (Chapter 2.2). Several of these are not straightforward to measure because of the small thickness of paper. Sometimes parameters can be approximated using engineering estimations (Mann et al., 1980; Baum et al., 1981).

In the homogenized model, the corrugated board is transformed into an equivalent homogenous layered structure (Fig. 3.21) using homogenization (Patel, 1996) and laminate theory (Agarwal and Broutman, 1990). The structural stiffnesses of the core are given by the following effective engineering constants:

$$E_x^{\text{eff}} \cong 0, E_y^{\text{eff}} \cong \alpha_{\text{take-up}} \cdot \frac{d_{\text{liner}}}{h_{\text{core}}} \cdot E_y^{\text{fluting}}, \nu_{xy} \cong 0, G_{xy} \cong 0, \tag{3.3}$$

where $\alpha_{\text{take-up}}$ is a geometric take-up factor (the length ratio of the fluting paper web before and after fluting), E_y^{fluting} is the elastic modulus of the fluting material in CD, d_{liner} is the thickness of the linerboard, and h_{core} is the height of the fluting waves (Fig. 3.2). The 4-node quadrilateral isoparametric shell elements consist of three plies. Each ply has its own effective elastic moduli, Poisson ratios, and shear moduli. The panel is divided into 2,500 elements.

Allansson and Svärd (2001) assumed that the linear elastic expression for in-plane stress (cf. Eq. 2.4) applies to both the refined model and the homogenized model. Both models predict global buckling of the panel. Figure 3.22 shows the load-deformation response of the models along with experimental results. It was then assumed that the corrugated board failure is triggered by failure of the linerboard facing on the concave side. The Tsai-Wu failure criterion (Tsai and Wu, 1971; Eq. 2.8) was used to predict failure. Figure 3.23 shows the resulting Tsai-Wu stress index f^{TW} for both the detailed and homogenized models. One can see that both models give similar stress distributions. However, the refined model reveals local buckling of the face sheets between the fluting crests. Failure was predicted at a corner of the panel, consistent with the experimental result in Figure 3.14.

The refined model is well-suited for the analysis of complex failure mechanisms that typically occur when the constituent papers are thin. It can predict the influences of paper thicknesses and elastic properties. The homogenized model is an efficient

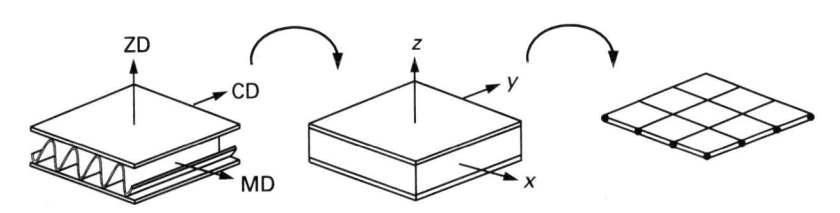

Fig. 3.21 Homogenization of the corrugated core.

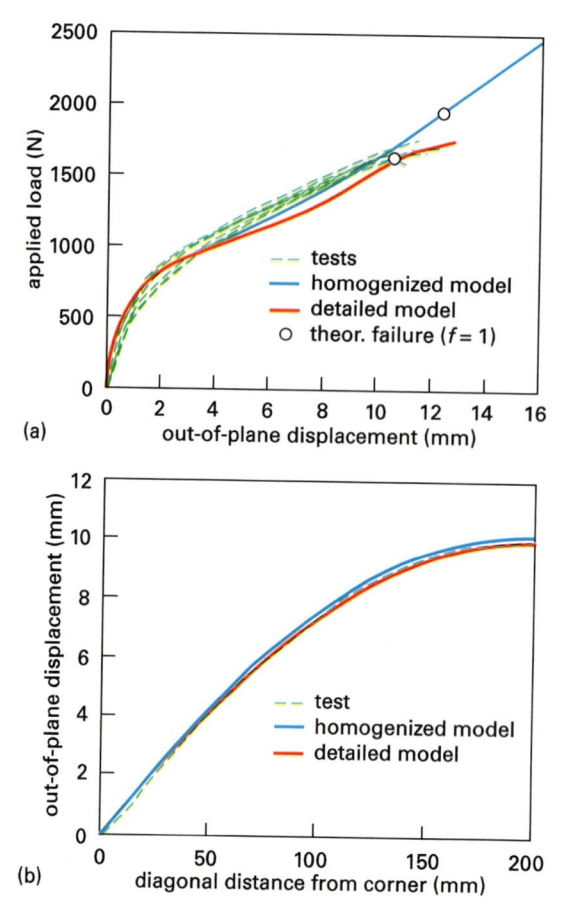

Fig. 3.22 Applied load against out-of-plane displacement (a), and out-of-plane displacement against position along the diagonal (b) of the panel in Figure 3.19 (Allansson and Svärd, 2001). Reproduced with permission from the first author.

alternative when the amount of local buckling is limited. Such a model predicts the ultimate failure with satisfying accuracy (Nordstrand, 2004), but it cannot be used to study the influence of the fluting properties. The Tsai-Wu stress index plot (Fig. 3.23) is useful for the box design. For example, cut-outs like hand-holds and ventilation holes should be positioned in areas where the Tsai-Wu stress index is small.

Carton board has a simpler structure than corrugated board and is well-suited for finite element analysis. Beldie et al. (2001) analyzed a carton board box under compressive loading using a three-layer laminate model. The outer layers of the board were thin and stiff, while the middle layer was thick and of low stiffness. Each layer was assumed to be elastic-plastic orthotropic material, described by Hill's orthotropic yield criterion (Hill, 1950). The resulting stress distribution (Fig. 3.24) shows that failure occurs at the upper corners, consistent with experimental observations, but the finite element model overestimates both the stiffness and ultimate strength of the box (Fig. 3.25).

3.4 Strength of boxes | 47

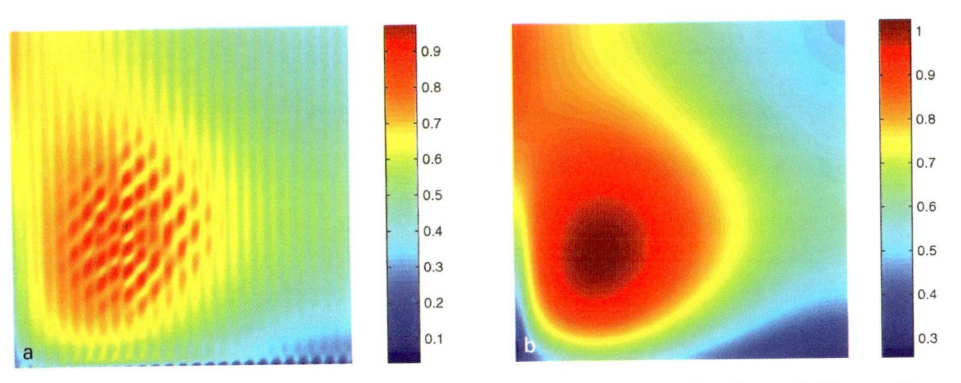

Fig. 3.23 Tsai-Wu stress index f^{TW} on the concave side of the panel in Figure 3.19 according to the refined model (a) and homogenized model (b) (Allansson and Svärd, 2001). The applied load and the out-of-plane displacement in the top right hand corner (center of the panel) are 1630 N and 10.6 mm in (a) and 1950 N and 12 mm in (b). Reproduced with permission from the first author.

Fig. 3.24 Stress field calculated for a carton board package under compressive load (Beldie et al., 2001). Reproduced with permission from John Wiley & Sons.

3.4.4 Long-term loading

One of the most significant environmental factors that make paper packaging weaker is exposure to high humidity or water. This is due to the hygroscopic nature of paper. In corrugated packaging, water not only makes paper softer (Chapter 2.3.2) but also weakens the water-based glue joints keeping the box panels together. An increase in ambient humidity has a dramatic effect on the strength of a box. Exposure to 85% RH may reduce the strength of a corrugated board box by 40% compared to exposure to 50% RH. Box performance may be enhanced by using moisture-resistant

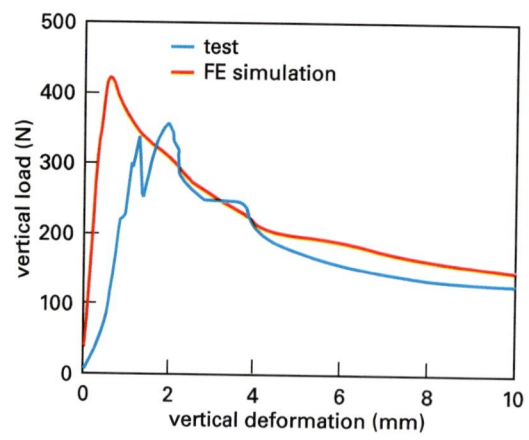

Fig. 3.25 Vertical load versus vertical deformation of the carton board package in Figure 3.24, according a measurement (blue) and finite element calculation (red; Beldie et al., 2001). Reproduced with permission from John Wiley & Sons.

papers, waterproof adhesives, and wax or plastic coatings that make the board surface water repellent. The effects of moisture are especially important when loading occurs over long times, as is typical in transportation and storage. Corrugated packaging sometimes fails during transportation and storage.

The gradual deformation of paper when loaded over long times is called *creep*. If loading is maintained, creep continues until the package breaks. The worst case is when humidity varies cyclically (Bronkhurst and Riedemann, 1994). In this case, the creep is accelerated (Fig. 3.26). Consequently, if the ambient conditions vary in a warehouse, stacked boxes may collapse prematurely. For engineering purposes, therefore, it is important to be able to determine the allowable load that does not cause collapse within the foreseen storage time. To characterize box behavior, one uses the concept of lifetime, which is the time that it takes for a box to fail under relatively high static loads. Extrapolation to low loads (and high lifetimes) then gives an estimate for the allowable load. Direct measurements would be impractical because of the long time needed. Further discussion of the creep properties of paper materials and the lifetime evaluation of boxes is provided in Chapter 7.

A typical example of the creep problems is corrugated board trays that are widely used for fresh produce (Fig. 3.27a). The produce is often stored in high humidity for some days. As a result, the tray bottom sags and the tray material loses strength. Figure 3.27b shows the deformation of a corrugated board tray due to a uniformly distributed load of 7 kg. When the load is applied, the bottom bends elastically ≈ 3 mm. During the next 72 hours, the total deformation increases to 5.3 mm.

3.5 Summary

Packaging is an important application area for paper materials. Corrugated board is the most popular material for secondary packaging of a wide variety of products.

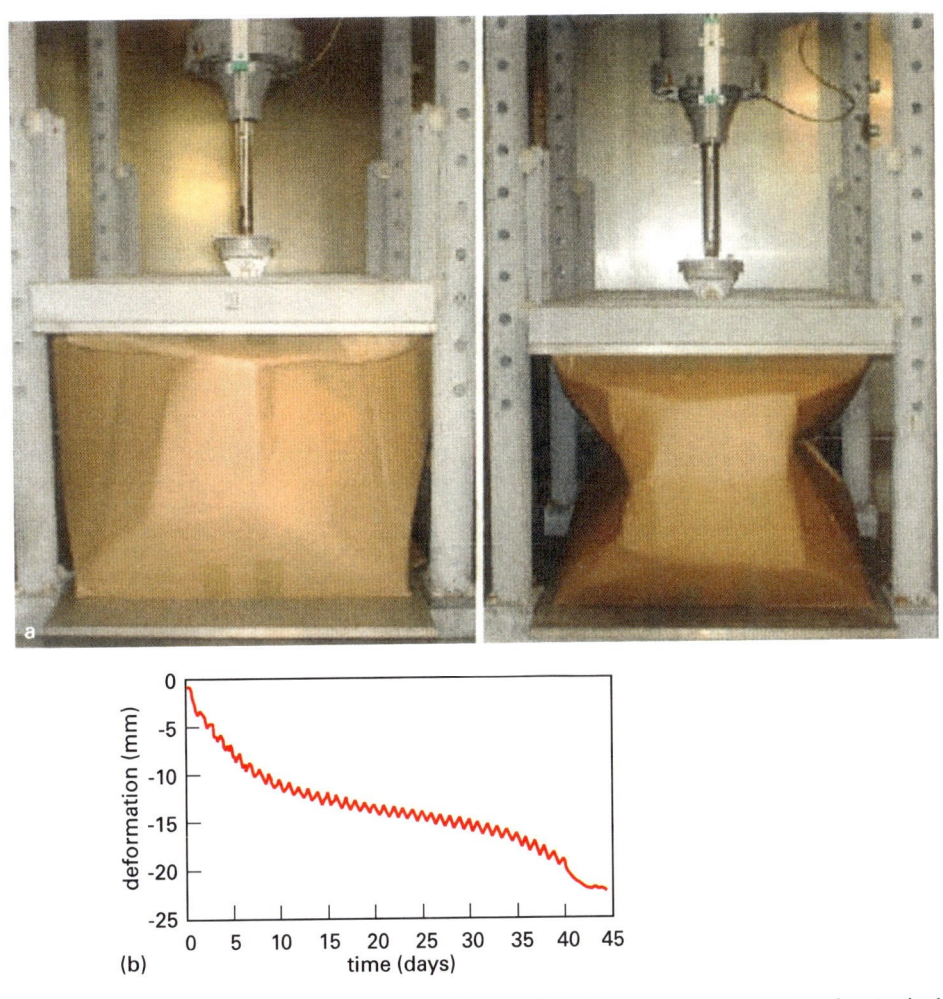

Fig. 3.26 Examples of corrugated board boxes failed in a creep test (a), and a typical deformation – time curve (b) of a box under constant load when humidity varies cyclically between 50% and 90% RH.

Carton board is used mainly for primary packaging. The performance of a corrugated board box is governed by its dimensions and the mechanical properties of its constituent papers.

Compressively loaded corrugated board boxes usually fail after the panels have buckled. The failure is caused by local buckles that develop in the face sheets on the concave side of a panel and coalesce into wrinkles. The wrinkles run diagonally from the corners toward the center of a panel. The semi-empirical McKee formula can be used to estimate the compression strength of regular slotted containers (RSC). The formula does not apply boxes for shallow boxes, asymmetric boxes, boxes of different length and width, or boxes containing holes or perforations, where other analysis

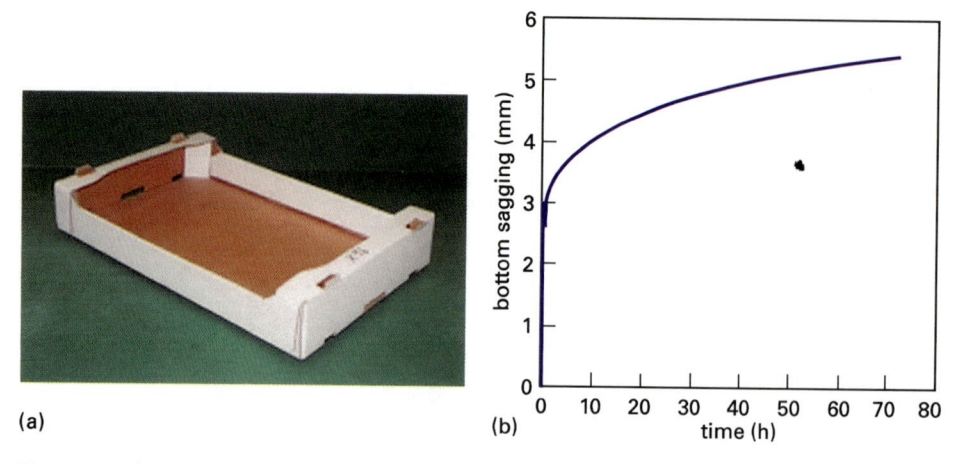

Fig. 3.27 A 266 × 400 mm² corrugated board tray (a), and the sagging of its bottom (displacement of the bottom center) when loaded with 7 kg of lead beads at 90% RH and 23°C. Thickness of the corrugated board was 4.14 mm.

methods are needed. Compared to corrugated boxes, carton board boxes exhibit similar stress distributions under compressive loading, but the failure is controlled by the quality of the corners, discussed in Chapter 4.

In practice, the most critical situation is packaging failure at some point during transportation or storage. The long-time behavior of packages is difficult to study, and relatively little work has been published. During long-time loading the deformation gradually accumulates over time. The deformation is strongly influenced by variations in humidity. The long-time creep behavior of paper materials and paper-based packages is discussed in Chapter 7.

References

Agarwal, B.D., and Broutman, L.J. (1990). Analysis and performance of fiber composites (New York, NY, USA: John Wiley & Sons).

Allansson, A., and Svärd, B. (2001). Stability and collapse of corrugated board. Master's thesis. Report TVSM-5102, Division of Structural Mechanics, Lund University, Sweden.

Bathe, K. J. (1982). Finite Element Procedures in Engineering Analysis (Englewood Cliffs, NJ, USA: Prentice-Hall).

Baum, G.A., Brennan, D.C., and Habeger, C.C. (1981). Orthotropic elastic constants of paper. Tappi 64(8), 97–101.

Beldie, L., Sandberg, G., and Sandberg, L. (2001). Paperboard packages exposed to static loads—finite element modelling and experiments. Packaging Tech. Sci. 14, 171–178.

Bronkhurst, C. A., and Riedemann, J. R. (1994). The creep deformation behaviour of corrugated containers in a cyclic moisture environment. In: Proceedings of the Symposium on Moisture Induced Creep Behaviour of Paper and Board, Stockholm, Sweden, December 5–7, pp. 249–273.

References

Cavlin, S. I., and Edholm, B. (1988). Converting cracks in corrugated board: Effect of liner and fluting properties. Packaging Tech. Sci. 1, 25–34.

Confederation of European Paper Industries (CEPI). (2011). 250 Avenue Louise Brussels, Belgium.

Cox, H. L. (1933). The buckling of thin plates in compression. Technical Report of the Aeronautical Research Committee. Report 1554.

European Carton Makers Association (ECMA). (2011). ECMA Code of Folding Carton Design Styles (The Hague, Netherlands: ECMA). http://www.ecma.org.

Federation of Corrugated Board Manufacturers (FEFCO). (2011). FEFCO-Code. (Brussels, Belgium: FEFCO). http://www.fefco.org.

Fellers, C., de Ruvo, A., Htun, M., Carlsson, L., Engman, C., Lundberg, R. (1983). Carton board—profitable use of pulps and processes. (Stockholm, Sweden: Swedish Forest Products Research Laboratory).

Gilchrist, A. C., Suhling, J. C., and Urbanik, T. J. (1999). Nonlinear finite element modeling of corrugated board. Mechanics of cellulose materials. ASME AMD 231; 101–106.

Grangård, H. (1970). Compression of board cartons part 1: Correlation between actual tests and empirical equations. Sv. Papperstidn. 73(16).

Hägglund, R., and Isaksson, P. (2008). Mechanical analysis of folding induced failure in corrugated board: A theoretical and experimental comparison. J. of Comp. Mat. 4(9), 889–908.

Hill, R. (1950). Mathematical Theory of Plasticity (Oxford, UK: Clarendon Press).

Isaksson, P., and Hägglund, R. (2005). A mechanical model of damage and delamination in corrugated board during folding. Engineering Fracture Mechanics 72(15), 2299–2315.

Kellicutt, K. Q. (1963). Effect of contents and load bearing surface on compressive strength and stacking life of corrugated containers. Tappi 46(1), 151–154.

Kuusipalo, J. (2008). Paper and Paperboard Converting, 2nd Edition (Helsinki: Finnish Paper Engineers' Association).

Maltenfort, G. G. (1956). Compression strength of corrugated containers. Fiber Containers 41(7).

Mann, R. W., Baum, G. A., and Habeger, C. C. (1980). Determination of all nine ortotropic elastic constants for machine-made paper. Tappi 63(2), 163–166.

McKee, R. C., and Gander, J. W. (1957). Top-load compression. Tappi 40(1), 57–64.

McKee, R. C., Gander, J. W., and Wachuta, J. R. (1963). Compression strength formula for corrugated boxes. Paperboard Packaging 48, 149–159.

Nordstrand, T. (2004). Analysis and testing of corrugated board panels into the post-buckling regime. Compos. Struct. 63(2), 189–199.

Nyman, U. (2004). Continuum mechanics modeling of corrugated board. Doctoral Thesis. Lund University, Sweden.

Patel, P. (1996). Biaxial failure of corrugated board. Licentiate thesis. Division of Engineering Logistics, Lund University, Sweden.

PIRA. (2011). Cleeve Road Leatherhead Surrey, United Kingdom. http://www.pira-international.com.

Pommier, J., Poustis, C., Bending, J., Fourcade, J., and Morlier, P. (1991). Determination of the critical load of a corrugated box subjected to vertical compression by finite element method. In: Proceedings of the 1991 International Paper Physics Conference, Kona, Hawaii, pp. 437–447.

Rahman, A. (1997). Finite element buckling analysis of corrugated fiberboard panels. In: Proceedings of the 1997 joint ASME/ASCE/SES summer meeting, Mechanics of Cellulosic Materials, June 29–July 2, Evanston, Illinois, pp. 87–92.

Thakkar, B., Gooren, L.G.J., Peerlings, R.H.J., and Geers, M.G.D. (2008). Experimental and numerical investigation of creasing in corrugated paperboard. Phil. Mag. 88, 3299–3310.

Tsai, S. W., and Wu, E. M. (1971). A general theory of strength for anisotropic materials. J. Comp. Mater. 5, 58–80.

Urbanik, T. J., and Saliklis, E. P. (2003). Finite element corroboration of buckling phenomena observed in corrugated boxes. Wood and Fiber Science 35(3), 322–333.

Westerlind, B. S., and Carlsson, L. A. (1992). Compressive response of corrugated board. Tappi 75(7), 145–154.

4 Behavior of corners in carton board boxes

Mikael Nygårds

4.1 Introduction

The preceding chapter discussed the manufacture and performance of boxes, with a focus on the corrugated board that is used in secondary packaging. In this chapter we consider the mechanics related to carton board boxes used as primary packaging. The process of converting carton board into packages is in essence the same as that used with corrugated boards, but the difference in the board structure means that the critical properties of a carton board are somewhat different from those of a corrugated board. A typical corrugated board box fails through a coalescence of small buckles that form between the fluting crests after the box face has buckled. This mechanism is absent in a carton board box because the board is homogeneous (there are no flutes). Instead, the strength of corners is critical for the strength of carton board boxes (Fig. 3.24; Beldie et al., 2001).

Carton board packages are often used to protect and carry liquids, frozen food, cigarettes, and other products that are sensitive to moisture, temperature, and air. Cracks in the carton board must be avoided because they can lead to leakage or oxidization of the product. In addition, the converting and filling of carton board packages are fast processes, and it is thus important that the board material is highly homogeneous and performs in the same way all the time.

In the process of converting carton board, making the corners is a critical phase. The corners must be sharp and form no cracks if folded by 90 degrees. The generally poor folding behavior of uncreased carton boards is easy to observe if one tries to fold a carton board sheet by hand. It is practically impossible to achieve a straight and sharp fold line. In contrast, it is easy to fold thin paper sheets sharply. In an untreated board, the tensile stresses at the outer edge of the corner may break the board there (Fig. 4.1a). Alternatively, the inner edge of the corner may break in compression. However, if the board is delaminated into layers before folding, the material on the inside can buckle inwards (Fig. 4.1b). This reduces the thickness of the layer that is bent, and the tensile stress on the outer edges decreases. A controlled amount of delamination in the corner area can be used to improve the converting behavior of a carton board.

The desired deformation of a carton board in the box-making process is shown schematically in Figure 4.2. The creasing phase creates delamination of the carton board structure. Figure 4.3 demonstrates the effect of creasing in reality. In this example, the uncreased board forms a sharp bend and ruptures in the inside of the corner; a carton board box made in this way would not protect the contents from external conditions. Even the creasing operation can break the board if the crease is too steep. In the optimal case, fibers are only pulled apart in the out-of-plane

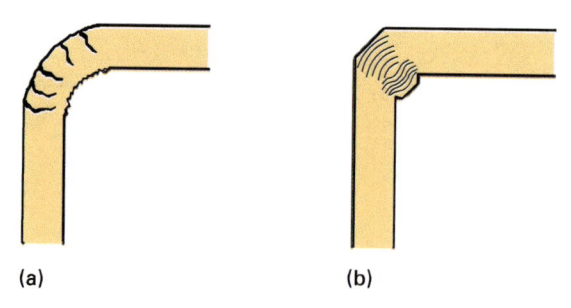

Fig. 4.1 Cracking in a folded corner of uncreased (a) and creased (b) carton board. Uncreased carton board tends to crack in the outer surface. In a creased carton board the inner surface bulges out after planar cracking or delamination.

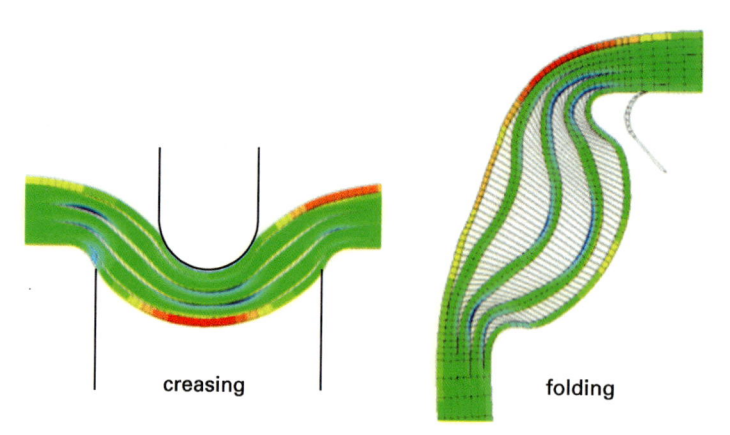

Fig. 4.2 Principle of creasing and folding.

direction and not at all in the in-plane direction. This is important for preserving the mechanical strength of the corner.

It is possible to achieve good delamination in the creasing phase by designing the structure of the carton board in alternative ways (Fig. 4.4). Carton boards are usually made by making several plies simultaneously in the forming section of the paper machine, before the wet web enters the press section (Fig. 2.23). If high thickness is needed, up to five plies can be used. Ordinarily, the middle plies have lower density than the outer plies. This gives an "I-beam" structure that has high bending stiffness and easy delamination in the middle plies. Typical mechanical properties of a three-ply carton board are shown in Table 4.1. Alternatively, all the plies can be similar in properties, but the interfaces between them are made weak. This gives a carton board that has very precisely set delamination planes.

In summary, the mechanical properties related to creasing and folding are crucial for the performance of a carton board in packaging. The strength of the filled carton board package is actually of lesser importance compared to the performance in the packaging process. Usually it is the corrugated secondary package where strength

Tab. 4.1 Experimentally determined properties of different plies in a three-ply carton board (Nygårds, 2008).

	Thickness (mm)	Elastic modulus (MPa)		Tensile strength (MPa)		Shear modulus (MPa)		Shear strength (MPa)
		E_x	E_y	σ_x^{max}	σ_y^{max}	G_{xz}	G_{yz}	τ_{xz}^{max}, τ_{yz}^{max}
Bottom	0.090	8760	3030	85	40	50	78	5
Middle	0.210	3110	1210	30	18	23	21	1
Top ply	0.090	5920	2670	50	30	73	85	5

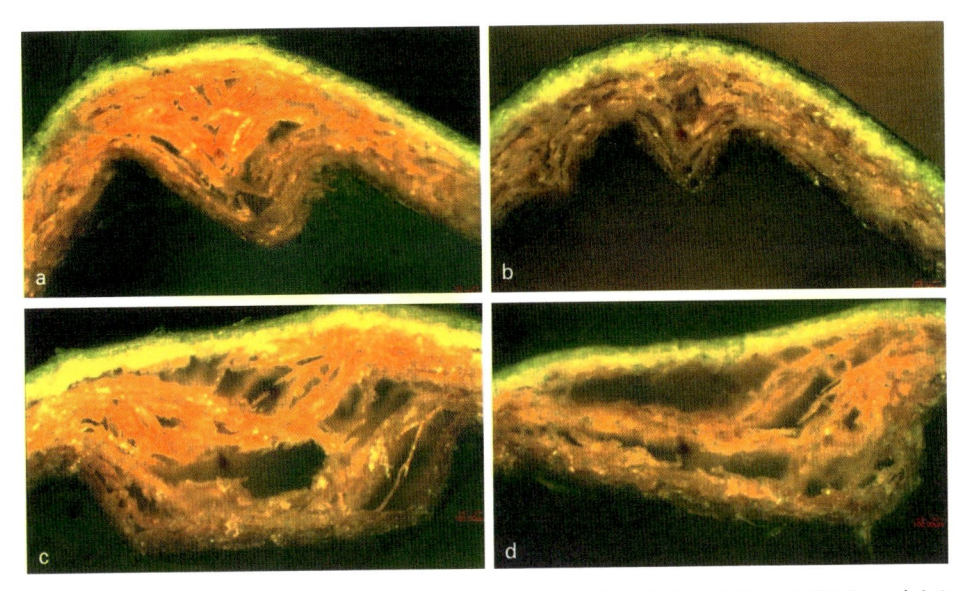

Fig. 4.3 Cross-sections of folded uncreased carton board in MD (top left) and CD (top right) and creased carton board in MD (bottom left) and CD (bottom right). Photos courtesy of Hui Huang.

is needed during transportation and storage (see Chapter 3.3). In this chapter, we discuss the mechanisms that control the creasing and folding behavior of a carton board.

4.2 Folding of a multiply carton board

The folding performance of carton boards can be tested using the set-up in Figure 4.5. One end of the board is fixed on a rotating pneumatic clamp, while a load cell measures the force P needed to bend the sample. Because one end of the beam has a

4 Behavior of corners in carton board boxes

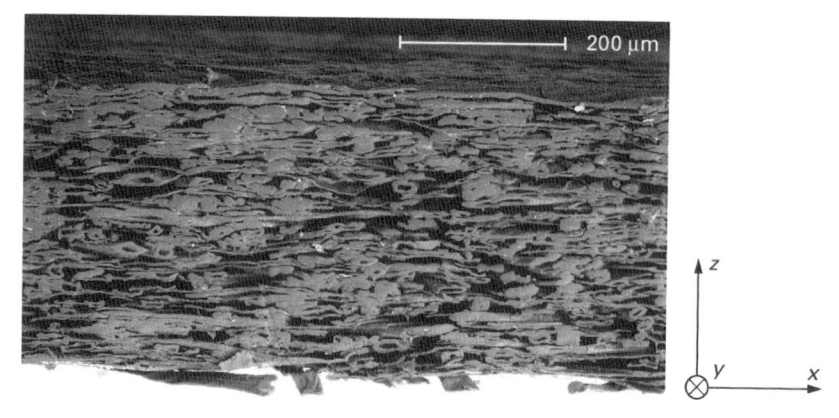

Fig. 4.4 Cross section of a carton board, showing the fiber network.

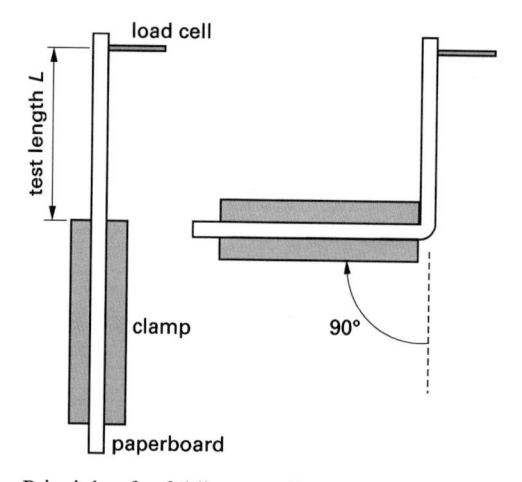

Fig. 4.5 Principle of a folding test (Lorentzen and Wettre, Sweden). One end of the carton board is fixed in a rotating pneumatic clamp, while a load cell measures the force needed to fold the sample.

rigid support, while the other is loaded by a point contact load, the stress state inside the specimen varies along the free length. The highest stresses occur at the rigid clamps. According to Timoshenko (1934), the in-plane stress is

$$\sigma_x = \frac{PL}{I} z = \frac{12PL}{wd^3} z, \tag{4.1}$$

and shear stress

$$\tau_{xz} = \frac{P}{2I}\left(\frac{d^2}{4} - z^2\right) = \frac{3P}{2wd}\left(1 - 4\frac{z^2}{d^2}\right). \tag{4.2}$$

4.2 Folding of a multiply carton board

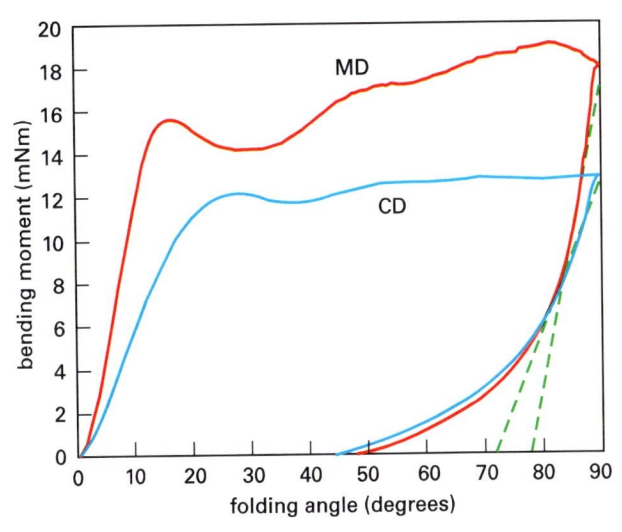

Fig. 4.6 Bending moment against folding angle in MD and CD for a folding–unfolding cycle of uncreased carton board defined in Tab. 4.1. The straight lines in the unloading phase indicate the slope detected in the initial bending phase.

Here, I is the moment of inertia of the specimen,

$$I = \frac{wd^3}{12}, \tag{4.3}$$

L the length, d the thickness, w the width, and z the ZD distance from the center line of the specimen. Usually the shear stresses are small and can be ignored; however, paper materials are highly anisotropic, and the shear stresses are important to consider.

Figure 4.6 shows the measurement result when the three-ply uncreased carton board defined in Table 4.1 is folded by 90° and released both in MD and in CD. The result is expressed as the bending moment $M_b = PL$. An initial linear region can be used to calculate the bending stiffness $S_b = EI$. The difference between MD and CD behavior arises from the anisotropy of paper (Chapter 2.2.2). In both directions the bending moment is linear almost all the way to a maximum value that is followed by a regime of fairly constant bending moments. The irregular curve shape in this region is due to slippage between the carton board and load cell.

The initial unloading stiffnesses in Figure 4.6 are close to the initial loading stiffnesses, indicating that the in-plane properties of the carton board are not affected even though a permanent fold of ca. 40° is created. This means that the average degree of inter-fiber bonding is not changed enough to reduce the in-plane elastic modulus. However, some inter-fiber bonds are probably broken by the shear stresses (Eq. 4.2). To see this, consider a simple experiment. Take a copy paper sheet, fold it by 180° and then unfold. This creates a local damage along a straight line. If you make a tensile test, the sheet will most likely not fail along the fold line. However, if you instead tear the paper, it is easiest to tear it along the fold line. This shows that the paper is slightly weaker along the fold line.

4 Behavior of corners in carton board boxes

Consider again the elastic stresses (Eqs. 4.1 and 4.2) during folding. The maximum bending moment in Figure 4.6 gives the stress distributions shown in Figure 4.7a. The normal stresses are much larger than the shear stresses. If the stress values in different plies are divided by the measured strength values of the carton board (Tab. 4.1), the renormalized normal and shear stresses are closer to each other (Fig. 4.7b). The renormalized values of axial stresses are still larger than those of the shear stresses. Furthermore, the compressive stress values should actually be divided by the compressive strength, not the tensile strength. As illustrated in Figure 2.14, the compressive strength values are much smaller than the tensile strength values. One infers that the uncreased carton board fails in compression in the inside edge of the fold line. This is qualitatively consistent with Figure 4.3. There is a small caveat in the

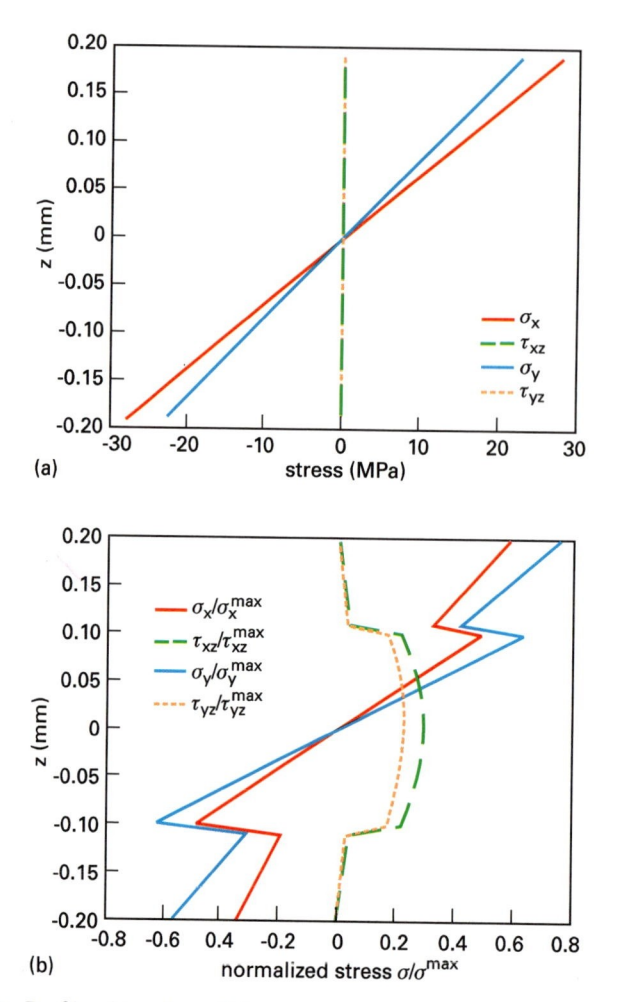

Fig. 4.7 Profile of tensile and shear stresses across board thickness (a) and the same stresses when divided by tensile strength and shear strength (Tab. 4.1) measured separately for different plies of the carton board. The stress profiles have been calculated for the maximum bending moment in Figure 4.6.

argument in that the stress distributions in Figure 4.7 were calculated assuming ideal linearly elastic behavior. In reality, plastic deformations in the different plies will influence the stress distributions and may influence the mode of failure that would be caused by the folding.

Stress distributions through the thickness of a paper board depend on the board thickness. From Equations 4.1 and 4.2 one can see that the shear stress component increases relative to the axial stress component if thickness d is increased. The ratio $\tau_{xz}^{max}/\sigma_{x}^{max} = 0$, if $d = 0$ mm, and increases linearly to $\tau_{xz}^{max}/\sigma_{x}^{max} = 1.5$ for $d = 1$ mm. This shows that the deformations caused by folding are different in thin paper grades (newsprint, office paper) and thick carton boards. Thin grades can be folded easily with no visible cracking (Barbier et al., 2002). Thick carton boards usually break when folded. Breaking usually occurs in the inside edge of the fold, and it is practically impossible to achieve a straight folding line. A box corner of this type would have low strength because an irregular folding line can easily trigger cracking, and the folding cracks themselves reduce the strength of a box corner.

4.3 Creasing

Creasing is used to prevent folding cracks and irregular fold lines. The resulting folding performance depends greatly on the creasing operation. In the creasing operation the ruler is pressed into the carton board that rests against the die (Fig. 4.8), and a permanent indentation is created. High shear stresses and compressive out-of-plane stresses arise in the carton board. The shear stresses break bonds between the board plies and between the fibers and then deform fibers so that several delamination planes are created in the carton board. However, delamination occurs only in a narrow zone, where the bending stiffness is therefore dramatically reduced.

The force-displacement curve of a creasing operation in Figure 4.9 is calculated from a finite element model. The vertical displacement of the ruler is defined to be zero when it is in plane with the die surface. The area under the curve gives the amount of energy consumed to deform the board irreversibly. Different deformation mechanisms act during the operation. At first a uniform shear stress field is created between the ruler and the die. The shear stress breaks inter-fiber and inter-ply bonds. Next, the carton board is compressed under the ruler. The compression prevents

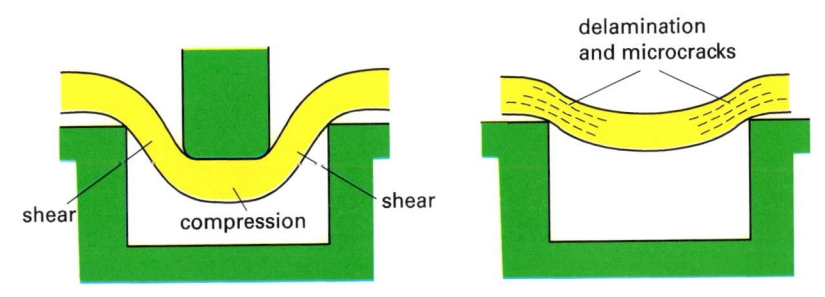

Fig. 4.8 Generation of shear-induced delamination and micro cracks in the creasing operation.

4 Behavior of corners in carton board boxes

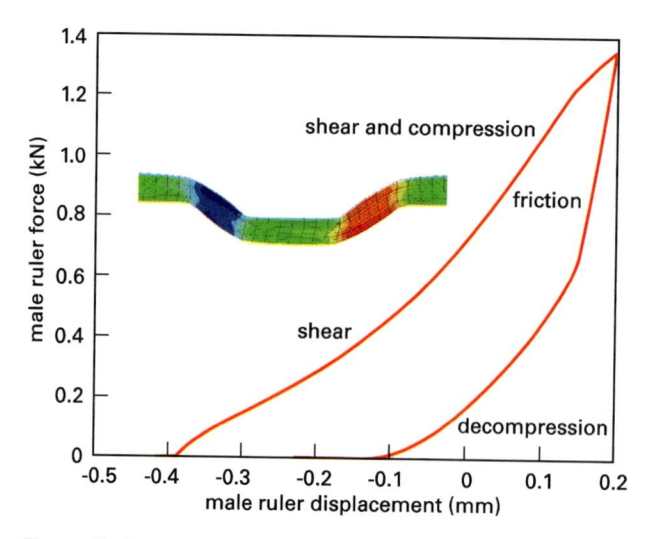

Fig. 4.9 Force-displacement curve calculated using FEM for the creasing of a 32-mm wide carton board specimen. The inset shows the shear strain field at peak load.

delamination from propagating under the ruler because friction between fibers within the board increases. After the elastic recovery one can see that the carton board is permanently deformed as force goes to zero before the ruler has returned to its original position. The shape of the creasing curve, and hence the amount of energy consumed, depends on the structure of the carton board and on the friction between it and the surfaces of the ruler and the die.

The shear induced delamination of the carton board reduces the bending stiffness of the creased carton board and enhances buckling of the carton board during folding. Due to the local damage in the creased area, the damaged region will behave similar to a hinge, which reduces the folding resistance. The folding properties do depend on the properties of the carton board, the geometry of the male ruler and female die, and on the creasing depth. The influence of creasing on folding behavior (bending moment as a function of folding angle) is shown in Figure 4.10 for the same 0.4 mm thick carton board that was used in Figure 4.6. Different crease depths from $d_{crease} = -0.15$ to 0.15 mm are compared. One can see that the initial bending stiffness S_b and the maximum bending moment M_b decreased when the crease depth d_{crease} was increased. In other words, a deeper crease makes a box corner easier to fold. However, it can also be seen from Figure 4.10 that roughly half of the 90° folding angle recovers when the bending moment is removed. This "spring-back" is unaffected by creasing. If box corners are folded by only 90° in the box forming, then the box tends to be rounded.

In order to facilitate a comparison of the different crease depths, we define a damage parameter D through both the reduction in the initial bending stiffness S_b,

$$D = \frac{\Delta S_b^{creased}}{S_b^{uncreased}}, \tag{4.4}$$

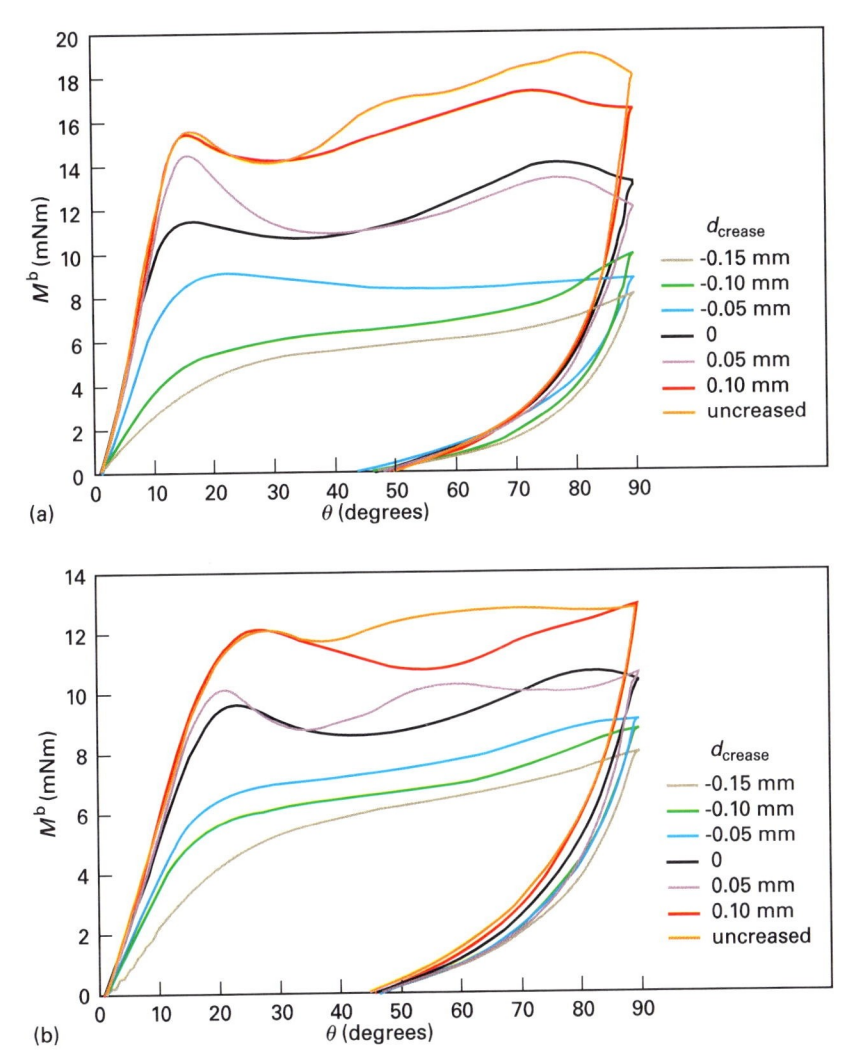

Fig. 4.10 Bending moment against folding angle for the same 0.4-mm carton board as in Figure 4.6 in MD (a) and CD (b) for a range of crease depths d_{crease} using a ruler of width 0.6 mm and die of width 1.2 mm (Nygårds, 2008).

and the maximum bending moment M_b,

$$D = \frac{\Delta M_b^{creased}}{M_b^{uncreased}}. \tag{4.5}$$

In an uncreased and hence undamaged material $D = 0$, and in a completely failed material $D = 1$. As seen in Figure 4.11, the two definitions of D give the same result. Also, there is essentially no difference in the damage evolution between MD and CD folding behavior. Hence, the degree of damage creased in a creasing process seems to

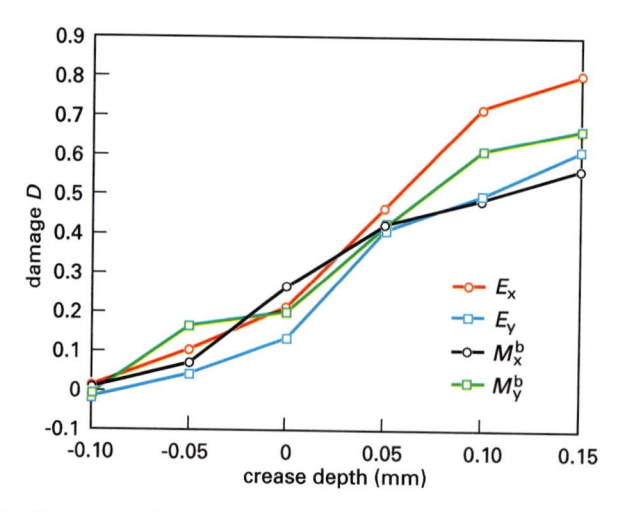

Fig. 4.11 Damage against crease depth, calculated from Equations 4.4 and 4.5.

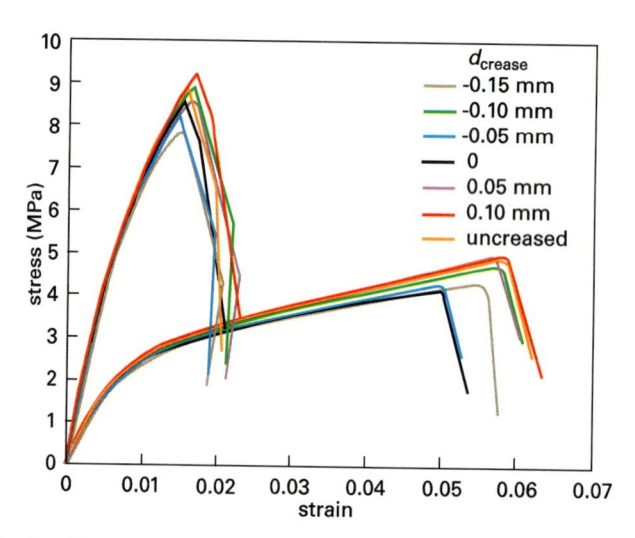

Fig. 4.12 In-plane stress–strain curves in MD and CD for the same 0.4-mm carton board as in Figure 4.10, measured after the creasing to different depths d_{crease}.

be independent of the anisotropy in board properties and controlled mainly by the creasing process.

If the carton board cracks during creasing, the in-plane strength is reduced. The crease depth then needs to be reduced, but at the same time, even the delamination decreases. The in-plane stress–strain curves of the samples in Figure 4.10 are shown in Figure 4.12. They show no change as a result of creasing to different depths, indicating that no cracking happened during creasing. In fact, none of the specimens tested failed in the region of the creases, clearly demonstrating that creasing did not influence the in-plane tensile properties of the carton board.

4.4 Important material properties

Carton board that is converted into boxes should fold easily but still maintain strength and stiffness along the folded corner. This is necessary for good package performance because box corners carry a great deal of compressive loads imposed on the box. Ideally, the carton board delamination in the creasing operation should be confined accurately and reproducibly along the crease line. In-plane cracking is not allowed in the folding operation. The desired behavior can be achieved and optimized by adjusting the material properties. The ZD properties especially govern the bending stiffness and performance in converting operations such as creasing, folding, and printing. Therefore, understanding how these can be changed is essential for successful product development.

The out-of-plane mechanical properties of a carton board can be characterized with respect to compression, tension, and shear. The ZD stress–strain curve in compression is generally described by an exponential curve because the material is porous. In contrast, the most dominant characteristic in the ZD tension and shear loading is the softening behavior (Fig. 2.15), which is caused by a local cohesive failure in planes where inter-fiber bonds break and fibers bridge the planar crack surfaces (Fig. 4.3). The shear failure mode is important in the creasing and folding operations.

From the preceding discussion of creasing we can conclude that good creasing behavior is controlled by two factors. First, low shear strength is needed so that delamination occurs instead of in-plane cracking. Second, the delamination needs to occur at well-defined planes. This can be achieved with multi-ply sheet structures. It is important especially to have a weak interface close to the outer surface of the fold because this makes it easier for the inner surface to bulge inwards. The effect of delamination close to the outer surface can be seen in Figure 4.3. In the creased samples bulging involves the whole structure except the layers closest to the outer surface. The uncreased samples show no delamination, and fibers are sharply into the inner corner, leading to cracking and strength losses.

The out-of-plane shear behavior of a carton board can be determined using the notched shear test (NST; Fig. 4.13; Nygårds et al., 2009). Shear failure occurs along the plane at which the two notches meet. If the position of this plane is varied, one can determine the variation of shear strength through the material thickness. In the example shown in Figure 4.13, the carton board has weak interfaces close to the surfaces, indicated by the low shear strength values. It should be favorable to have both low average shear strength and well-defined minima in the profile. However, the random scatter in the local shear stress values also can be larger than the differences in the mean values. Carton boards like that normally perform badly in converting because the delamination planes are scattered at random, and thus, the folding behavior varies at random.

4.5 Final remarks

We have discussed the behavior of box corners, which relies on creasing and folding of carton board. This is a fairly well-defined boundary value problem, but it also

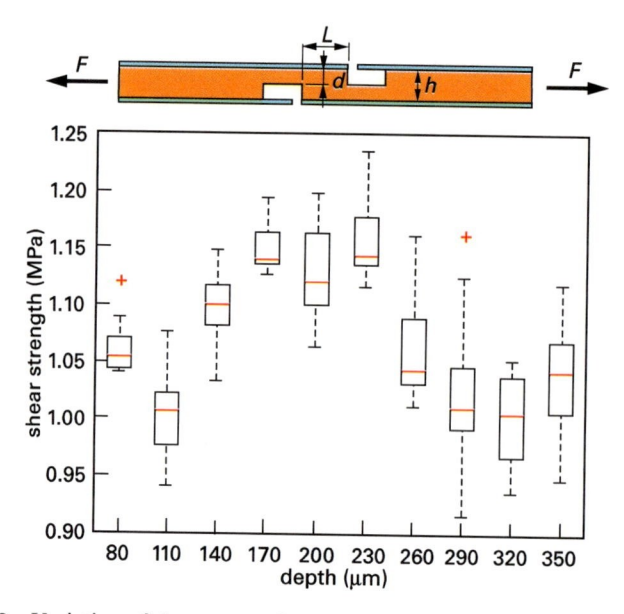

Fig. 4.13 Variation of shear strength across the thickness of the same 0.4-mm carton board as in Figure 4.6, measured using the notched shear test shown on the top (Nygårds, 2008).

involves many different deformation and fracture mechanisms. We have presented a qualitative description of the main mechanisms of creasing and folding, but we have not considered the effect of corners on the box strength. A thorough study of the strength of a box with a given geometry would require knowledge of not only the panel properties (Chapter 3.4.3; Beldie et al., 2001) but also the strength properties of the corners, which, in turn, are sensitive to the delamination of the creases. In principle, the latter effect can be studied using fracture mechanics (Chapter 5), although little work on this has been published.

The mechanical properties of a carton board and the behavior of box corners must also depend on the loading rate, moisture content, and temperature of the carton board. The moisture content also influences the thickness of carton boards, which, in turn, influences the creasing, folding, and buckling behaviors. The general effects of moisture content, temperature, and loading rate are discussed in Chapters 7 and 9.

References

Barbier, C., Larsson, P.-L., and Östlund, S. (2002). Experimental observation of damage at folding of coated papers. Nordic Pulp and Pap. Res. J. 17(1), 34–38.

Beldie, L., Sandberg G., and Sandberg L. (2001). Carton board packages exposed to static loads—finite element modelling and experiments. Packaging Tech. and Sci. 14, 171–178.

Nygårds, M. (2008). Experimental techniques for characterization of elastic-plastic material properties in carton board. Nordic Pulp and Pap. Res. J. 23(4), 432–437.

Nygårds, M., Fellers, C., and Östlund, S. (2009). Development of the notched shear test. In: Advances in Pulp and Paper Research, Transactions of the 14th Fundamental Research Symposium, S.J. I'Anson, ed. (Oxford, UK: PPFRC), pp. 877–897.

Timoshenko, S. (1934). Theory of Elasticity, Engineering Societies Monographs, H. W. Craver, ed. (New York, NY, USA: McGraw-Hill).

5 Fracture properties

Sören Östlund and Petri Mäkelä

5.1 Introduction

Defects reduce the strength of paper structures considerably. In a typical office paper sheet a 10 mm long cut on the edge reduces the failure load to one-half and the failure elongation to one-third of the original (Fig. 5.1). Although somewhat artificial, this example demonstrates the impact defects have on the strength of paper products. The strength reduction comes from the fact that when a sheet containing a defect is loaded, stress concentrations are induced in the vicinity of the defect (Fig. 5.2). A sharp defect gives rise to the most severe stress concentration, and therefore, sharp defects are most critical for the load-carrying capacity of a structure.

At small enough external stresses, a defect is stable and the structure does not break macroscopically. If the external stress level is increased, damage will accumulate in the neighborhood of the defect and eventually lead to rupture of the structure. Accordingly, defects will affect the performance of paper products. It is also possible that ruptures are triggered by factors other than clear defects, such as the nonuniformity of the paper material at different length scales, as is discussed in Chapter 8. This chapter concentrates on the effect of macroscopic crack-like defects.

Fracture mechanics is a discipline within solid mechanics that deals with the strength of structures containing defects. Knowledge of fracture mechanics is of fundamental interest when preventing failures of paper in converting and end-use, as well as when designing the material properties for applications where controlled failure is needed, such as in the creasing and folding operations (Chapter 4), or the tear lines in packages.

Three principal factors influence the strength of a notched structure made of a specific material (Fig. 5.3): the external loading, the geometry of the structure (including the size and shape of defects), and the fracture toughness, or the capability of the material to sustain high local stresses. The objective of fracture mechanics is to relate these factors so that if two of them are known, the third can be calculated.

The general deformation of a defect under loading is described by the displacements of the two crack surfaces relative to one another. The relative displacement can be separated into the principal fracture modes I, II and III defined in Figure 5.4. In practice, defects influence the integrity of structures first and foremost under tensile loading, and this has been the main focus in the discipline of fracture mechanics.

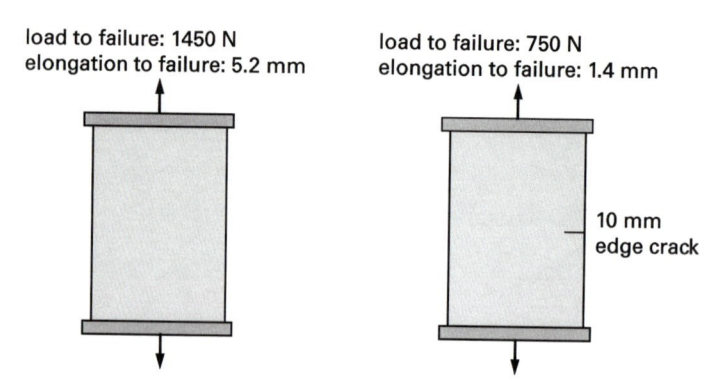

Fig. 5.1 Comparison of the tensile load and elongation to failure of an ordinary A4 office paper sheet (210 mm wide and 297 mm long) with and without a 10 mm long edge cut.

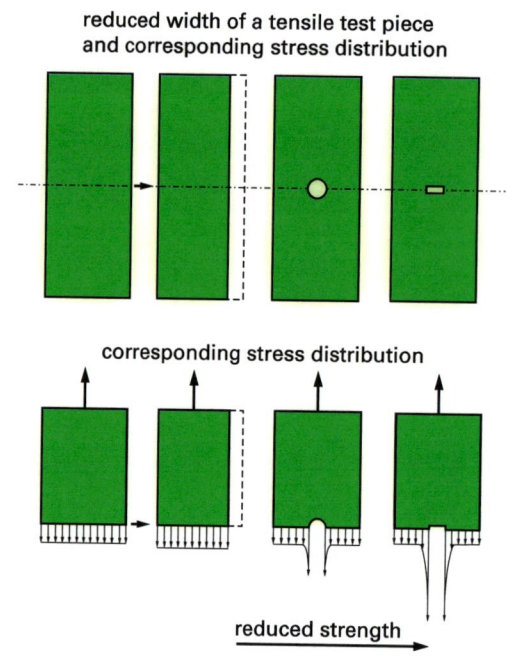

Fig. 5.2 Change in the tensile stress distribution in a paper sheet because of a width reduction, a circular hole, or a cut line.

However, the stability of structures under compressive loading also is sensitive to the existence of defects. In the case of paperboard packaging (Chapter 3), crack-like defects can cause mode II transverse shear failure, particularly if they occur close to the edges of box panels. Especially in carton board boxes, the quality of creasing (Chapter 4) can have large impact on the structural performance.

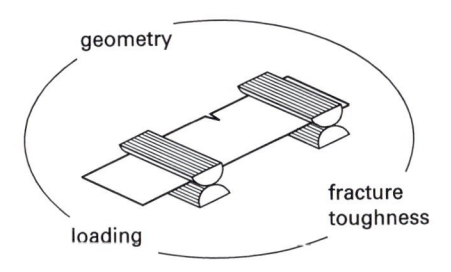

Fig. 5.3 Summary of the principal factors that influence the strength of notched structures. The size of the notch is part of the definition of geometry. Reproduced from Mäkelä (2002) with permission from Svenska Pappers- och Cellulosaingeniörsföreningen (SPCI).

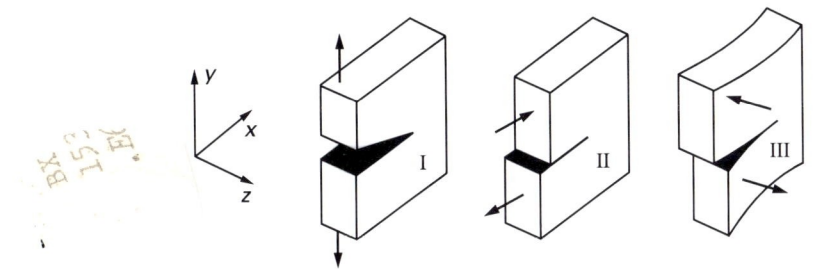

Fig. 5.4 Principal fracture modes I, II, and III in terms of the relative displacement of crack surfaces.

5.2 Examples of practical applications of fracture mechanics

Before entering into the specific theories of fracture mechanics for paper materials, we briefly discuss a few applications where fracture mechanics are important for conceptual understanding and problem solving. It is appropriate to consider in-plane failure and out-of-plane failure (also known as delamination) separately.

5.2.1 Mode I failure under in-plane tension

This is the dominating mode of failure in many paper applications and has been a subject of considerable interest in the literature. Web breaks (Fig. 5.5) sometimes happen because of macroscopic defects, such as rigid fiber bundles, dirt spots, wrinkles or holes in the web, or edge cuts caused by careless handling of the paper roll. Fracture mechanics conveniently analyzes their effects. Improving the insensitivity of the paper material against defects undoubtedly increases web strength.

To assess the structural integrity of a paper web, it is necessary to know the stresses close to the crack tip and the fracture toughness of the material. The crack tip stress state depends on the crack length, crack position, and web tension (Fig. 5.6). Its

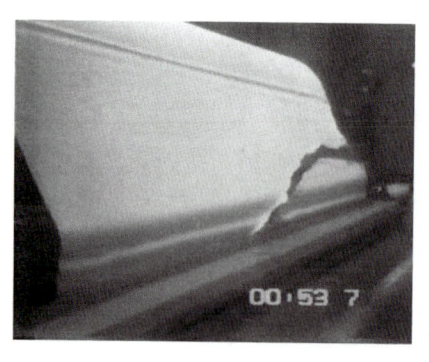

Fig. 5.5 Paper web break. Photo courtesy of Albany Nordiskafilt AB.

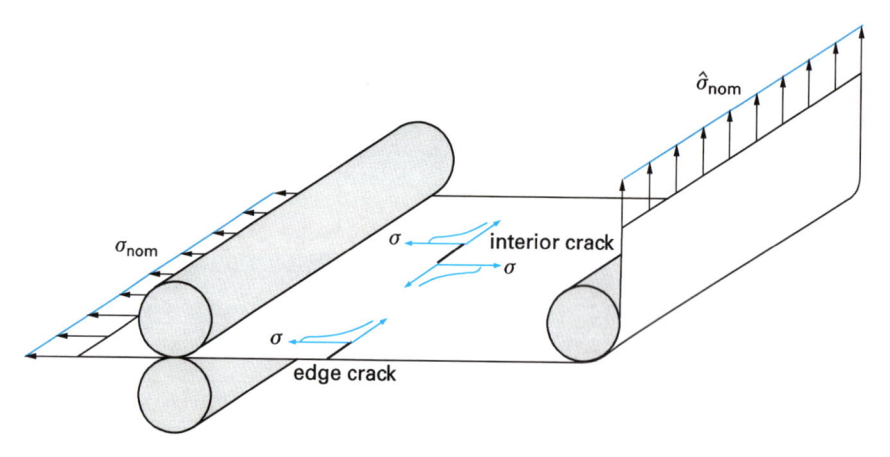

Fig. 5.6 Schematic definition of the fracture mechanics problem of web breaks caused by defects.

explicit mathematical form is described in Section 5.3. It is important to realize that the problem statement in Figure 5.6 contains several simplifying assumptions that should be quantified in order to accurately predict the real performance of a paper web. These include

– Out-of-plane deformation
– Loading perpendicular to the plane of the web
– Skew web tension

The web is, in general, not flat, and, particularly in the crack tip region, tensile buckling will affect the stress state. If the crack is not oriented exactly in the cross-machine direction CD, the crack tip is loaded in the opening mode I and shear mode II (Fig. 5.4). However, CD cracks (as in Fig. 5.6) are, in general, the most severe, and they are loaded primarily in mode I. When passing a pressurized turner bar, the crack tip is loaded also perpendicular to the plane of the web by the air pressure (Fig. 5.7). There are also several factors that can cause a skew web or otherwise

5.2 Examples of practical applications of fracture mechanics

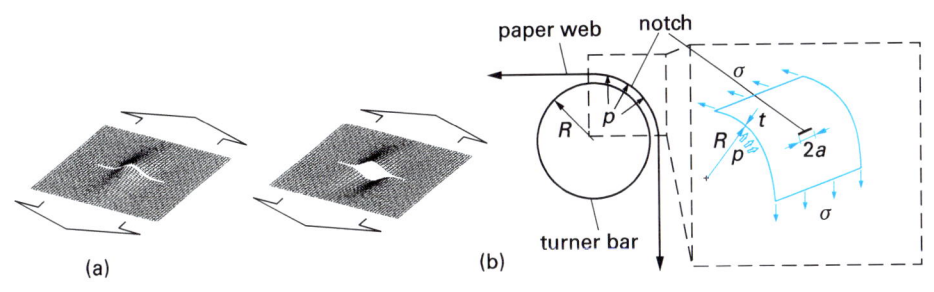

Fig. 5.7 Examples of factors that affect crack tip stresses in a paper web: tension-induced buckling (a; adapted from Skjetne, 2006), and bulging over a pressurized turner (b; after Nordhagen, 2009). Reproduced with permission from the authors.

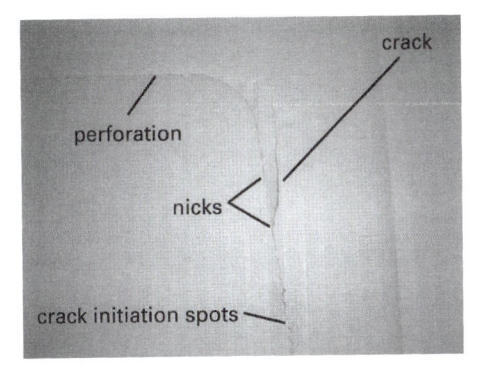

Fig. 5.8 Cracks created in the perforation operation of a corrugated board.

nonuniform web tension profile. Nonuniformity can arise from misaligned turner bars, frictional effects, and web tension profiles created on the paper machine, and it can severely affect crack tip loading.

The influence of loading factors on crack tip stresses in a printing press has been analyzed by Nordhagen (2009). The results indicate that even in normal operating conditions, the nonideal features can increase crack tip stress to levels that are comparable to the stresses that cause rupture in laboratory testing. Thus, it is not enough to compare the nominal web tension and the failure stress in laboratory testing. Damage tolerance methodology is a practical way to use fracture mechanics in the analysis of web breaks. One assumes the existence of a certain crack, and then uses a detailed stress analysis to investigate the integrity of the cracked structure.

In many paper products cut-outs and perforations are applied to facilitate converting and end-use. Examples include die-cutting, ventilation holes and hand holes in corrugated board containers, and perforations to enable easy opening of carton board packages. Stress concentrations are created at the tips of the cut-outs and perforations. Figure 5.8 shows a crack triggered when die-cutting perforation in a corrugated container and crack initiation spots (short cracks) along the straight cut. Good performance requires that the perforation should be neither too weak nor too

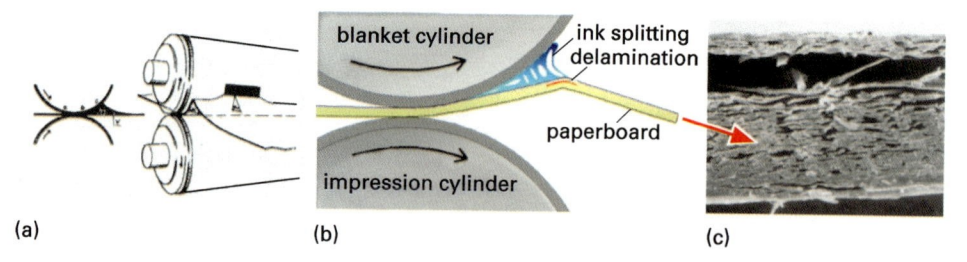

Fig. 5.9 Delamination problems in offset printing: Nonuniform adhesion (a), ink tack (b), and delamination in a paperboard (c).

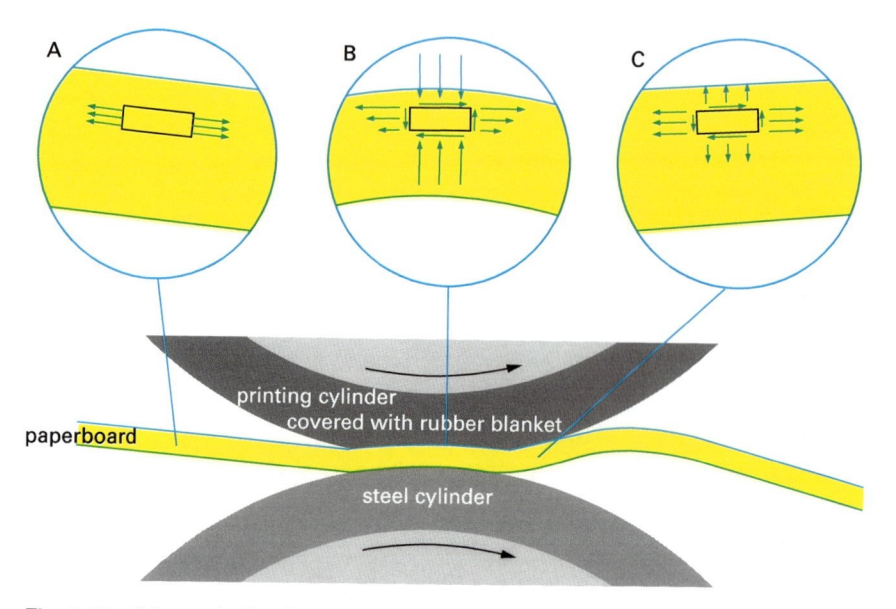

Fig. 5.10 Schematic distribution of MD and ZD stresses in a paperboard in different locations relative to an offset printing nip.

strong. Fracture mechanics is a convenient tool for such optimization. However, if the material volume in front of a cut is too small, crack tip stress fields cannot fully develop, and standard fracture mechanics may not apply.

5.2.2 Out-of-plane delamination

Because fibers are predominantly aligned in the plane of the sheet, paper is considerably more compliant and weak in the out-of-plane direction than in the in-plane directions (Chapter 2.3.3). There are important cases where delamination determines the performance of paper products.

In offset printing of paper, delamination is a well-known problem that reduces production efficiency (Fig. 5.9). Loading in a printing nip (Chapter 10) can create cracks. When the web exits the nip, a crack may grow into visually unacceptable delamination due to ink tackiness. The complex multi-axial stress state of a material element passing through a nip varies, as shown in Figure 5.10. Before the nip, pure tension applies (inset A in Fig. 5.10). Inside the nip, the rubber blanket causes bending, shear stresses, and compressive ZD stresses (inset B). Shear stresses can also arise from asymmetry. On the exit side, ink tack can cause ZD tensile stresses (inset C). The real behavior is quite complex, and detailed mathematical analysis is needed to evaluate the stress components.

Failure analysis can be used to determine if a certain combined stress state creates macroscopic delamination or if typical existing microscopic defects can grow to unacceptable size. In the case of carton boards, the layered structure has to be accounted for because it influences the ZD variation of the normal and shear strengths. The location of an eventual delamination cannot be known in advance, although it tends to occur close to the printing surface. Full three-dimensional analysis is often necessary, as is the case in Figure 5.9a, for example, where ink tack is nonuniform.

Another case where delamination is vital is creasing and folding (Chapter 4). The shear mechanism that creates delamination in the creasing operation is shown in Figure 4.8. A sharp crease line reduces the bending moment locally, which is necessary for achieving straight fold lines and forming a box with flat panels. A successful creasing operation requires that the material delaminates in a controlled manner, which conflicts with the requirement that delamination must be avoided in offset printing. Therefore, optimization is needed. Fracture mechanics can be used for quantitative analysis and conceptual understanding that can identify the key parameters (Fig. 4.7).

Delamination can also play a role in the converting of corrugated boards, and it can occur particularly in the die-cutting of multi-ply materials. Also, when manufacturing the sine-wave shaped fluting medium, delamination damage can decrease the transverse shear stiffness of the paper, which, in general, reduces the stacking strength of the box.

5.3 Crack tip modeling in paper materials

This section describes the theoretical foundation for crack tip modeling in paper materials. We do not cover the entire theory of continuum fracture mechanics, or review fracture mechanics literature for paper, but depict the most important crack tip models for the analysis of paper materials and products. For a more detailed presentation, see Anderson (2005) and Mäkelä (2002). Breakthrough in the application of fracture mechanics to paper materials came when Swinehart and Broek (1995) and Wellmar et al. (2000) showed that the failure in notched paper sheets can be predicted from measured material properties. These studies shared the use of a robust fracture mechanics theory combined with numerical methods to determine the failure point of notched paper samples.

5.3.1 Characteristic length scale and the basis of crack tip modeling

Most mathematical theories of fracture mechanics describe the effects of sharp cracks where the material cohesion is lost completely. The real nature of cracks depends on

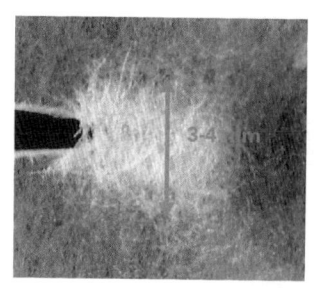

Fig. 5.11 Distribution of microscopic bond failures in a laboratory sheet made of chemical pulp, just before crack propagation starts. Damage is made visible with silicone oil impregnation (Niskanen et al., 2001). Reproduced with permission from The Pulp and Paper Fundamental Research Society (www.ppfrs.org).

the microscopic fracture mechanisms. In paper materials, inter-fiber bond failure, the related fiber pull-out, and fiber breakage dominate. Of these, bond failure is usually the prevalent mechanism in ordinary paper materials (Chapter 11.5). Tens of bonds must fail to release a fiber, and the order of hundreds of fibers must either be released or broken when a crack perpendicular to the paper plane propagates by one fiber length. It follows that the failure of paper cannot be triggered by a single microscopic event. Before macroscopic rupture can occur, microscopic damage is typically accumulated across several millimeters, a distance comparable to fiber length (Fig. 5.11). In engineering fracture mechanics, this area of microscopic damage, which surrounds the crack tip where stress concentrations result in extensive, irreversible deformation and damage, is known as the fracture process zone (FPZ).

As shown in Figure 5.11, the apparent fracture process zones in paper can be diffuse and extend well ahead of the assumed crack tip location. These features are ignored in the continuum mechanics models considered in this chapter, but the microscopic fracture process in paper is discussed in more detail in Chapter 11.5. In similarity to other materials, the dispersion of crack boundaries does not limit the applicability of fracture mechanics in paper materials. As demonstrated later, fracture mechanics can be applied to paper even when macroscopic, clearly visible cracks are absent.

Many fracture mechanics models can be applied to paper materials and products depending on the problem and objectives of the analysis, but it is best to use the simplest possible model that has predictive capability. As shown below, in certain cases a model based on elastic-plastic material properties and a point-wise fracture criterion is sufficient, while in other cases the size of the failure process zone must be accounted for. The latter is particularly true for delamination problems. The choice of model is largely determined by the size of the FPZ in comparison to characteristic dimensions, such as crack length and in-plane dimensions of the product. The point-wise fracture criterion requires that the FPZ is proportionally small, whereas this is not required if the extension of FPZ is included in the model used. The size of the FPZ is usually not known, and the accuracy of the chosen crack tip modeling approach must be verified by comparing the predictions either with experiments or with predictions from a more advanced model that has already been proven reliable.

The fracture criterion from a specific crack tip model must be transferable from laboratory testing to real structures, which, in fracture mechanics, is called *autonomy*.

Autonomy can be investigated by testing different test pieces in laboratory. It has been observed (Mäkelä and Östlund, 2007) that fracture mechanics predictions for paper can be valid, and autonomy can prevail even though the predicted stress fields are not consistent with experimental observations.

5.3.2 Linear elastic fracture mechanics LEFM

The most simple fracture mechanics model is based on the assumptions that damage and nonlinear deformation are confined in a failure process zone at the crack tip that is small in comparison to all characteristic in-plane dimensions. This is known as small-scale yielding, and the FPZ is mathematically treated as a point. The crack tip stress fields can then be determined using a linear elastic material model. The corresponding theory is called linear elastic fracture mechanics (LEFM). Irwin (1957) showed that in mode I loading, the multi-axial stress and strain fields close to the crack tip are characterized by one single parameter, that is, the stress intensity factor K. The nonzero stress components are singular according to

$$\sigma_{ij} = \frac{K}{\sqrt{2\pi r}} f_{ij}(\varphi), \tag{5.1}$$

where the coordinates r and φ are defined in Figure 5.12. Notice that here we define the x- and y-axes relative to the crack orientation, not the MD and CD of paper as elsewhere in this book (cf. Chapter 2.2.1). Details such as the angular functions, $f_{ij}(\varphi)$, the displacement and strain fields, and the effects of orthotropic material properties (of particular importance in the case of paper) are available in the fracture mechanics literature (see, e.g., Sih and Liebowitz, 1968), but Equation 5.1 is sufficient for the present purpose.

Equation 5.1 is the first term in a series expansion, and its validity is therefore restricted to the so-called singularity-dominated zone near the crack tip. Further away from the crack tip, higher order terms should be included. In reality, singular stresses do not exist because no real material can carry infinite loads, and Equation 5.1 is, therefore, invalid in the fracture process zone. Figure 5.13 gives a schematic picture of real crack opening stress σ_{yy} in a linear elastic material.

The stress intensity factor K is useful because one can determine it from measurements with notched specimens. Closed-form solutions for K are available for some simple geometries and loading types, but in most situations numerical methods are

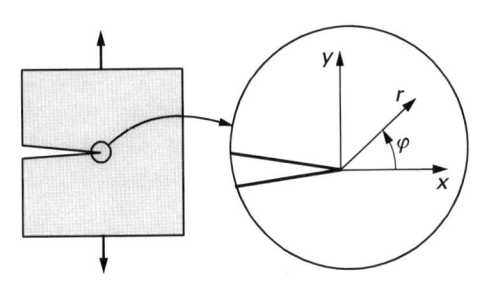

Fig. 5.12 Crack tip coordinates used in Equation 5.1.

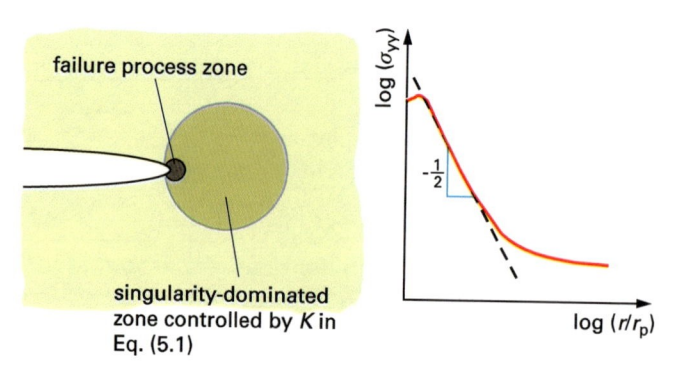

Fig. 5.13 Qualitative distribution of the crack opening stress σ_{yy} in a linear elastic material. The characteristic length scale r_p may be chosen to be the size of the fracture process zone.

required for evaluation of K. The numerically obtained relations between K and external stresses for finite specimens are usually expressed in the form of

$$K = \sigma\sqrt{\pi a}f(\text{geometrical parameters}), \tag{5.2}$$

where σ is the remote stress, a is the crack size (e.g., diameter or half length), and f is a function of the geometry and characteristic dimensions of the structure. The factor $\sigma\sqrt{\pi a}$ originates from the closed-form solution of a mode I center crack in an infinite plate. Experiments with notched specimens can be used to evaluate the value of K at which crack growth begins. In LEFM, this critical value K_c is taken to be a material property called *fracture toughness*. If K_c is known, one can calculate from Equation 5.2 the critical stress state that initiates crack growth. Because K_c is a material property in LEFM, the theory predicts that, independent of geometric dimensions and load type, crack growth and, therefore, macroscopic fracture begins when K calculated from Equation 5.2 reaches K_c. Thus, the LEFM fracture criterion is

$$K = K_c. \tag{5.3}$$

The first studies on the defect-sensitivity of paper materials were based on the LEFM theory, but later investigations have shown that LEFM is accurate only for large cracks in large paper structures.

5.3.3 Nonlinear fracture mechanics using J-integral

If the material behavior in the crack tip region deviates distinctly from linear elasticity, more complex fracture mechanics models are needed. This is the case for paper materials. The most well-known crack tip model for nonlinear materials assumes a parabolic nonlinear elastic material model. The uniaxial version of the stress-strain expression is

$$\varepsilon = \frac{\sigma}{E} + \left(\frac{\sigma}{E_0}\right)^n, \tag{5.4}$$

where n is a strain-hardening exponent and E_0 is called the strain-hardening modulus. This leads to the following stress field in the vicinity of the crack tip:

$$\sigma_{ij} = \alpha \cdot \left(\frac{J}{r}\right)^{\frac{1}{n+1}} \cdot f_{ij}(n,\varphi). \tag{5.5}$$

The scalar multiplier α depends on the material and stress state (plane strain or plane stress). This crack tip solution is known as the HRR-field, after Hutchinson (1968) and Rice and Rosengren (1968). They showed that the crack tip conditions in a nonlinear elastic material are characterized by a path-independent contour integral known as the J-integral, which was introduced by Rice (1968).

The HRR-field predicts that (a) the stresses in the vicinity of a crack tip are singular and governed by the stress–strain behavior of the material, (b) the J-integral determines the amplitude of the stresses, and (c) the stresses have a uniform distribution. Therefore, in nonlinear elastic materials the stresses close to a crack tip are characterized by J, in analogy with K in the case of linear elastic materials. In fact, LEFM is a special case of nonlinear fracture mechanics. If the material is linear elastic and plane stress conditions apply, then $J = K^2/E$, which means that the HRR-model can be used to estimate the length scale outside which the LEFM model (small-scale yielding approximation) applies. A schematic illustration of this is shown in Figure 5.14. In addition to the HRR-field, other single-parameter characterizations of the crack tip stress fields that use J are possible. The necessary conditions for the existence of a J-dominated zone are that the stresses and strains scale with J/r and are distributed uniformly inside this zone.

In analogy with LEFM, experiments on notched specimens can be used to determine the critical value, or nonlinear fracture toughness J_c, that is needed for crack growth to start. Hence the criterion for crack growth is expressed as

$$J = J_c. \tag{5.6}$$

Nonlinear fracture mechanics was introduced for paper materials by Uesaka et al. (1979). Initially, the research on fracture mechanics of paper materials was focusing on the experimental methods to determine the fracture toughness J_c and its dependence on the papermaking parameters. However, J_c itself cannot be used to compare

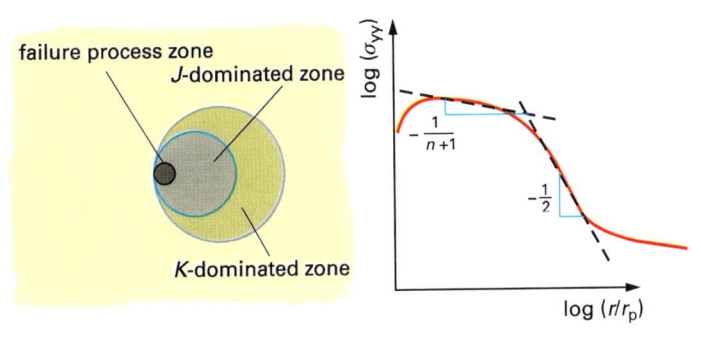

Fig. 5.14 Qualitative distribution of the crack opening stress σ_{yy} in a nonlinear elastic material. The characteristic length scale r_p may be chosen to be the size of the fracture process zone.

the defect-sensitivity of paper materials because the relationship between J and external loading, and therefore also the critical value J_c, depends also on the material behavior (n and E_0 in Eq. 5.4). Apart from a few exceptions, numerical methods are therefore needed to determine the relationship between J and the external loading. The J-integral evaluation is in fact implemented in most of the commercially available finite element codes that are intended for nonlinear fracture mechanics analyses.

Nonlinear elastic materials have a unique relationship between strains and stresses, while in elastic-plastic materials the stress state depends on both the strain state and the strain history. The latter is the case for paper (Chapter 2.3). However, there is no difference between the two materials if no unloading takes place. Thus, the previously discussed nonlinear elastic treatment applies also to elastic-plastic materials as long as the material does not recoil because of damage creation prior to crack growth. The nonlinear elastic approximation of an elastic-plastic material is called the *deformation theory of plasticity*.

In the case of multi-axial stresses and deformations, the nonlinear elastic treatment of elastic-plastic materials would also require proportional loading. In fact, a deformation theory relationship is generally not identical to the corresponding elastic-plastic relation, even under the special case of strictly increasing uniaxial loading, but the differences are small. Therefore, the J-integral together with the concept of deformation theory of plasticity is valid far beyond the limits of LEFM.

Figure 5.15 summarizes the practical use of nonlinear fracture mechanics to evaluate the defect sensitivity of paper materials with the following steps:

1. Determine the material behavior (Eq. 5.4) from tensile testing on the usual standard test pieces.
2. Determine the nonlinear fracture toughness J_c from tensile testing on notched rectangular test pieces. The finite element method is needed in the evaluation.
3. Predict the failure of products or structures of interest from Equation 5.6. The calculation of J requires finite element analysis, and the values of J differ depending on the type of loading.

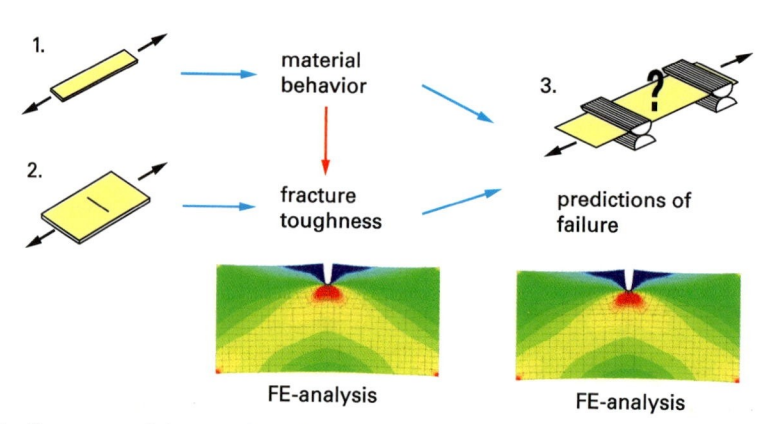

Fig. 5.15 Summary of the use of nonlinear fracture mechanics to predict failure.

5.3.4 Cohesive zone models

The cohesive zone model was originally suggested as a correction of LEFM for large-scale plasticity in metals. It was further developed for the analysis of damage evolution in concrete and later in paper (Tryding, 1996). In the cohesive zone modeling, damage is assumed to be confined to a narrow zone ahead of the crack tip. The explicit damage behavior is replaced by a cohesive stress-widening curve. For mode I loading, this curve is a function of only the crack widening u_w (Fig. 5.16). This simple model description of damage is uniaxial so that only the material properties parallel to the external load are considered.

The cohesive stress-widening curve describes the stress–strain behavior of the material after the point of maximum stress is passed, in other words, the post-peak behavior. Short specimens have to be used to observe this behavior because standard test specimens of paper fail immediately and completely when the maximum stress is reached. The system becomes unstable because the recoil in elongation outside the fracture process zone must be equal the elongation inside the damaged zone, and the latter easily makes up a very large strain increment if the damaged zone is narrow compared to specimen length. Thus, a short enough specimen length removes the instability, and the post-peak stress-elongation curve can be detected (Hillerborg et al., 1976). In order to determine the cohesive stress-widening curve, one assumes that the total elongation of a test piece (Fig. 5.17) can be decomposed into a sum of uniform elongation in the undamaged material and widening u_w in the damaged zone

$$u = \varepsilon L + u_w, \tag{5.7}$$

where ε denotes the strain in the undamaged material and L is the unstrained length of the test piece. In the post-peak region ε can be expressed as

$$\varepsilon = \frac{u_0}{L} - \frac{\sigma_0 - \sigma}{E}. \tag{5.8}$$

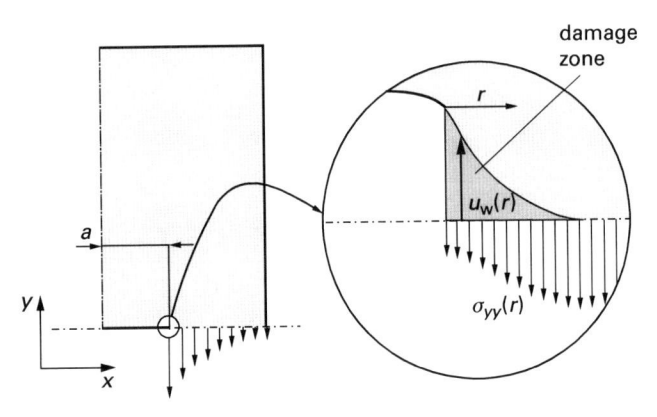

Fig. 5.16 Crack tip modeling using the cohesive zone model. The cohesive stress-widening curve is given by $\sigma(u_w)$.

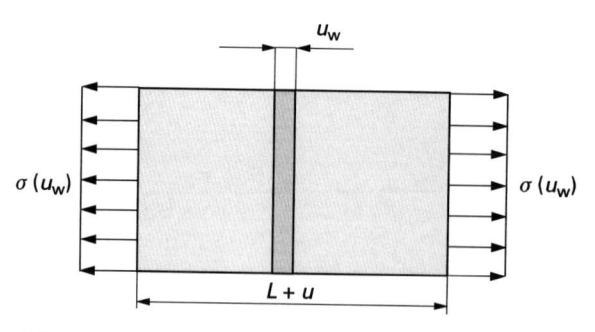

Fig. 5.17 Elongation of a test piece with a narrow damage zone, used to determine the post-peak cohesive stress. Reproduced from Mäkelä (2002) with permission from Svenska Pappers- och Cellulosaingeniörsföreningen (SPCI).

Here σ_0 and u_0 are the peak-stress (also referred to as the cohesive stress) and the corresponding elongation (Fig. 5.18), and E is the elastic modulus. Combining Equations 5.7 and 5.8, one obtains the cohesive stress-widening relation

$$u_w = u - u_0 + \frac{\sigma_0 - \sigma}{E} L. \tag{5.9}$$

The third term of this expression accounts for the elastic recovery (unloading) of the structure outside the damaged zone. The main assumptions behind Equation 5.9 are that this recovery occurs far from the damaged zone and that the irreversible elongation in the damage zone consists only of post-peak damage growth. Observe that the measurement does not require notched structures.

Measured cohesive stress-widening relations can be included in a material model to evaluate the defect-sensitivity of paper structures. A separate fracture criterion is not needed. When the calculated local strain exceeds the peak value u_0, the softening of the material is given by Equation 5.9. One can then see when the overall load-carrying capacity of the structure begins to decrease. This leads to instability if it is the external load that is increased, but not necessarily if the external deformation is increased. In the latter case one has to determine if the system is stable or unstable against further infinitesimal increases in deformation.

Usually the cohesive zone is not fully developed before a typical paper structure with defects becomes unstable. This means that some of the theoretical fracture energy

$$W_f = \int_0^{u_w^{max}} \sigma(u_w) \, du_w \tag{5.10}$$

is *not* consumed before the instability occurs.

The cohesive crack model is not limited to situations where a macroscopic defect exists. Any situation that brings the local stress above σ_0 will create a damaged region that may subsequently transform into a growing crack due to the localized softening of the material. Practical use of the cohesive crack model requires the following three steps:

1. Determine the stress–strain properties of the material from the usual standard specimens.

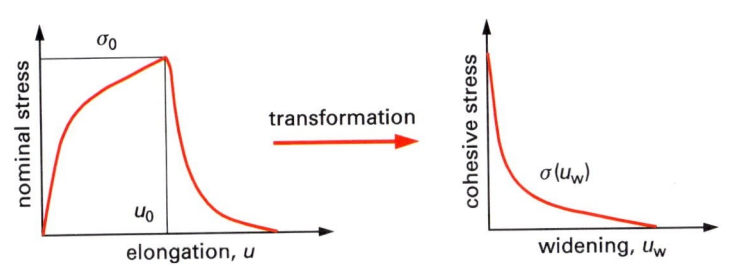

Fig. 5.18 Transformation of a stable load-elongation curve into a cohesive stress-widening curve. Reproduced from Mäkelä (2002) with permission from Svenska Pappers- och Cellulosaingeniörsföreningen (SPCI).

2. Determine the cohesive stress-widening curve from short test pieces.
3. Calculate what load or elongation is needed for instability.

The last step requires finite element analysis with the cohesive crack model included.

In Figure 5.19 three different crack tip models are compared in the case of MD and CD loading of center-cracked copy paper specimens. The analysis includes LEFM and the J-integral method, and cohesive zone modeling with an incremental elastic-plastic material behavior. The comparison with experiments shows that, in general, the linear elasticity assumption yields poor accuracy. The nonlinear J-integral method gives accurate predictions for medium-size and large cracks. Cohesive zone modeling with parabolic strain-hardening gives accurate predictions for the load and strain at failure for all the crack sizes.

5.3.5 Continuum damage mechanics modeling of paper

In paper materials the microscopic fracture mechanisms are fiber–fiber bond failure and fiber breakage. Continuum damage mechanics describe the effect of microscopic fractures at the macroscopic level without making any assumptions about the nature of the microscopic processes (Kachanov, 1958). The foundation of this theory is a damage parameter that characterizes the average degradation of the material. In the simplest case, a scalar damage parameter D ranges from $D = 0$ (undamaged material) to $D = 1$ (completely ruptured material). Several damage mechanisms could also be treated simultaneously, and anisotropy in the damage development could be included by treating D as a higher order tensor, but that will not be considered here.

There are different theories regarding the effect of D on the constitutive parameters, such as the elastic energy equivalence used by Isaksson, Gradin, and Kulachenko (2005). In their analysis of isotropic paper sheets, the change in elastic energy is described by a change in the stress tensor σ_{ij},

$$\sigma_{ij} = (1-D)\hat{\sigma}_{ij}, \qquad (5.11)$$

where the effective stress tensor $\hat{\sigma}_{ij}$ represents the behavior of the undamaged material. The elastic stiffness tensor C_{ijkl} of the damaged material is related to the stiffness tensor C^0_{ijkl} of the undamaged material by

$$C_{ijkl} = (1-D)^2 C^0_{ijkl} \qquad (5.12)$$

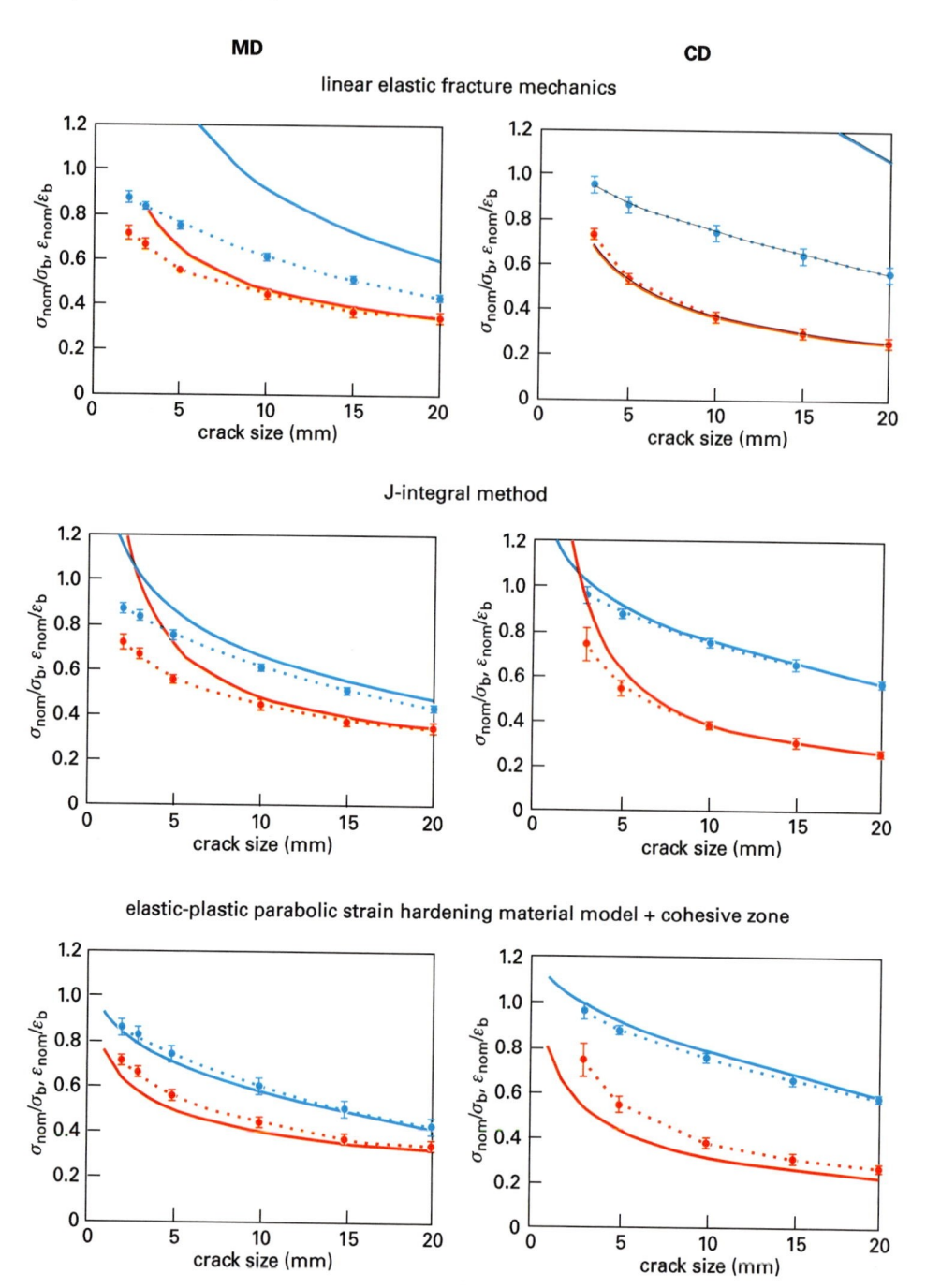

Fig. 5.19 Comparison between numerically predicted (solid curves) and experimentally determined (dashed-dotted curves) critical stress (blue curves) and strain (red curves) at failure for center-cracked copy paper test pieces in MD and CD (left and right, respectively) using three different fracture mechanics models. The results are normalized with respect to stress and strain at break for a noncracked paper test piece of the same dimension as in the experimental study (100 mm by 50 mm).

The driving force for damage evolution at a point is the damage energy release rate Y given by, $Y = \partial \psi / \partial D$, where ψ is the complementary elastic stress energy density. If the material is linear elastic in its undamaged state, then

$$Y = (1-D)^3 \sigma_{ij} \left[C_{ijkl}^0 \right]^{-1} \sigma_{kl}. \tag{5.13}$$

Therefore, when Y reaches a critical value Y_0 at a point, damage will start growing at that point.

Because damage in paper materials is intrinsically nonlocal (Fig. 5.12), the damage accumulation at a point is affected not only by the local value of Y, but also by stresses in the local neighborhood. To address this, Isaksson et al. (2005) introduced a nonlocal damage driving force characterized by a length scale that turns out to be of the order of mean fiber length. To complete the model, Isaksson et al. (2005) proposed the damage evolution law

$$D = 1 - e^{-k(Y-Y_0)^m} \text{ for } Y = H, \tag{5.14}$$

where k and m are parameters that can be determined from experiments, and H is a threshold value that Y has reached during the loading history representing a damage hardening behavior. Figure 5.20 shows a typical comparison of the model with a measured stress–strain curve and damage evolution in a short and wide paper specimen. The cumulative number of acoustic emission events is used as the experimental measure of damage. In the calibration and application of continuum damage mechanics models it is necessary to use the finite element method.

5.3.6 Delamination of paper materials

The importance of out-of-plane mechanical properties was discussed in Section 5.2.2. Considerable efforts have been put into the development of robust and reliable methods for experimental determination of the ZD normal and shear stiffnesses, and the ZD strength properties of paper materials. These are not simple tasks, and

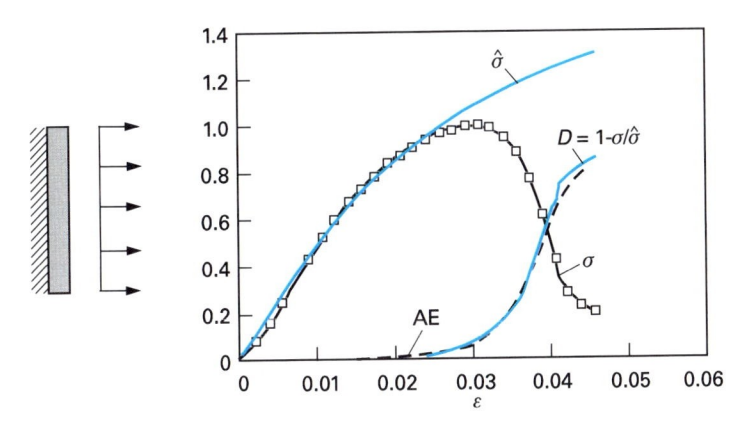

Fig. 5.20 Measured and modeled stress–strain and damage curves of a short and wide paper specimen, shown on left. Acoustic emission (AE) events are used to measure damage accumulation. Reprinted from Isaksson et al. (2005) with permission from Elsevier.

even fracture mechanics–based test methods, such as the double cantilever beam (Anderson, 2005), have been suggested for the purpose. However, Girlanda et al. (2005) have shown that crack tip stress fields cannot fully develop in the ZD because of the heterogeneous structure and low thickness of paper materials.

Figure 5.21 shows one approach to model the ZD deformation behavior of a multi-ply carton board. The undamaged material is described with a continuum material model and the interfaces between plies with a traction-displacement relation similar to the cohesive zone model in Section 5.3.4. An advanced anisotropic elastic-plastic model is needed to describe the undamaged material behavior. The interface model works without macroscopic defects, which is useful in the present case. A structural integrity analysis based on the existence of defects could also be used. The interface model described here could be applied at any ZD position, not only at the interfaces between the carton board plies.

The interface model describes how two material surfaces get separated or delaminate during ZD loading. Delamination is assumed to initiate due to a combination of out-of-plane normal (mode I) and shear loadings (mode II and III). The displacements of the two opposing surfaces relative to one another are assumed to consist of elastic and damage components,

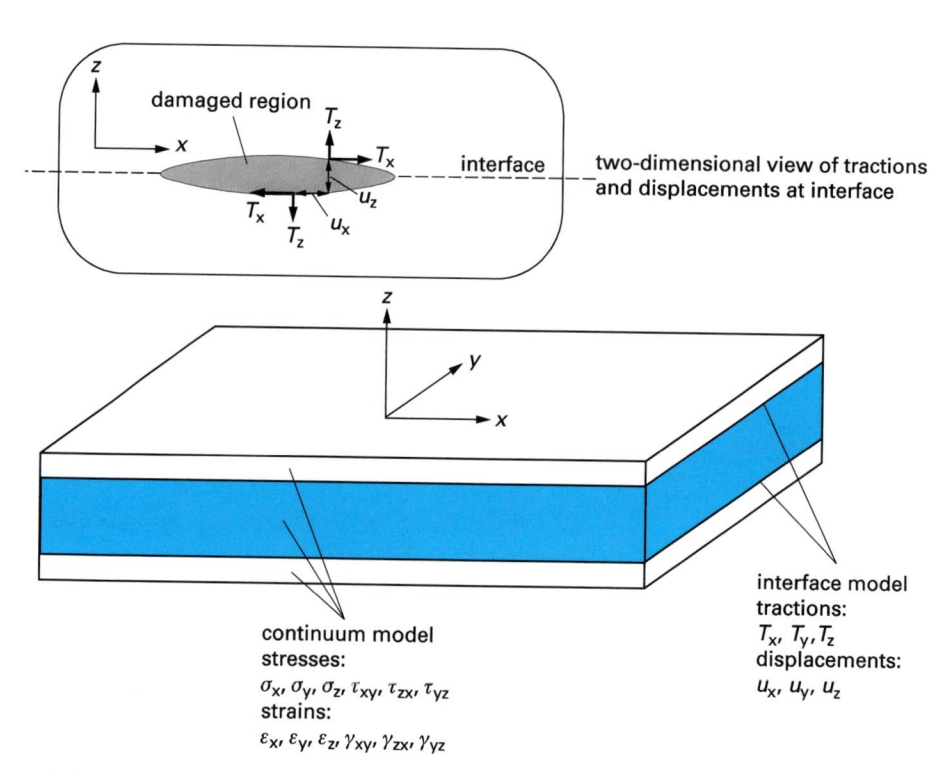

$$u_i = u_i^e + u_i^d, \tag{5.15}$$

Fig. 5.21 Combination of a continuum model and an interface model for a three-ply carton board.

where i represents the three directions x, y, and z in the local coordinate system. This assumption means that plastic deformations can happen only inside the undamaged plies. The relevant stress components are σ_z, τ_{zx}, and τ_{yz}. The total displacement grows in increments

$$\Delta u_i = \Delta u_i^e + \Delta u_i^d, \tag{5.16}$$

and we assume that the corresponding traction increments are proportional to the change in the elastic displacement, or

$$\Delta T_i = C_i \left(\Delta u_i - \Delta u_i^d \right) \text{ (no summation)}. \tag{5.17}$$

Here C_i are the components of the instantaneous interface stiffness in the i-direction. Experimental data (Stenberg, 2002) demonstrates that damage affects the interface properties. In the model this is accounted for by *reductions* in the interface strength components σ_i^{max} and stiffnesses C_i. Interface failure caused by the damage is then described by a failure criterion analogous to the Tsai-Wu criterion (Section 2.4), stating that local failure occurs when

$$f^{\text{interface}} \left(\sigma_i^{max}, C_i \right) = 0, \tag{5.18}$$

In the analysis of the creasing of a carton board, Nygårds et al. (2009) used the expression

$$f^{\text{interface}} = \left(\frac{T_x}{\sigma_x^{max}} \right)^2 + \left(\frac{T_y}{\sigma_y^{max}} \right)^2 + \frac{T_z}{\sigma_z^{max}} - 1. \tag{5.19}$$

No initial crack is needed in the model. Initiation of damage is modeled in a manner equivalent to plastic deformation in metals, and the integration of the previous equations follows a procedure analogous to the one used in the theory of plasticity (Ottosen and Ristinmaa, 2005). In the case of pure shear stresses ($T_z = 0$), an associated damage rule will result in normal dilatation, that is, an increase in paper thickness. This has also been observed in experiments. For details on the application of an interface model of the type described here, see Nygårds et al. (2009).

Although the principle of the model is straight-forward, its experimental calibration is challenging. Firstly, low thickness means that high precision is needed when measuring the ZD displacements, and secondly, there is no well-established method to separate the nonlinear part of the ZD deformation into the continuum and damage components that are required by the model.

5.4 Compressive failure

Compression properties are important in the performance of boxes under stacking load, as discussed in Chapter 3. The measurement of the compressive failure of paper materials is challenging because thin sheets buckle easily. Macroscopic buckling is the most important compressive failure mechanism in many products. Different approaches to avoid buckling have been proposed (Fellers, 1986), including short span testing, using tubular specimens, or preventing out-of-plane deformations with narrow blades.

As demonstrated in Figure 2.14, the compressive stress–strain behavior of paper is different from the tensile behavior. The compressive strength is only about 30% of the tensile strength. Compressive loading of a paper specimen to levels close to failure will not affect the strength obtained in a subsequent tensile test, whereas tensile loading to levels close to failure leads to a pronounced reduction in the compressive strength. Thus, the microscopic mechanisms of failure in compression are different from the mechanisms in tension.

The microscopic failure of paper in compression is caused by a structural instability (not rupture) of the fiber network. This happens through either the buckling of free fiber segments (Chapter 11.2) or the shear dislocations in fiber walls (Fig. 5.22). The first happens primarily in low density sheets, and the latter in medium and high density sheets. The microscopic buckling and shear dislocations in the fiber network change the distribution of compressive stresses. When compressive load is increased, more fiber segments buckle and shear dislocations increase until the whole sheet becomes structurally unstable. The macroscopic compressive failure is often associated with a shear slip dislocation that shows also delamination (Fig. 5.23).

The microscopic failure mechanism in compression explains why prior tensile loading affects compressive strength but prior compressive loading does not affect tensile strength. Inter-fiber bond failures in tensile loading increase the mean length of free

Fig. 5.22 Microscopic compressive failures in paper: buckling of fiber segments (left) and shear dislocations in fiber walls (right). Courtesy of Christer Fellers.

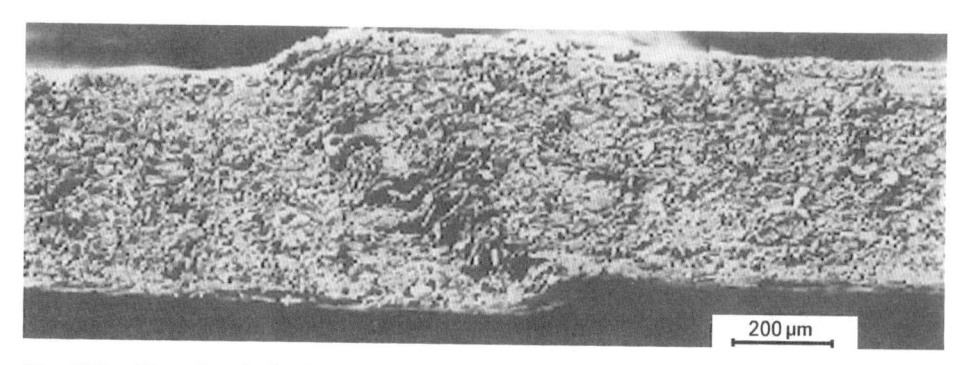

Fig. 5.23 Shear band slip failure in compression, also showing delamination. Courtesy of Christer Fellers.

fiber segments (Chapter 11.2.3). As a result, the buckling threshold of fiber segments decreases. In the same way, if the creasing or folding of a carton board creates excessive delamination outside the fold area, then the compressive strength of a box decreases.

5.5 Summary

Let us return to the problem of tensile breaks caused by defects. Figure 5.24 demonstrates the application of the J-integral method of fracture mechanics to four different paper grades: testliner (a special type of linerboard), fluting, sack paper, and newsprint. The material properties were determined using standard tensile testing. The fracture toughness of each material was determined using 50 mm wide test pieces, clamping length 100 mm, and 20 mm wide center notches. Antibuckling guides prevented out-of-plane buckling in the vicinity of the notches. The experiments were used to determine the parameters of a nonlinear fracture mechanics model based on the isotropic deformation theory of plasticity of Section 5.3.3. Reasonable agreement was found between experiments and model predictions for the elongation and force at failure of very large paper sheets (800–1,000 mm wide and 1,800 mm long) that contained edge cracks of different sizes.

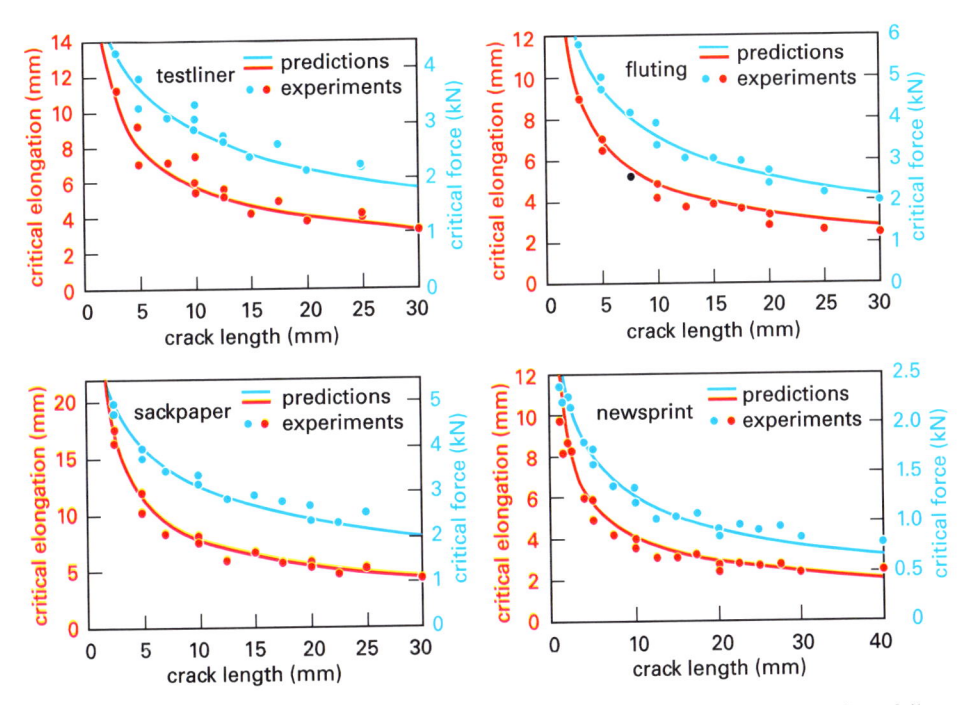

Fig. 5.24 Comparison of measured and calculated force (blue) and elongation (red) at failure plotted against the length of edge crack in 1,800 mm long and 800–1,000 mm wide paper sheets of testliner, fluting, sackpaper, and newsprint. Calculation used nonlinear fracture mechanics based on isotropic deformation theory of plasticity (Mäkelä et al., 2009). Reproduced with permission from Svenska Pappers- och Cellulosaingeniörsföreningen (SPCI).

The goal of this chapter was to describe the use of continuum fracture mechanics in the analysis of paper materials and products. The choice of the appropriate model depends on the size of the fracture process zone (FPZ) in relation to the structural dimensions. If it can be assumed that the FPZ is proportionately small, stress and strain fields are singular at the crack tip. The results in Figure 5.24 demonstrate that, in this case, nonlinear fracture mechanics based on the J-integral predicts well the mode I in-plane failure of notched paper structures. If the nonzero size of the FPZ cannot be ignored, then the cohesive zone model based on the cohesive stress-widening curve gives excellent predictions for mode I in-plane failure. Models based on the cohesive stress-widening curve are also applicable for out-of-plane failure of paper materials where the concept of crack tip singularity is not necessarily applicable.

Fracture mechanics modeling can thus be used to predict the failure point of notched structures as well as for damage tolerance analysis of structures that contain defects. The latter approach has not been fully exploited in the performance analysis of paper products, in contrast to its widespread use in other branches of engineering. Special complications with paper materials arise from the effects of moisture, creep and relaxation, loading dynamics, and the heterogeneity of paper properties. These aspects of paper materials are further explored in the next chapters.

References

Anderson, T. L. (2005). Fracture Mechanics: Fundamentals and Applications, 3rd Edition (Boca Raton, FL, USA: CRC Press, Taylor & Francis Group).

Fellers, C. (1986). The significance of structure for the compression behavior of paperboard. In: Paper – Structures and Properties, J. A. Bristow and P. Kolseth, eds. (New York, NY, USA: Marcel Dekker).

Girlanda, O., Hallbäck, N., Östlund, S., and Tryding, J. (2005). Defect sensitivity and strength of paperboard in out-of-plane tension and shear. J. Pulp Pap. Sci. 31, 100–104.

Hillerborg, A., Modéer, M., and Petersson, P.-E. (1976). Analysis of crack formation and crack growth in concrete by means of fracture mechanics and finite elements. Cement Concrete Res. 6, 773–782.

Hutchinson, J. W. (1968). Singular behavior at the end of a tensile crack tip in a hardening material. J. Mech. Phys. Sol. 16, 13–31.

Irwin, G. R. (1957). Analysis of stresses and strains near the end of crack traversing a plate. J. Appl. Mech. 24, 361–364.

Isaksson, P., Gradin, P. A., and Kulachenko, A. (2005). The onset and progression of damage in isotropic paper sheets. Int. J. Solids Struct. 43, 713–726.

Kachanov, L. M. (1958). Time of the rupture process under creep conditions. Izv. Akad. Nauk. SSSR. Otd Tekhn Nauk (In Russian), 26–31.

Mäkelä, P. (2002). On the fracture mechanics of paper. Nord. Pulp Pap. Res. J. 7, 254–274.

Mäkelä, P., and Östlund, S. (2007). Cohesive crack modelling of paper materials. In: Proceedings of the 61st Appita Annual Conference and Exhibition, Gold Coast, Australia, pp. 357–364.

Mäkelä, P., Nordhagen, H., and Gregersen, Ö. W. (2009). Validation of isotropic deformation theory of plasticity for fracture mechanics analysis of paper materials. Nord. Pulp Pap. Res. J. 24, 388–394.

Niskanen, K., Kettunen, H., and Yu, Y. (2001). Damage width: A measure of the size of fracture process zone. In: The Science of Papermaking, Transactions of the 12th fundamental research symposium, Oxford, September 2001, C. F. Baker, ed. (UK: The Pulp and Paper Fundamental Research Society).

Nordhagen, H. (2009). Development of fracture mechanics to study end use of paper webs. Doctoral thesis. Norwegian University of Science and Technology, Trondheim, Norway.

Nygårds, M., Just, M., and Tryding, J. (2009). Experimental and numerical studies of creasing of paperboard. Int. J. Solids Struct. 46, 2493–2505.

Ottosen, N. S., and Ristinmaa, M. (2005). The Mechanics of Constitutive Modeling (Oxford, UK: Elsevier).

Rice, J. R. (1968). A path independent integral and the approximate analysis of strain concentration by notches and cracks. J. Appl. Mech. 35, 379–386.

Rice, J. R., and Rosengren, G. F. (1968). Plane strain deformation near a crack tip in a power-hardening material. J. Mech. Phys. Sol. 16, 1–12.

Sih, G. C., and Liebowitz, H. (1968). Mathematical theories of brittle fracture. In: Fracture, Vol. 2, H. Liebowitz, ed. (New York, NY, USA: Academic Press).

Skjetne, B. (2006). Numerical studies of brittle-elastic fracture in random nedia. Doctoral thesis. Norwegian University of Science and Technology, Trondheim, Norway.

Stenberg, N. (2002). Out-of-plane shear of paperboard under high compressive loads. J. Pulp Pap. Sci. 17, 387–394.

Swinehart, D., and Broek, D. (1995). Tenacity and fracture toughness of paper and board. J. Pulp Paper Sci. 21, J389–J397.

Tryding, J. (1996). In-plane fracture of paper. Doctoral thesis. Division of Structural Mechanics, Lund University, Lund, Sweden.

Uesaka, T., Okaniwa, H., Murakami, K., and Imamura, R. (1979). Tearing resistance of paper and its characterization. J. Japan Tappi 33, 403–409.

Wellmar, P., Gregersen, Ø. W., and Fellers, C. (2000). Predictions of crack growth initiation in paper structures using a *J*-integral criterion. Nord. Pulp Pap. Res. J. 15, 4–11.

Part II
Dynamic stability

6 Web dynamics in paper transport systems

Tetsu Uesaka

6.1 Introduction

Paper and board are manufactured and converted in web form. The smooth, trouble-free transport of the paper web is essential not only in papermaking but also in printing and converting. Examples of web transport systems can be seen in paper machines (Fig. 6.1), corrugators, coaters, printing machines (Fig. 6.2), label machines, and so forth. They typically consist of (1) sections called open draw, where the web is running under certain tension without any supporting fabric; (2) sections called closed draw, where the web is supported by one or two fabrics; and (3) the nips or contact points, where the web is contacted by two rolls under some compression, by one roll, or by a roll and fabric. The web is transported over a number of rolls, some of which drive other rolls, and others simply change the direction of the web. The relative speed difference between two roll nips is called *draw*. The draw is an important parameter of web transport problems.

When considering practical problems in web transport systems, one may need to pay attention to phenomena at two distinct length scales: (1) web behavior between rolls (meter or sub-meter scale) and (2) within the nip (sub-centimeter scales). This chapter focuses on problems at the first length scale. Chapter 10 deals with the second aspect.

The primary motivation for investigating the dynamics of web transport systems comes from practical issues, such as web breaks on paper machines and printing presses and web instabilities (e.g., wrinkles that lead to creasing and web breaks). Already the literature from early days shows that the way in which the web is run in a transport system has enormous effects on the frequency of web breaks and wrinkles (Grant, 1967; Larsson, 1984; Page and Bruce, 1985; Uesaka, 2005; Uesaka and Ferahi, 1999). In the field of printing, operators recognized for years that some press units are easy to run and others are difficult. It is also not exaggeration to say that the history of paper machine development is filled with a series of attempts to stabilize the web transport systems (Paulapuro, 2008).

Typical problems encountered in practical situations are (1) tension variations due to changes in draw; (2) large transverse vibrations of web due to aerodynamic effects and system resonances; (3) time-varying or spatially nonuniform adhesion on press rolls and dryer rolls; (4) instabilities associated with web property variations and hygroexpansion or shrinkage; and (5) web tension nonuniformity in the cross machine direction (often called "baggy edges"). Instead of trying to cover the whole area of web handling, this chapter focuses on webs running in open draw sections of printing machines and the wet end of paper machines. Winding and reeling mechanics or paper roll deformations in the finishing section of paper machine are not discussed.

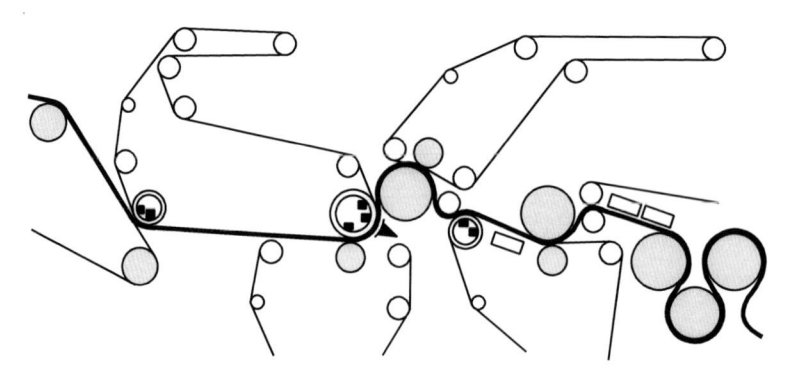

Fig. 6.1 Web transport in the press section of a paper machine.

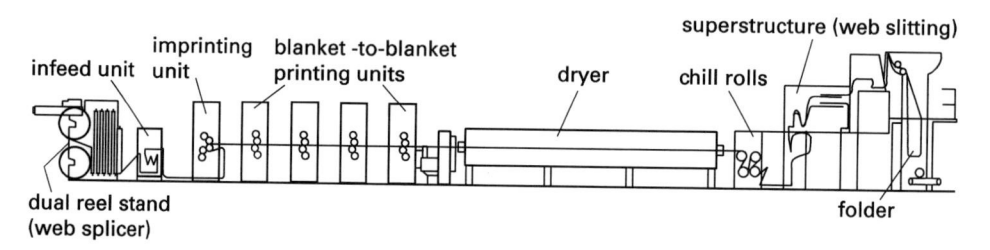

Fig. 6.2 Web transport system in a commercial web heatset offset press (Source: © Helmut Kipphan, Handbook of Print Media, Springer, 2001). Reproduced with permission.

The reader is referred to excellent textbooks on these matters (Good, 2007; Jokio, 1999; Paulapuro, 2008; Roisum, 1994, 1998).

6.2 Dynamics of web transport

6.2.1 Basic formulation of web transport problems

When analyzing the dynamics of web transport, one needs the conservation equations for mass and momentum. The applications to specific situations are solved either analytically or numerically. This section introduces the basic formulation of the problems in preparation for the subsequent analyses.

Figure 6.3 shows a portion of web moving in a web transport system. In general, the web is a three-dimensional object. We consider an infinitesimally small volume element of the web. The conservation law of mass states that the rate of mass change is equal to the difference between the mass entering and leaving the volume element (Fung, 1965):

$$\frac{\partial \rho}{\partial t} + \frac{\partial \rho v_i}{\partial x_i} = 0, \tag{6.1}$$

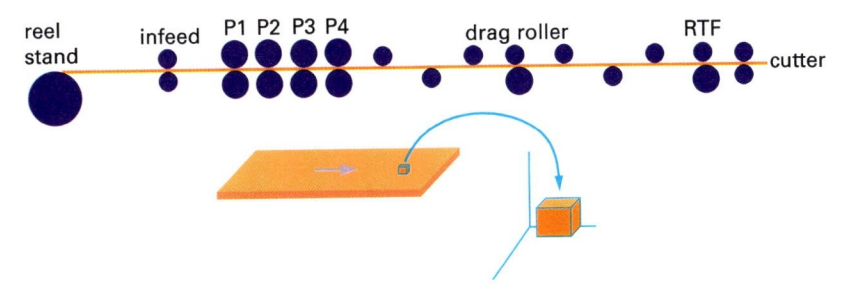

Fig. 6.3 A volume element of a web moving through the web transport system of a printing press.

where ρ is density and v_i is velocity in the i-direction. The first term represents the rate of density change, and the second term represents the net mass flow per unit volume. In general, both density ρ and velocity v_i depend on the position and time. In web dynamics, density changes are mainly caused by web strain that is spatially nonuniform and time-dependent, for example, a sudden tension surge.

The second important law is the conservation of momentum (or Newton's second law) (Fung, 1965):

$$\rho\left(\frac{\partial v_i}{\partial t} + v_j \frac{\partial v_i}{\partial x_j}\right) = \frac{\partial \sigma_{ij}}{\partial x_j} + B_i, \tag{6.2}$$

where σ_{ij} is a stress tensor component, and B_i is a body force component (e.g., gravity). The left side represents the rate of momentum change for the moving unit volume element. The right side gives the sum of forces acting on the volume element, consisting of surface forces and the body forces acting on the unit volume.

The general conservation laws (Eqs. 6.1 and 6.2) must be solved together with the boundary conditions and stress–strain relationship to obtain the three-dimensional motion of the web and the distribution of stress and strain in the web. Typical boundary conditions in web transport systems arise from (1) roll contacts: velocity, contact forces, and web-roll adhesion forces; and (2) interaction with the surrounding air: drag force, air friction, and inertial force due to air motion. In the following sections we apply equations 6.1 and 6.2 to extract the specific characteristics of web transport in the printing press and paper machine.

6.2.2 The case of an axially moving web

An axially moving web can be simplified by assuming that all the field variables are functions of only the position in the machine direction (MD), or **x**-direction of the web, and that the stress–strain relationship is linear elastic:

$$v = (v_x, 0, 0) \tag{6.3}$$

$$\sigma_x = C_x \varepsilon_x,$$

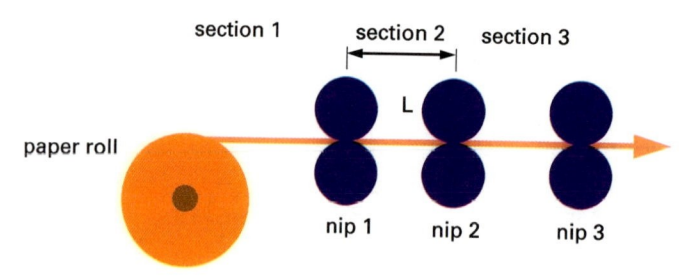

Fig. 6.4 Notation of web sections and nips in an axially moving web. L is the length of the open span. (Hristopulos and Uesaka, 2002). Reproduced with permission from Pulp and Paper Technical Association of Canada (PAPTAC).

where v_x is the speed and ε_x the strain of the web in the machine direction, and C_x is the corresponding tensile stiffness. When considering a web section between two nips (Fig. 6.4), it is convenient to take ε and σ to represent the strain and stress *increments* created in the section. Then, the mass conservation and momentum conservation equations can be rewritten as (Hristopulos and Uesaka, 2002),

$$\frac{\partial \varepsilon}{\partial t} - \frac{\partial ((1-\varepsilon)v)}{\partial x} = 0, \tag{6.4}$$

$$\frac{\partial ((1-\varepsilon)v)}{\partial t} + \frac{\partial ((1-\varepsilon)v^2)}{\partial x} = c^2 \frac{\partial \varepsilon}{\partial x}, \tag{6.5}$$

where the subscripts for strain and velocity are omitted for convenience so that $v = v_x$ and $\varepsilon = \varepsilon_x$; $c^2 = C_x/\rho$ represents the speed of longitudinal waves in the MD.

We first consider *the steady state case* so that the field variables ε and v are independent of time. Equations 6.4 and 6.5 then give

$$(c_0^2 - c^2)\frac{\partial \varepsilon}{\partial x} = 0, \tag{6.6}$$

where $c_0 = v(1-\varepsilon)$ is a constant according to Equation 6.4, and

$$c_0^2 = v^2(1-\varepsilon)^2 \ll c^2 = \frac{C_x}{\rho}. \tag{6.7}$$

Therefore, the only possible solution for Equation 6.6 is that strain and speed are constant over each open draw section and instead change discontinuously at the nips, as shown in Figure 6.5. The total strain in a certain section is the sum of all the strain increments created up to that section.

An important implication of the general steady-state solution is that paper is strained at very high rate in the nips. The corresponding stress–strain behavior cannot be detected with the standard tests. For example, suppose that a printing press is running at 10 m/s with a 0.02 m/s speed difference between two rolls, and a typical nip length is 1 cm. Then the strain rate of the web in the nip is 200%/s, or 1,200 times higher than the standard strain rate in tensile testing. One should instead measure

6.2 Dynamics of web transport

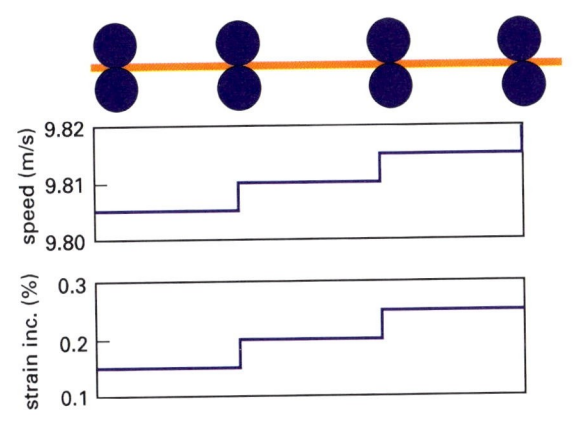

Fig. 6.5 Speed and strain changes in a steady state web transport system, from Hristopulos and Uesaka (2002). Reproduced with permission from Pulp and Paper Technical Association of Canada (PAPTAC).

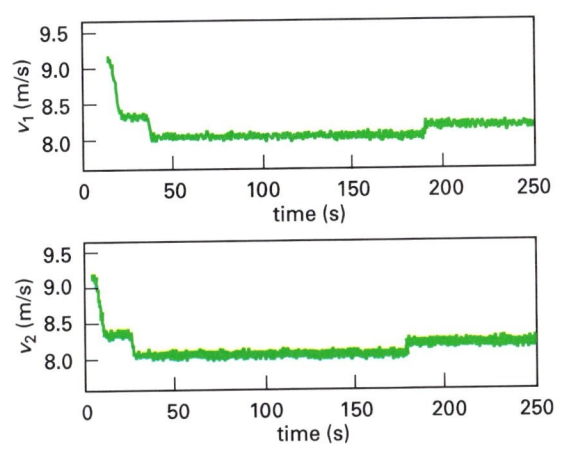

Fig. 6.6 Web speeds at nip 1 and nip 2 (Fig 6.4) in a printing press, measured using double-beam laser Doppler sensors, from Hristopulos and Uesaka (2002). Reproduced with permission from Pulp and Paper Technical Association of Canada (PAPTAC).

the response to a stepwise strain increment and subsequent stress relaxation (Chapter 7).

Draw is the parameter that governs the running condition of the web. The draw Δ_{12} is the *relative* speed difference of two nips, 1 and 2, that follow one another (Fig. 6.4). Because web speed is constant between nips in the steady state, nip speed sets the web speed before the nip, and draw is equal to the strain increment created at the nip:

$$\overline{\Delta}_{12} \equiv \frac{\overline{v}_2 - \overline{v}_1}{\overline{v}_1} \cong \overline{\varepsilon}_2 - \overline{\varepsilon}_1. \tag{6.8}$$

We use the notation where a bar over a variable indicates a steady state value. Equation 6.8 does not hold in nonsteady states, as shown later.

The draw is normally calculated from the rotational speeds of the rolls, but also it can be measured directly from web speeds (Fig. 6.6), for example, using double-beam laser Doppler sensors. A typical draw variation measured in a U.S. pressroom is shown in Figure 6.7 (Deng et al., 2007). It shows very large variation in the draw during the normal operation of the printing press. It is not uncommon that the peak values exceeded the nominal strain-to-failure of the paper without causing a web break. The questions therefore arise: Is the measured draw variation actually equal to the variation of paper strain (or web tension)? If so, why doesn't the web break all the time? In order to answer these questions we need to discuss the general case where velocities and draw are fluctuating.

In the general nonsteady state we have to go back to Equations 6.4 and 6.5. When we integrate both sides of Equation 6.4 over an open draw section $0 < x < L$ and use the mean value theorem, we obtain

$$L\frac{\partial \varepsilon(\xi,t)}{\partial t} = v_2(t) - v_1(t) + \varepsilon_1(t)v_1(t) - \varepsilon_2(t)v_2(t), \quad 0 < \xi \leq L. \tag{6.9}$$

In order to apply this equation to the conditions typical in printing houses, we consider low-frequency draw fluctuations so that their wavelengths, say $\lambda : c/f$ (where f is the frequency), are much larger than the open span length L. Equivalently, the fluctuation frequencies are much lower than a *critical* frequency f_c of the open span,

$$f \ll f_c \equiv \frac{c}{L}. \tag{6.10}$$

Under this condition, the variation of strain depends little on the position coordinate ξ, or $\varepsilon(\xi,t) = \varepsilon(t)$ so that Equation 6.9 can be rewritten as (Hristopulos and Uesaka, 2002)

$$L\frac{d\varepsilon(t)}{dt} = v_2(t) - v_1(t) + \varepsilon_1(t)v_1(t) - \varepsilon_2(t)v_2(t). \tag{6.11}$$

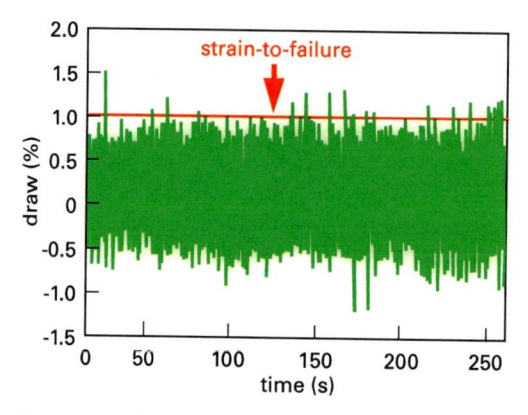

Fig. 6.7 Draw variation calculated from Figure 6.6, from Hristopulos and Uesaka (2002). The red line indicates typical level of strain-to-failure for newsprint. Reproduced with permission from Pulp and Paper Technical Association of Canada (PAPTAC).

This equation is the same as in the standard model of web dynamics (Reid and Shin, 1991). Numerical solutions to Equation 6.11 also agreed closely with those obtained from the partial differential equations in the theory of an elastic string (Brown, 1999).

The important point here is that Equation 6.11 is valid under the condition of Equation 6.10. For example, if the speed of longitudinal waves in a printing paper is 3 km/s and the open span is 1 m, then the critical frequency $f_c = 3$ kHz. Typical frequencies that are encountered in the pressroom are typically well below 60 Hz (Deng, Ferahi, and Uesaka, 2007; Uesaka, 2005); therefore, Equation 6.10 holds very well. In contrast, Equation 6.11 may not capture phenomena related to pulse-like disturbances, such as a wrinkle in the web or a splice that joins the ends of two paper webs, higher-speed operations (a coating machine or paper machine), or wet paper webs. These phenomena require full solution of Equations 6.4 and 6.5.

Although Equation 6.11 is nonlinear, we can linearize and solve it by considering small speed fluctuations $\delta v \ll v$ around the steady state:

$$\Delta \varepsilon_{12} = \overline{\Delta_{12}} + \frac{1}{L}\int_0^t \delta v(s) \exp\{-f_k(t-s)\}ds, \tag{6.12}$$

where $\Delta \varepsilon_{12}$ is the strain increment, $\overline{\Delta_{12}}$ is the nominal mean draw in the open draw section, and f_k is a characteristic frequency defined by

$$f_k = \frac{\overline{v_2}}{L}, \tag{6.13}$$

where $\overline{v_2}$ is the mean speed of the second nip (Fig. 6.4). Figure 6.8 illustrates the calculated strain response for two frequencies of draw oscillation. In the low-frequency case (1 Hz, upper graph), the strain follows almost perfectly the draw, whereas in the

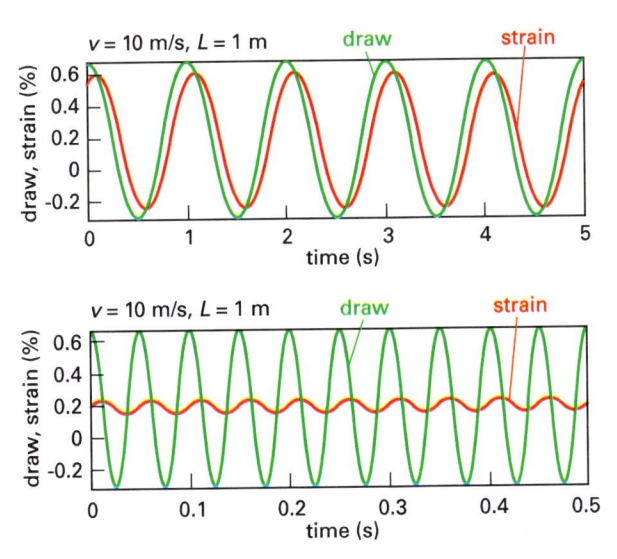

Fig. 6.8 Calculated strain response for two frequencies of draw oscillations, $f = 1$ Hz (upper graph) and $f = 20$ Hz (lower graph), from Hristopulos and Uesaka (2002). Reproduced with permission from Pulp and Paper Technical Association of Canada (PAPTAC).

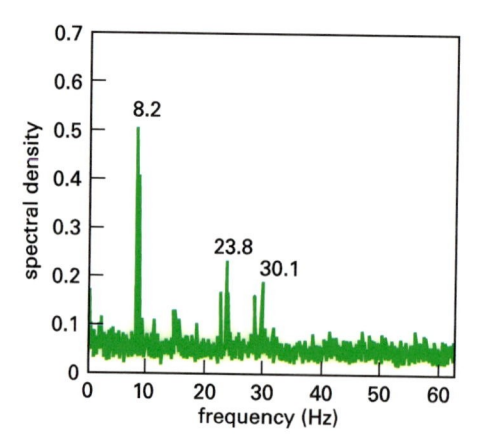

Fig. 6.9 Spectral density distribution determined from Figure 6.7 (Hristopulos and Uesaka, 2002). Reproduced with permission from Pulp and Paper Technical Association of Canada (PAPTAC).

high-frequency case (20 Hz, lower graph), the strain variation is greatly attenuated with a significant phase delay. In this specific example, the characteristic frequency of the open draw section is $f_k = 10$ Hz. This means that if the frequency of a draw disturbance exceeds 10 Hz, then the resulting strain variation is strongly attenuated. This is also expected from Equation 6.12; the open draw section behaves like a low-pass filter of the disturbances.

Figure 6.9 shows the spectral density distribution determined from Figure 6.7. Except white noise and a peak at 8.2 Hz, most of the draw variations occurred well above the characteristic frequency $f_k \approx 10$ Hz of the open draw section, and they do not contribute to the strain variation. However, at higher machine speeds or shorter open spans, the characteristic frequency would be higher, and the disturbances at higher frequencies would start contributing to the strain variation, and thus to the tension variation. Therefore, it is important to eliminate any sources of vibrations in the rotational components of web transport systems. Otherwise, draw variations contribute to web breaks in printing houses (Deng et al., 2008).

6.2.3 Moving thread problem

The obvious extension to the previous *axially-moving* web model is to add another degree of freedom, allowing web motion in the *transverse* direction perpendicular to the web surface. In other words, the web moves in the MD–ZD plane but maintains the one-dimensional geometry like a thread (Fig. 6.10). Although the extension seems straightforward, the conservation equations (6.1 and 6.2) now become a set of fully coupled, nonlinear differential equations that are difficult to solve analytically or even numerically. (An excellent review of the one-dimensional thread models has been given by Ahrens et al., 2004). Therefore, many researchers have made simplifying assumptions about the web displacements (e.g., an arc shape) and the continuity equation (strain variations along web length) (Pakarinen et al., 1993; Kurki et al.,

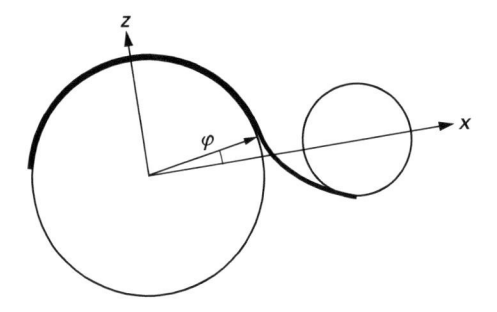

Fig. 6.10 Open draw section as an example of the moving thread problem, from Edvardsson and Uesaka (2010). The web is moving from left to right, and φ indicates the take-off-angle that represents the peeling position. Reproduced with permission from American Society of Mechanical Engineers (ASME).

1995, 1997). Another difficulty is that the boundary conditions, such as web adhesion force and contact forces on the rolls, are not known a priori because they are part of the solution. Further, when considering wet web behavior on the paper machine, the stress–strain relations must represent the visco-elastic-plastic nature of wet paper, which introduces another source of time-dependency. The time-dependent nature of wet paper has been investigated intensively in the past (Hauptmann and Cutshall, 1977; Kurki et al., 2004; Kouko, et al., 2006; Kouko, et al., 2007; Kouko and Sorvari, 2008). The main challenge is how to estimate visco-elastic parameters that appear in mechanical models (e.g., Maxwell, Kelvin, and combined models). The parameters are highly dependent on time scale, solids content, temperature, and furnish compositions because the models are phenomenological. The previously named literature covers a large set of such examples.

Published studies have accumulated important insights on the web dynamics in the open draw sections in the wet-end of paper machine. For example, Pakarinen and coworkers (1993) estimated how different force components contribute to the total force applied on the web in the steady state. The forces they considered were the pressure difference between top and bottom surfaces of the web, gravity, centrifugal force, and air friction. They showed that at the speed of 900 m/min, the pressure difference gave the biggest contribution (68% of the total force), whereas at 1,500 m/min, the centrifugal force was dominant (84%). In nonsteady states, tension spikes were predicted for the cases when the disturbances are created either by external pressure or by adhesion strength (Pakarinen et al., 1993; Kurki et al., 1995) or adhesion strength (Kurki et al., 1997).

Recently a new numerical approach was proposed to overcome the previously discussed difficulties in solving the moving thread problem with variable boundary conditions (Edvardsson and Uesaka, 2009, 2010). The idea is to represent the continuous thread in terms of a series of interconnected *particles* (Fig. 6.11). For each particle i the equation of motion is

$$\frac{d\boldsymbol{p}_i}{dt} = \sum_j \boldsymbol{F}_{ij}^{\text{int}} + \sum_j \boldsymbol{F}_{ij}^{\text{ext}}, \tag{6.14}$$

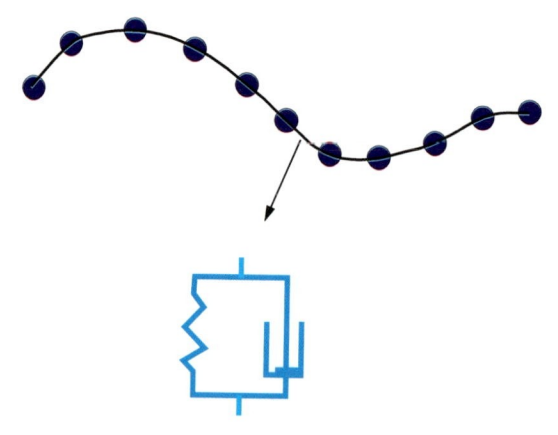

Fig. 6.11 Particle model where the moving thread is represented by a series of particles that interact with elastic and viscous forces.

where F_{ij}^{int} are the internal forces from interconnected particles, and F_{ij}^{ext} are the external forces from the pressure of surrounding air, adhesion forces, friction, and so forth. The momentum p_i is given by

$$p_i = m_i \frac{dx_i}{dt}, \tag{6.15}$$

where m_i and x_i are the mass and position of the particle i. The interaction forces F_{ij}^{int} can represent various stress–strain relationships (elastic, visco-elastic, and visco-plastic) during the stretching and bending deformations of the thread (Edvardsson and Uesaka, 2010).

If the number of particles is increased, the discrete particle model converges precisely to the continuum solution of the partial differential equations. Figure 6.12 compares the solutions calculated from the particle approach and the continuum approach for the web strain variation along an open draw section. Interestingly, when a viscous term (Fig. 6.11) is included in F_{ij}^{int}, web strain does not show the step-wise change at the nip as it did in the elastic case (Fig. 6.5). Instead, web strain displays some delay that depends on the value of viscosity (Kurki et al., 1997).

Figure 6.13 shows how the take-off angle φ (defined in Fig. 6.10) of a wet paper web depends on the machine speed. During normal operation, φ is constant over time and fairly insensitive of the speed. At higher speeds, the web sticks to the roll surface a little longer before detaching, as seen from the fact that φ decreases. However, above a certain critical speed, 1,700 m/min in this case, the open web section becomes unstable, and the take-off angle φ increases without a limit. Below the critical speed of 1,700 m/min, the web strain (Fig. 6.14) is precisely equal to the 3% draw that was applied, whereas above 1,700 m/min, the maximum strain in the open draw section grows without any sign of saturation. The obvious consequence is a web break.

The existence of a critical speed has been speculated among papermakers for many years, and it is always implied in paper machine design. However, the analysis of Edvardsson and Uesaka (2009) showed what factors actually control the critical speed. For a given draw, tensile stiffness of the web is the most influential factor. If

6.2 Dynamics of web transport | 103

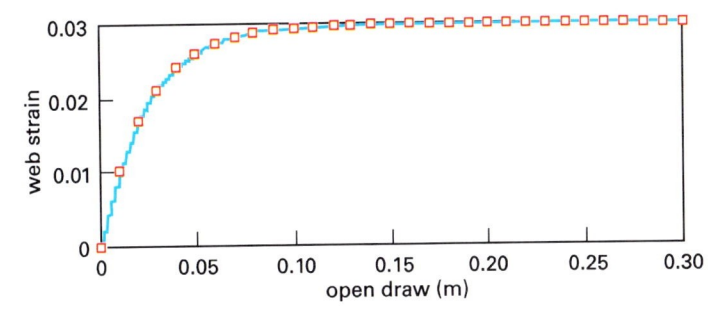

Fig. 6.12 Web strain evolution along an open draw section, from Edvardsson and Uesaka (2010). Reproduced with permission from American Society of Mechanical Engineers (ASME).

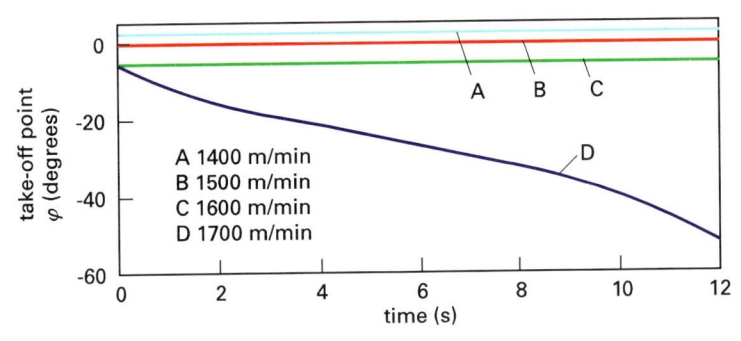

Fig. 6.13 Effect of web speed on the time-evolution of the take-off position, from Edvardsson and Uesaka (2009). Reproduced with permission from The Pulp and Paper Fundamental Research Society (www.ppfrs.org).

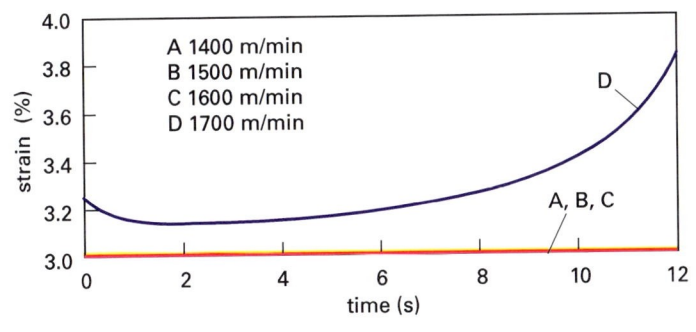

Fig. 6.14 Effect of web speed on the time-evolution of the maximum strain in an open draw section, from Edvardsson and Uesaka (2009). The initial fall in the curve D is due to a numerical artifact. Reproduced with permission from The Pulp and Paper Fundamental Research Society (www.ppfrs.org).

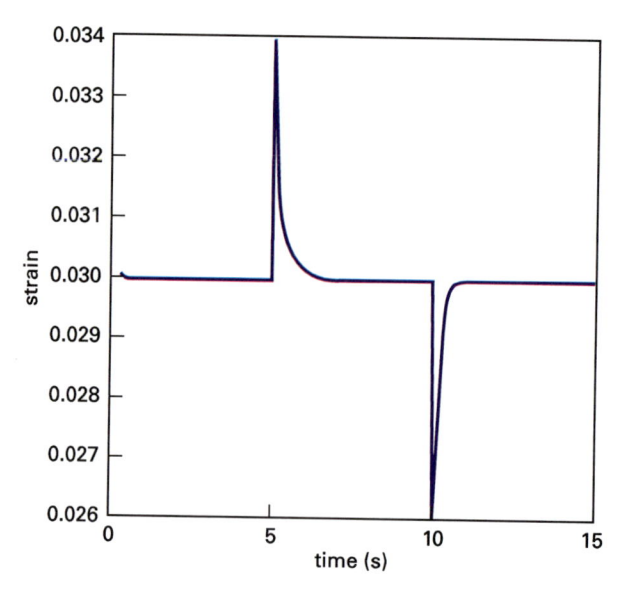

Fig. 6.15 Response of web strain to a step-wise reduction by 15% in web stiffness (at $t = 5$ s) and a step-wise return to the original level (at $t = 10$ s), from Edvardsson and Uesaka (2010). Web speed 1,400 m/min. Reproduced with permission from American Society of Mechanical Engineers (ASME).

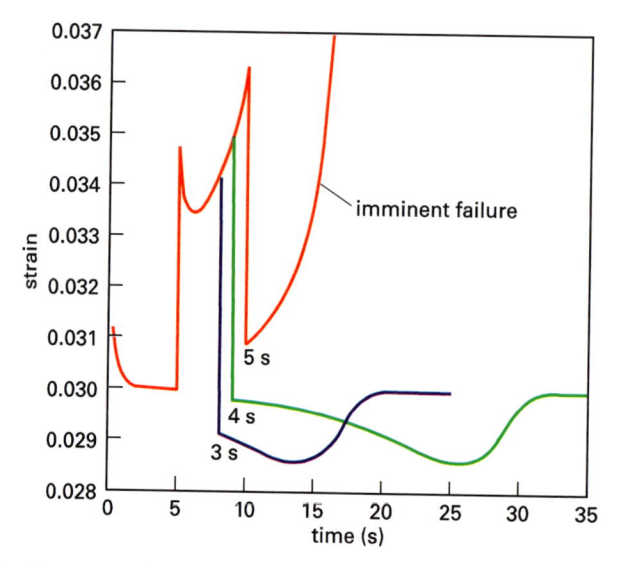

Fig. 6.16 Response of web strain to a step-wise reduction by 15% in web stiffness, lasting from 3–5 s, from Edvardsson and Uesaka (2010). Web speed 1,600 m/min. Reproduced with permission from American Society of Mechanical Engineers (ASME).

the stiffness decreases, the critical speed decreases as well. This implies that achieving high speeds on a paper machine with open draws in the wet end requires that one increase the stiffness of the wet web. Because the tensile stiffness is very sensitive to moisture content (Fig. 2.7), one must then increase the dryness of the web.

Figure 6.15 shows the effect of a stiffness pulse at the speed of 1,400 m/min. The stiffness was decreased by 15% for a 5 second period. This created a strain disturbance at the times when the stiffness was changed. However, at higher speeds, for example, the speed of 1,600 m/min shown in Figure 6.16, the strain behaved in a much more complex way. For the cases of short strain pulses, the web strain eventually returned to its original value. For a longer strain pulse (5s), the web strain no longer returned to a steady state but instead continued to grow. In practice, fluctuations are always present in the wet conditions, such as moisture content after the forming section, roll-paper adhesion, fiber orientation, basis weight, and so forth. It is conceivable that they can affect web stability through the stiffness, but no direct evidence currently exists indicating that stiffness variations trigger web breaks.

6.2.4 Fluttering of a two-dimensional web

The realistic representation of a paper web is a two-dimensional planar object or curved surface when the web is not flat. The practical problems are diverse, but one important issue is vibration, or what papermakers often call the flutter. In mechanics, the word *flutter* is reserved for a special kind of vibration, but in this chapter we follow the casual papermaking usage. As machine speeds increase and web stability requirements become more stringent, fluttering in the dryer section has drawn attention as the potential culprit of web breaks. Fluttering in the open draws of a double-felted dryer is a notorious example. Another is the instability of the web in the air pockets of a single-felted dryer. Complex air flows inside the air pocket produce pressure differences that detach the paper web from the felt, causing instabilities (Pakarinen et al., 1995).

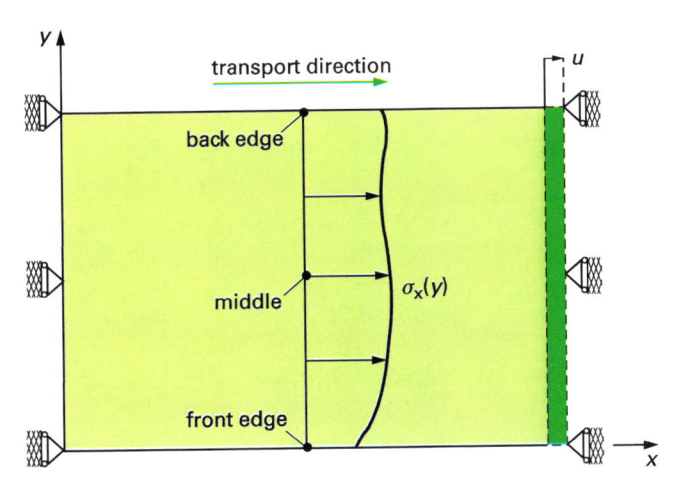

Fig. 6.17 Boundary conditions used in the study of web vibrations in the presence of web tension nonuniformity. Reprinted from Kulachenko et al. (2007) with permission from Elsevier.

$f_1 = 6.30$Hz $f_2 = 7.27$Hz

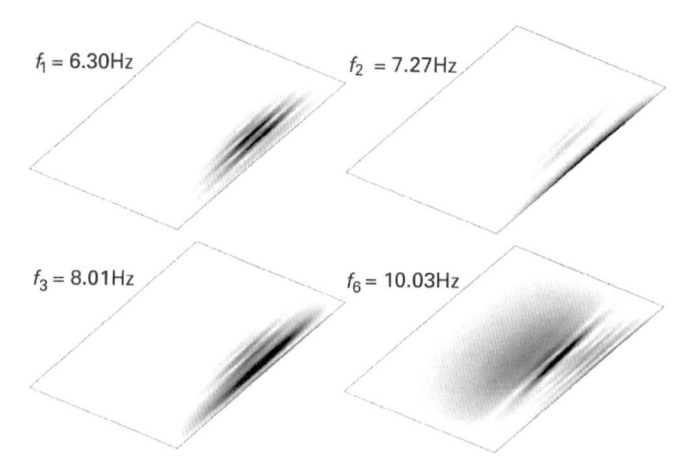

$f_3 = 8.01$Hz $f_6 = 10.03$Hz

Fig. 6.18 Natural frequencies of vibration without air in an 1-m long and 1-m wide open draw section. Reprinted from Kulachenko et al. (2007) with permission from Elsevier.

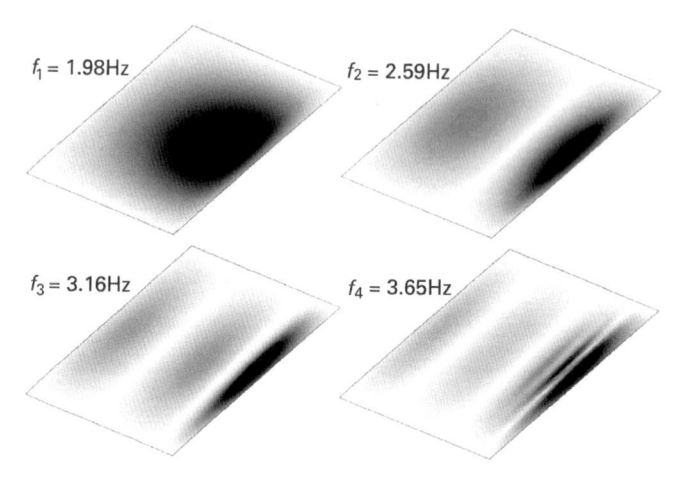

$f_1 = 1.98$Hz $f_2 = 2.59$Hz

$f_3 = 3.16$Hz $f_4 = 3.65$Hz

Fig. 6.19 Natural frequencies of vibration with air an 1-m long and 1-m wide open draw section. Reprinted from Kulachenko et al. (2007) with permission from Elsevier.

The one-dimensional moving thread models discussed in the previous section have been used to analyze flutter problems both with air (Pramila, 1986) and without air (Mujumdar and Douglas, 1976). The air surrounding the paper web has an enormous effect on the vibration characteristics. The natural frequencies predicted with air were only 15%–30% of those without air. In the equation of motion of the paper web, this effect can be approximately described by adding the mass of surrounding air to the mass of the web (Pramila, 1986). Another important factor influencing web flutter, and one of the most important, unsolved problems in papermaking, is the CD nonuniformity of web tension (Chapter 2.5.2). Linna and his colleagues (Linna and Lindquist, 1987; Linna et al., 1991, 1995, 2002) and Parola and his colleagues (Parola and Beletski, 1999; Parola et al., 2000a, 2000b) discuss the development and

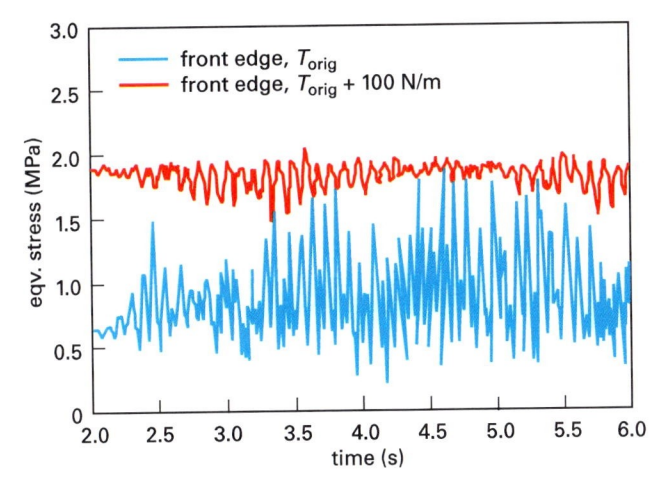

Fig. 6.20 Effect of increasing web tension on the amplitude of stress variations at one edge of a paper web. Reprinted from Kulachenko et al. (2007) with permission from Elsevier.

effects of web tension nonuniformity. In this case the extension of the analysis into two dimensions is essential.

Kulachenko et al. (2007) proposed a nonlinear finite element model for a running web that is coupled with the surrounding air through interacting air elements. The paper web is represented by shell elements, and the web speed is kept constant. However, most importantly, the web has a nonuniform CD profile $\sigma_x(y)$ of the web tension (Fig. 6.17). The predicted natural frequencies without air and with air are shown in Figures 6.18 and 6.19. As discussed previously, the presence of air depresses greatly the natural frequencies. Both figures show the famous edge flutter associated with web tension nonuniformity. The presence of air increases complexity of the vibration patterns. In both cases, the nonuniformity of web tension had a dramatic effect on the edge flutter. Web speed had virtually no effect on the natural frequencies.

When excessive edge fluttering occurs, it is common to increase the web tension (through draw) to reduce the flutter in an effort to prevent web breaks and wrinkles. Figure 6.20 illustrates what happens to the stress variations. Notice that stress is expressed as the equivalent "von Mises" stress (Fung, 1965), dominated by the tensile stress in the present case. Increasing the web tension indeed damped the edge fluttering. However, now all stress values have exceeded the highest peak stresses seen before the web tension was increased. Therefore, increasing web tension does not help to reduce the stress levels, except to reduce the probability of web wrinkling.

6.3 Concluding remarks

This chapter discussed the basic characteristics of typical paper web transport systems. In a steady state, the system is characterized simply by the web speed and the draws between nips. Draw determines the strain increment (or the asymptotic strain increment in the case of viscous wet webs) and thus the tension increments of the web. The web tension is balanced by various forces acting on the web, including the inertial force, proportional to the speed squared, and the adhesion force.

In a nonsteady state, the web strain increment of an open draw section may differ from the draw if the frequency of variations is high. This happens because many open draw sections behave like a low-pass filter, attenuating significantly high frequency variations. The critical frequency of filtering depends on the length of the open draw section and the speed of the web, increasing with higher speeds and shorter spans.

The dynamic response in a typical paper web transport system is often dominated by relatively low frequency components. However, at the nips, the paper web is strained at a very high strain rate. This strain rate is much higher than what is used in standard laboratory tests. This raises the following question: are standard tests appropriate in determining the strength properties of paper? Another important aspect is the cross-directional nonuniformity of the web properties (tension, moisture, drying history, etc.) that impact web stability. Anecdotes in the field seem to abound, but serious analyses have only begun.

The issue of water application was not discussed in this chapter. The state of water applied to a moving web is complex. Water can absorb into the paper structure, but a considerable amount of water can also evaporate as the web moves. The latter effect is a strong function of the web speed (Gomer and Lindholm, 1991; Hansen, 1997). It is often said that in offset printing, the water in the fountain solution and ink induces hygroexpansion and alters mechanical properties of the web, thereby leading to the *misregistration* problem, where different colors are not in registry relative to one another. However, the real mechanisms of misregistration are more involved. Interested readers can find a good review of the subject in Lif, Östlund, and Fellers (2005) and can also consult Chapter 9.

From the point of analyzing web behavior, increasingly powerful methods, both numerical and experimental, have become available and are being utilized. In particular, analyzing the cross-directional nonuniformity problems is no longer a distant goal in the area of web dynamics.

References

Ahrens, F., Patterson, T., and Bloom, F. (2004). Mathematical modelling of web separation and dynamics of a web adhesion and drying simulator. Int J. Applied Mechanics and Engineering 9, 227–271.

Brown, J. L. (1999). Propagation of longitudinal tension in a slender moving web. In: Proceedings of the Fifth International Conference on Web Handling, Oklahoma State University, Stillwater, Oklahoma.

Deng, X., Ferahi, M., and Uesaka, T. (2007). Press room runnability: A Comprehensive analysis of press room and mill database. Pulp and Paper Canada 108, T39–T48.

Deng, X., Parent, F., Manfred, T., Aspler, J. S., and Hamel, J. (2008). Pressroom draw variations and its impacts on web breaks, Part 1. Newspaper press trials. In: Proceedings of 35th International Research Conference of IARIGAI, Valencia, Spain.

References

Edvardsson, S., and Uesaka, T. (2009). System stability of the open draw section and paper-machine runnability. In: Advances in Pulp and Paper Research, Trans. 14th Fund, Res. Symp. S. J. I'Anson, ed. (Oxford, UK: FRC), pp. 557–575.

Edvardsson, S., and Uesaka, T. (2010). System dynamics of the open-draw with web adhesion: Particle approach. J. Applied Mechanics 77, 1–11.

Fung, Y. C. (1965). Foundations of Solid Mechanics (Englewood Cliffs, NJ, USA: Prentice-Hall).

Gomer, M., and Lindholm, G. (1991). Hygroexpansion of newsprint as a result of water absorption in a printing press. In: Proceedings of 43rd TAGA Annual Conference, Rochester, New York.

Good, J. K. (2007). Winding: Machines, Mechanics and Measurements (Lancaster, PA, USA: Destech Publications).

Grant, J. P. (1967). A decade of suffering from web breaks in a metropolitan pressroom. Pulp and Paper Magazine of Canada 68, 143–149.

Hansen, Å. (1997). Water absorption and dimensional changes of newsprint during offset printing. In: Advances in Printing Science and Technology, W. H. Banks, ed. (London: Pentech Press), pp. 247–268.

Hauptmann, E. G., and Cutshall, K. A. (1977). Dynamic mechanical properties of wet paper webs. Tappi 60, 106–108.

Hristopulos, D., and Uesaka, T. (2002). A model of machine direction tension variations in paper web with runnability applications. J. Pulp and Paper Sci. 28, 389–394.

Jokio, M., ed. (1999). Papermaking: Part 3, Finishing (Helsinki, Finland: Fapet Oy).

Kipphan, H., ed. (2001). Handbook of Print Media: Technologies and Production Methods (Berlin: Springer).

Kouko, J., Kekko, P., and Kurki, M. (2006). Effect of strain rate on the strength properties of paper. Paper presented at the Progress in Paper Physics Seminar, Oxford, Ohio.

Kouko, J., Salminen, K., and Kurki, M. (2007). Laboratory scale measurement procedure for the runnability of a wet web on a paper machine, Part 2. Paperi ja Puu 89, 424–430.

Kouko, J., and Sorvari, J. (2008). Mechanical testing and viscoelastic strength modeling of wet paper. Paper presented at the Progress in Paper Physics Seminar, Otaniemi, Finland.

Kulachenko, A., Gradin, P., and Koivurova, H. (2007). Modelling the dynamical behaviour of a paper web. Part 2. Computers and Structures 85, 148–157.

Kurki, M., Juppi, K., Ryymin, R., Taskinen, P., and Pakarinen, P. (1995). On the web tension dynamics in an open draw. In: Proceedings of the Third International Conference on Web Handling, Oklahoma State University, Stillwater, Oklahoma.

Kurki, M., Vestola, J., Martikainen, P., and Pakarinen, P. (1997). The effect of web rheology and peeling on web transfer in open draw. In: Proceedings of the Fourth International Conference of Web Handling, Oklahoma State University, Stillwater, Oklahoma.

Kurki, M., Kekko, P., Kouko, J., and Saari, T. (2004). Laboratory scale measurement procedure of paper machine wet web runnability; Part 1. Paperi ja Puu 86, 256–262.

Larsson, L. O. (1984). What happens in the press to cause web breaks? Pulp and Paper Canada 85, T249–T253.

Lif, J. O., Östlund, S., and Fellers, C. (2005). In-plane hygro-viscoelasticity of paper at small deformations. Nordic Pulp and Paper Research J. 20, 139–149.

Linna, H. I., and Lindqvist, U. (1987). Runnability problems in web presses and paper machines. Paper presented at the Topical Themes in Newsprint – Printing Research, TFL International Multidisciplinary Symposium, Lidingö, Sweden (Swedish Newsprint Research Centre (TFL)), pp. 63–74.

Linna, H. I., Molianen, P., and Koskimies, J. (1991). Paper anisotropy and web tension profiles. Paper presented at the 1991 International Paper Physics Conference, Kona, Hawaii, TAPPI, vol. 1, pp. 327–339.

Linna, H. I., Kaljunen, T., Moilanen, P., Mahonen, A., and Parola, M. (1995). Long-term study of variations in the web tension profile. Paper presented at Advances in Printing Science and Technology, The 23rd Research Conference of IARIGAI (Paris: John Wiley and Sons), pp. 213–225.

Linna, H. I., Parola, M. J., and Virtanen, J. O. (2002). Improving productivity by measuring web tension profiles. Appita 55, 317–322.

Mujumdar, A. S., and Douglas, W.J.M. (1976). Analytical modeling of sheet flutter. Svensk Papperstidning 79, 187–192.

Page, D. H., and Bruce, A. (1985). Pressroom Runnability. Paper presented at the Newsprint and the Pressroom Conference (Chicago, IL, USA: CPPA and ANPA).

Pakarinen, P., Ryymin, R., Kurki, M., and Taskinen, P. (1993). On the dynamics of web transfer in an open draw. In: Proceedings of the 2nd International Conference on Web Handling, Oklahoma State University, Stillwater, Oklahoma.

Pakarinen, P., Juppi, K., and Karlsson, M. (1995). A study of air flows in single-tier pockets at high machine speeds. J. Pulp and Paper Sci. 21, J68–J73.

Parola, M. J., and Beletski, N. (1999). Tension across the paper web – A new important property. Paper presented at the 27th EUCEPA Conference (Grenoble, France: ATIP), pp. 293–298.

Parola, M. J., Kaljunen, T., and Vuorinen, S. (2000a). New methods for the analysis of the paper web performance on the press. Paper presented at Advances in Printing Science and Technology, The 27th Research Conference of IARIGAI (Graz, Austria: PIRA International), Vol. 27, pp. 203–218.

Parola, A., Sundell, H., Virtanen, J., and Lang, D. (2000b). Web tension profile and gravure press runnability. Pulp and Paper Canada 101, T35–T39.

Paulapuro, H., ed. (2008). Papermaking, Part I: Stock Preparation and Wet End, 2nd Edition (Helsinki, Finland: Paperi ja Puu).

Pramila, A. (1986). Sheet flutter and the interaction between sheet and air. Tappi 69, 70–74.

Reid, K. N., and Shin, K. H. (1991). Variable-gain control of longitudinal tension in a web transport system. In: Proceedings of the First International Conference on Web Handling, Oklahoma State University, Stillwater, Oklahoma.

Roisum, D. R. (1994). The Mechanics of Winding (Atlanta, GA, USA: TAPPI).

Roisum, D. R. (1998). The Mechanics of Web Handling (Atlanta, GA, USA: TAPPI).

Uesaka, T. (2005). Principal factors controlling web breaks in pressrooms – Quantitative evaluation. Appita Journal 58, 425–432.

Uesaka, T., and Ferahi, M. (1999). Principal factors controlling pressroom breaks. Paper presented at the TAPPI International Paper Physics Conference (San Diego, CA, USA: TAPPI Press), pp. 229–245.

7 Creep and relaxation

Douglas W. Coffin

7.1 Introduction

As one delves deeper into the mechanics of paper, one gains an appreciation that the passing of time itself is a major contributing factor to the response of paper to load. This time-dependence is inherently a response of the material and cannot simply be captured by incorporating changes in kinetic energy from the acceleration and deceleration of the material. The effects of time-dependence on the mechanical behavior of paper were briefly illustrated in Chapter 2.3.2. Sometimes these effects are negligible, and sometimes they cannot be ignored.

From an engineering standpoint, neglecting the time-dependent nature of paper will lead to reliable predictive equations and fundamental understanding. The fracture problems studied in Chapter 5 and the box corner problems presented in Chapter 4 do not require consideration of time-dependence to understand the important mechanisms and behaviors. An important aspect for consideration in dealing with the mechanics of paper is the dissipation of energy, and in many situations this could be effectively accounted for in terms of rate-independent plasticity. For the engineer, it may even be irrelevant as to the causes of energy dissipation as long as it has been accounted for in some fashion and the calculations provide a reasonable estimate for the process being modeled. However, if an approach that ignores time-dependence is unsatisfactory, the inclusion of time-dependence may be necessary.

The time-dependence of the mechanical response of paper extends across large time scales from microseconds to decades. At very short times, where relaxations have not started, paper will appear brittle and stiff. At long-time scales, paper will appear more ductile and compliant. The focus of phenomena covered in Chapters 4 and 5 is on relatively short-time scales, and time-dependence is of less interest. In some short-time scale phenomena, such as web dynamics, time-dependence is sometimes of interest because stress-relaxation affects the magnitude of stresses (Chapter 6.2.2). However, time-dependence really takes center stage in the long-time strength of stacked boxes (Chapter 3.4.4). Understanding the time-dependence of paper over long periods of time is the focus of this chapter.

To study time-dependence, we could title the chapter visco-elastic-plastic behavior, or even rheology, but as an introduction to the time-dependence of paper we focus on two phenomena: creep and stress relaxation. First, the essential elements that must be considered when treating the creep and relaxation behavior of paper are discussed. Then, the basic observed behavior of paper is presented and the effects of some relevant parameters are shown. Finally, a simple treatment of box lifetime is presented to underline that even in simple practical cases where creep is the main phenomena, other factors such as material variability and environmental conditions are of equal importance.

7.2 Relaxation and creep as phenomena

Creep and *relaxation* are the two customary phenomena associated with time-dependent behavior. They are opposite manifestations of the dissipation of energy from a material under some state of stress. The definitions are as follows:

- Creep: The increase in strain $\varepsilon(t)$ over time under a constant state of stress, $\sigma =$ constant $\forall\, t \geq 0$.
- Relaxation: The decrease in stress $\sigma(t)$ over time under a constant state of strain, $\varepsilon =$ constant $\forall\, t \geq 0$.

The state of stress or strain can be any combination of normal and shear components.

Typically creep and relaxation testing is performed under conditions of uni-axial tension or compression. For relaxation (Fig. 7.1b), a sample of initial length L is changed to length $L + u$ and held constant for a period of time. The load P is recorded as a function of time. For creep (Fig. 7.1c), a constant load is applied, and the change in length is recorded as a function of time. From a practical standpoint, one cannot control the local state of stress and strain in every region of the material, and for testing only the applied load or applied deformation is held constant.

The creep and relaxation are a result of redistribution of stresses at some length scale in the material. These could be molecular, micro-, or macro-level redistributions. Here we treat the phenomena as being valid as a continuum and ignore what leads to them. The role of structure on the global creep and relaxation response is addressed later in this chapter, but only regarding how it affects the effective continuum response.

The measured load P and displacement u are related to the stress σ and strain ε via integrating over the cross-section area A and the length of the sample L, or

$$P = \int_A \sigma\, dy\, dz \quad \text{and} \quad u = \int_0^L \varepsilon\, dx. \tag{7.1}$$

Assuming the stress and strain are uniformly distributed, we can consider the creep and relaxation test to correspond with the definition of creep and relaxation given

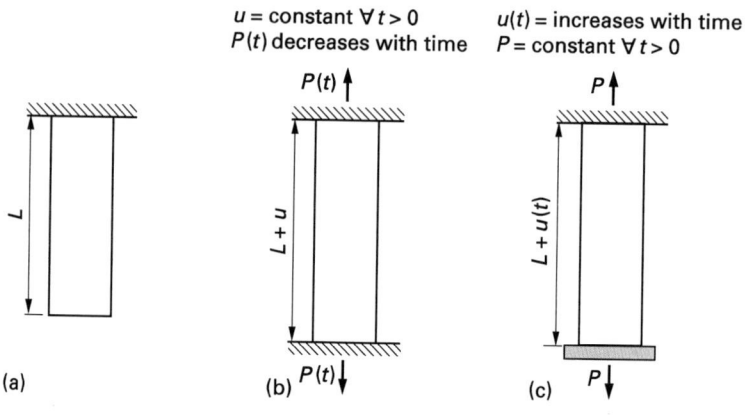

Fig. 7.1 Stress relaxation (b) and creep (c) of a strip of paper (a).

previously. As we progress in the discussion of creep and relaxation, it is important to remember that the measured quantities are P and u. Stresses and strains are likely to be distributed unevenly through the sheet in a manner that can change with the passage of time.

Creep and relaxation tests are often carried out at various levels of load or deformation (Fig. 7.2). This yields a family of N creep curves or relaxation curves. The creep curves can be written as

$$\varepsilon(t,\sigma_i) \forall t \leq 0, i = 1, 2 \ldots N, \tag{7.2}$$

and relaxation curves as

$$\sigma(t,\varepsilon_i) \forall t \leq 0, i = 1, 2 \ldots N. \tag{7.3}$$

These curves can be used to establish material characteristics or properties. The material response is characterized as creep compliance J and relaxation modulus G. These parameters are obtained by dividing Equations 7.2 and 7.3 with the constant stress or strain, respectively, leading to

$$J(t,\sigma) = \frac{\varepsilon(t,\sigma)}{\sigma} \tag{7.4}$$

and

$$G(t,\varepsilon) = \frac{\sigma(t,\varepsilon)}{\varepsilon}. \tag{7.5}$$

In general, the creep compliance and relaxation modulus can be functions of the stress and strain in different directions, temperature, and moisture content. If one wishes to account for time-dependence in engineering calculations, some type of model for Equations 7.4 and 7.5 must be used. Important considerations when developing specific relations are addressed in the next section.

For many materials and specific time and load domains, the creep compliance and relaxation modulus are found to be independent of the level of load and strain, respectively. In other words, when normalized by the initial loading, each family of curves in Figure 7.2 converges to one curve, either $J(t)$ or $G(t)$. At least this is an adequate approximation. The collapse implies that the material behavior is linear

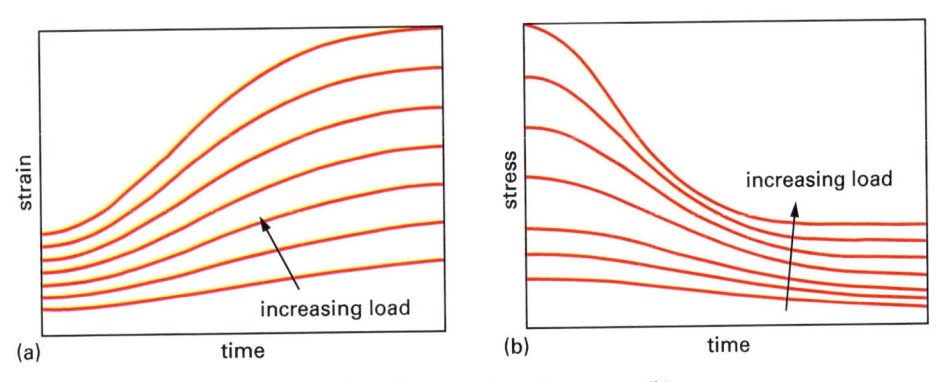

Fig. 7.2 Family of creep curves (a) and stress relaxation curves (b).

in that range of time and load. If linearity is not present, it may still be possible to develop systematic shifts so that a family of curves is transformed into one single so-called *master curve*. Even though the material response is nonlinear, expressions can be developed to account for the dependence on load level. Paper materials are inherently nonlinear, but it is possible to find ranges of load levels where the behavior can be given by linear expressions or nonlinear master curves.

7.3 Modeling of time-dependence

Both creep and stress relaxation are manifestations of time-dependent dissipative processes. These behaviors influence the response of paper during papermaking, converting, and end-use. Constitutive equations that capture this behavior must be developed for use in models. This section discusses some of the basic features of formulating a model for paper. The discussion is centered on uni-axial loading but could be generalized to multi-axial states of stress and strain. Readers requiring more in-depth knowledge can find many sources, for example, Ward and Sweeny (2004).

7.3.1 Linear behavior

In 1876 Boltzman proposed that *creep* is a function of the entire past loading history and that each step of loading makes an independent contribution to the final deformation (Ward and Sweeney, 2004). This is true for linear visco-elastic materials and allows for the description of general loading sequences using the creep compliance and stress relaxation modulus. These general equations can be written as

$$\varepsilon(t) = \int_{-\infty}^{t} J(t-t^*) \frac{d\sigma}{dt^*} dt^* \tag{7.6}$$

and

$$\sigma(t) = \int_{-\infty}^{t} G(t-t^*) \frac{d\varepsilon}{dt^*} dt^*. \tag{7.7}$$

For both of these equations to be simultaneously valid, the creep compliance and relaxation modulus must satisfy

$$\int_{0}^{t} G(t-t^*) J(t^*) dt^* = t. \tag{7.8}$$

The implication is that simply fitting a convenient empirical equation to data and substituting that into Equations 7.6 and 7.7 may not provide a good model. On the other hand, with the proper constitutive equation, time-dependent behavior can be captured regardless of the loading conditions.

A critical factor to include in a model is the degree to which the deformation is recoverable. If a portion of the energy imparted to the sample during a test is stored in the material, the deformation may be recoverable. If some energy is dissipated (say as heat) during the process, the deformation cannot be recoverable. Figure 7.3 illustrates the need to explore time-dependence beyond the simple creep test. The creep and creep-recovery in the two cases have the same initial linear-elastic response

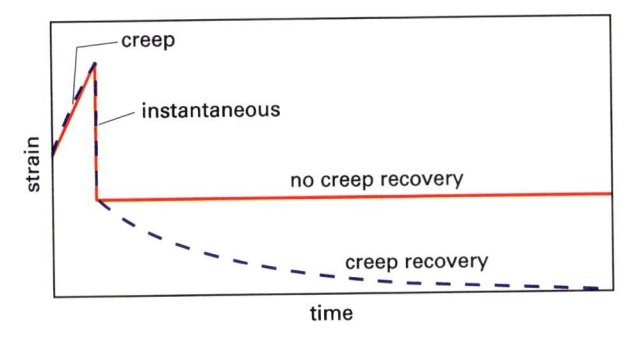

Fig. 7.3 Comparison of creep recovery for two cases with equal creep curves.

and essentially the same constant-rate creep response. After a time lapse, the load is removed and the initial elastic response is recovered. In one case, there is no further recovery of deformation because all the stored energy has been dissipated. In the other case, all the deformation is recovered because potential energy is stored in the material. Observe that much more time is needed for recovery than for creep. This is because the external force driving creep is greater than the internal force driving recovery.

This discussion has introduced the following three considerations that must be addressed when developing appropriate models for the time-dependent nature of paper.

1. Nonlinearity
2. Recoverability
3. Time-scale

7.3.2 Nonlinearity

Superposition works only for linear systems. A system and process described by some operator $F[f(t)]$ is linear, if $F[f+g] = F[f] + F[g]$ holds, and nonlinear otherwise. For creep compliance and stress relaxation the response is said to be nonlinear if, in addition to time, J and G are functions of stress or strain, or both stress and strain. When the material is linear, J and G are only functions of time. The degree of nonlinearity can be assessed several ways, but one must evaluate the response for a sufficient range of initial load levels to generate relevant data.

Isochronous curves provide a simple method to evaluate the degree of nonlinearity (Fig. 7.4). One transforms the measured creep or stress relaxation data into separate curves that display stress and strain at one constant time, hence the term *isochronous*. From Equations 7.4 and 7.5 one can see that, depending on whether creep or stress relaxation is considered, the slopes of the curves give either J or G for the time chosen. If the material is nonlinear, the isochronous curves are nonlinear.

At short times, the response is linear and corresponds to the highest slopes in Figure 7.4. For this example, the line corresponding to the longest time also indicates a linear response. This long-time linearity results because relaxations were delayed but elastic so that at long times only linearly elastic energy is stored in the material.

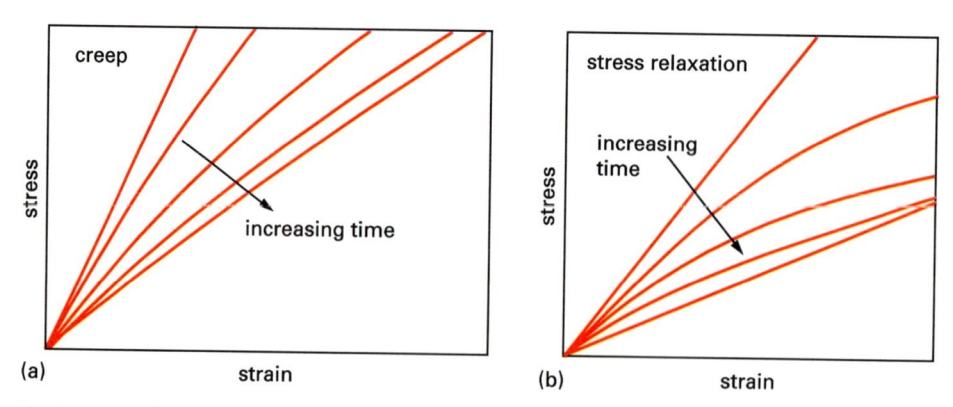

Fig. 7.4 Isochronous curves for nonlinear creep (a) and stress relaxation (b).

At intermediate times, where energy is actively being dissipated, the curves exhibit nonlinear dependence such that at higher load levels a disproportional amount of energy is dissipated.

As discussed by Ward and Sweeney (2004), Leaderman was the first to generalize Equations 7.6 and 7.7 to capture the nonlinear behavior of polymers, and then Schapery used thermodynamics of reversible systems, deriving the following constitutive equation for nonlinear materials,

$$\varepsilon(t) = g_1(\sigma) \int_{-\infty}^{t} D\big(\psi(t) - \psi(t^*)\big) \frac{d(g_2(\sigma))\sigma}{dt^*} dt^*. \tag{7.9}$$

Here,

$$\psi(t) = \int_0^t \frac{1}{a(\sigma(t^*))} dt^* \tag{7.10}$$

is a stress-dependent reduced time function, and $g_1(\sigma)$, $g_2(\sigma)$ and $a(\sigma)$ are functions of stress. For a creep test, Equation 7.9 reduces to

$$\varepsilon(t) = g_1(\sigma) \cdot g_2(\sigma) \cdot D\left(\frac{t}{a(\sigma)}\right). \tag{7.11}$$

Similarly, one could form a master creep curve by determining appropriate shifting functions to collapse the creep compliance curves for different load levels to one curve. The shifting parameters are basically the functions g_1, g_2, and a. It could be that nonlinearity does not arise from time-dependent behavior but rather from some other mechanism, such as rate-independent plasticity. If this were the case, the isochronous curves would still be nonlinear, and some shift would still be required to form a master creep curve.

7.3.3 Recoverability

Even if one were to determine all the parameters in Equations 7.9 and 7.10, one would still have to rely on fitting curves to experimental data. The ease of using Equation 7.9 depends on whether or not an empirical equation captures the recoverability of the strain. It is reasonable to split the total strain into recoverable and nonrecoverable components,

$$\varepsilon(t) = \varepsilon^{recov}(t) + \varepsilon^{nonrec}(t). \qquad (7.12)$$

The concept of recoverable and nonrecoverable strain is separate from that of nonlinearity. To be recoverable, the material must have potential energy stored in some manner such that when load is removed there is an internal driving force that facilitates recovery of strain. In the stress–strain discussion of Chapter 2, the permanent plastic deformation (strain at the point "∞" in Fig. 2.7) corresponds to the nonrecoverable strain here, and the total elastic strain (between the points "3" and "∞" in Fig. 2.7) corresponds to the recoverable strain.

Consider now a material that is subjected to a load cycle load defined as

$$\sigma(t) = \sigma_0 \cdot H(t) - \sigma_0 \cdot H(t - t_1), \qquad (7.13)$$

where $H(t)$ is the Heavyside step function, and t_1 is a sufficiently long time so that

$$\varepsilon(t > t_1) = \int_{-\infty}^{t} J(t - t^*) \frac{d\sigma}{dt^*} dt^* = \sigma_0 \cdot [J(t) - J(t - t_1)]. \qquad (7.14)$$

Creep is recovered if the strain given by Equation 7.14 goes to zero. In other words, creep is recovered if

$$J(t) = J(t - t_1). \qquad (7.15)$$

At longer times, this holds if $dJ/dt \to 0$, when $t \to \infty$ If, on the other hand,

$$\sigma_0 \cdot (J(t) - J(t - t_1)) = \varepsilon(t_1), \qquad (7.16)$$

then creep is not recovered. For linear materials, this occurs when $J(t) = C \cdot t$.

One modeling approach to account for recoverable and nonrecoverable behavior would be to specify that the creep compliance consists of two components,

$$\varepsilon(t) = \varepsilon^{rec}(t) + \varepsilon^{nonrec}(t) = \int_{-\infty}^{t} \left[J^{rec}(t - t^*) + J^{nonrec}(t - t^*) \right] \frac{d\sigma}{dt^*} dt^*. \qquad (7.17)$$

Similar arguments could be made for both stress relaxation and nonlinear creep equations. We note that separating recoverable and nonrecoverable creep requires both loading and unloading tests. The simple creep and relaxation curves by themselves are insufficient.

7.3.4 Time scales

Finally, we must consider the time scale of the loading event. When one models time-dependent behavior, there is a minimum time scale, t^{min}, below which time-dependence is ignored. This can be used as the time increment in the model, and the response at $t < t^{min}$ can be taken to be instantaneous. Thus, the total strain consists of an instantaneous component $\varepsilon^{instant}$ and a time-delayed component $\varepsilon^{delayed}(t)$:

$$\varepsilon(t) = \varepsilon^{instant} + \varepsilon^{delayed}(t). \qquad (7.18)$$

The instantaneous component of strain can be either elastic (recoverable) or plastic (nonrecoverable) deformation (cf. Chapter 2.3.1). The same holds for the time-dependent component $\varepsilon^{\text{delayed}}(t)$, as discussed in the preceding section. The separation point between the instantaneous and delayed responses corresponds to $t^{\min}$. For small, lower cut-off of times, the instantaneous elastic modulus typically increases, and the plastic deformation decreases if more of the total deformation is counted as delayed response. In a model, one could specify the minimum time, $t^{\min}$, so that all history before $t^{\min}$ is lumped into an instantaneous response.

Typically, there is also an upper limit of time of interest for the problem being addressed. Events that occur at times longer than the upper limit are not important. Consideration of this fact may help simplify the analysis because one can ignore very slow relaxations. Thus, the model should capture the behavior in the range of time scales of interest, and extrapolation to shorter or longer time scales should be applied only with the understanding that results would be more questionable.

Another way to consider time scales is the concept of a time constant τ as a material property governing a time-dependent response. The creep compliance can then be expressed as a sum of functions each active at a different time scale τ_i. For example,

$$J(t) = J_0 \cdot \delta(t) + \sum_{i=1}^{N} J_i \left(\frac{t}{\tau_i} \right), \tag{7.19}$$

where J_0 represents the material response at $t < \tau_1$.

To summarize the important characteristics of modeling time-dependent behavior of paper, any developed equations should address the following

1. Instantaneous elastic response to account for recoverable deformation that occurs at shorter time scales than the lower cut-off response.
2. Instantaneous plastic response to account for nonrecoverable deformation that occurs at shorter time scales than the lower cut-off response.
3. Recoverable time-dependent behavior for delayed-elastic behavior.
4. Nonrecoverable time-dependent response for a delayed-plastic or viscous response.
5. Consideration of longer time and load limits for the validity of the model.

A general expression for the creep compliance that could be used for modeling is

$$\varepsilon(t) = \frac{\sigma}{E} \delta(t) + \varepsilon_p \delta(t) + \sum_{i=1}^{N^{\text{rec}}} J_i^{\text{rec}} \left(\frac{t}{\tau_i^{\text{rec}}} \right) + \sum_{i=1}^{N^{\text{nonrec}}} J_i^{\text{nonrec}} \left(\frac{t}{\tau_i^{\text{nonrec}}} \right). \tag{7.20}$$

For creep and relaxation, there are typically time regimes where the response shows some characteristic behavior. Consider a typical creep response that exhibits a primary, a secondary, and a tertiary creep response. Often for polymers, the primary creep is predominantly recoverable, the secondary creep is mainly nonrecoverable, and the tertiary response is indicative of onset of failure or localization of stress in the material. Of course, the same mechanisms are likely to be active (to some extent) in multiple regimes. Separating out the responses is only an approximation, albeit a useful one. Figure 7.5 shows three cases where different mechanisms may contribute to the creep in different regimes.

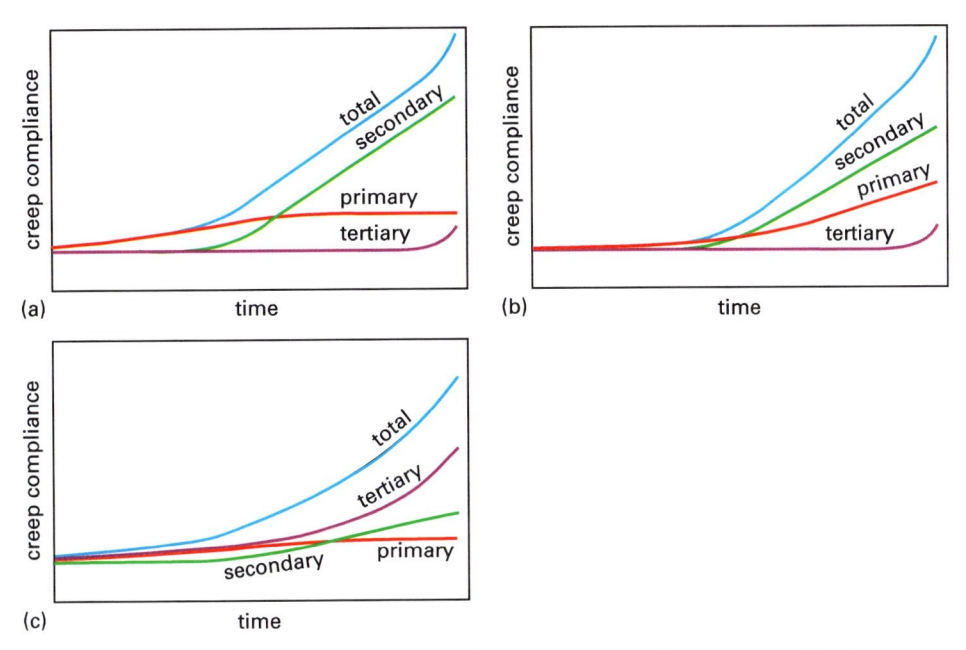

Fig. 7.5 Different possible mechanisms of primary creep, secondary creep, and tertiary creep.

7.4 Creep and relaxation properties of paper

With the overview given in the previous section, we are now ready to discuss the time-dependent nature of paper. The behavior is quite complex, and to account for all effects would require many parameters and extensive testing. Not only is the global loading response of paper a combination of elastic and inelastic behavior at various time scales, but also the effects of temperature and moisture content and loading history need to be addressed. Furthermore, the effects of network structure are coupled to the effects of the fibers that constitute the paper (Chapter 11). All this together gives the response measured in the laboratory or in the field.

Although one should develop a model that captures all these effects, it would likely be more cumbersome than instructive. The other extreme in modeling is to separate all effects and look at the general response to each parameter. The problem with the second approach is that one will almost always be able to find contradictory experimental results because of the couplings between so many parameters. Nevertheless, the second approach is more instructive and is adopted here. Naturally, not all papers will behave exactly as described below. The challenge is to place the results in the right context and provide explanations for why the observed behavior is different than the general view. Almost all the results in this section can be found in the literature, and the reader is referred to Salmén and Hagen (2002), Coffin (2005), and Ketoja (2008) for more specific information.

7.4.1 Creep

The discussion of creep begins with the basic response of paper to a constant uniaxial tensile load. The tensile creep response of paper is similar to that of many polymers. Figure 7.6 provides experimental data obtained by Brezinski (1956) for the creep of handsheets. The response shown in the figure is typical of many papers and illustrates several important aspects of the behavior. The dimensionless creep compliance is plotted both as a function of logarithmic time (Fig. 7.6a) and linear time (Fig. 7.6b), the latter to more clearly demonstrate the rapid decay of creep rate that occurs.

The important observations are as follows:

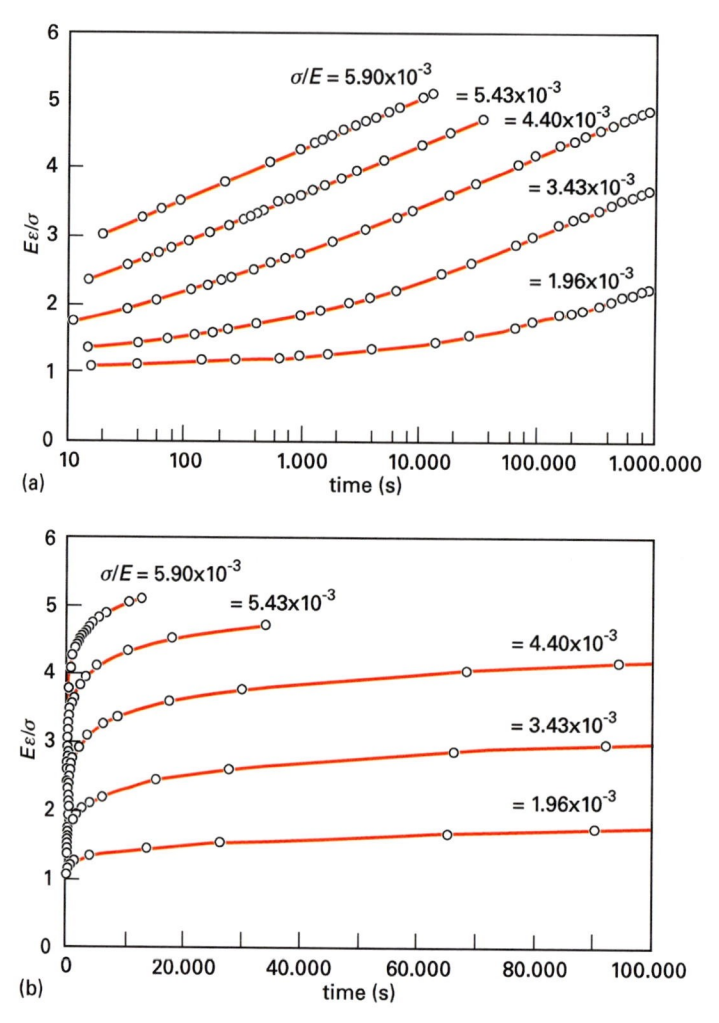

Fig. 7.6 Representation of the data of Brezinski (1956) as the product of the initial modulus and creep compliance for logarithmic time scale (a) and linear time scale, but for a shorter time range (b).

1. The rate of creep decays with time (exponential decay in the rate of creep).
2. Increased load results in disproportionately larger creep compliance (nonlinearity).
3. At high loads or long times, the rate of change in creep compliance is constant when expressed with respect to logarithmic time.

The second observation indicates that the creep compliance is a function of stress level and that linear visco-elasticity alone is inadequate to capture the total strain versus time response. Observation 3 suggests that a master creep curve can easily be formed from the data, and thus, an equation could be constructed to describe the creep behavior for a range of times and load levels. In fact, a simple linear shift of the logarithmic time is all that is required for many papers, as shown in Figure 7.7. In that case, the shift required to form the master curve of creep is given by

$$J(\sigma_2, t) = J\left(\sigma_1, t \cdot e^{\lambda(\sigma_1 - \sigma_2)}\right), \tag{7.21}$$

where λ is the slope of the shift factor versus stress curve. This equation illustrates how the creep compliance of paper is a strongly nonlinear function of load. As the load acts to shift the compliance curve in time, higher load levels are equivalent to allowing longer time for creep. One interpretation is that the creep in paper is a load activated phenomenon. In other words, the characteristic time constants in Equations 7.19 and 7.20 can be written as

$$\tau = \tau_0 e^{\lambda \sigma} \tag{7.22}$$

where τ_0 is the characteristic relaxation time at zero load.

An alternative approach to combining the creep curves is to shift the curves shown in Figure 7.6a vertically. This would work in any regime where the creep is linear with the logarithmic time. For high loads, significant inelastic deformation occurs at times lower than those measured. This deformation can be taken as an instantaneous plastic component. The vertically shifted curves would superimpose, and the remaining creep response could be explained with a linear visco-elastic model. At low loads,

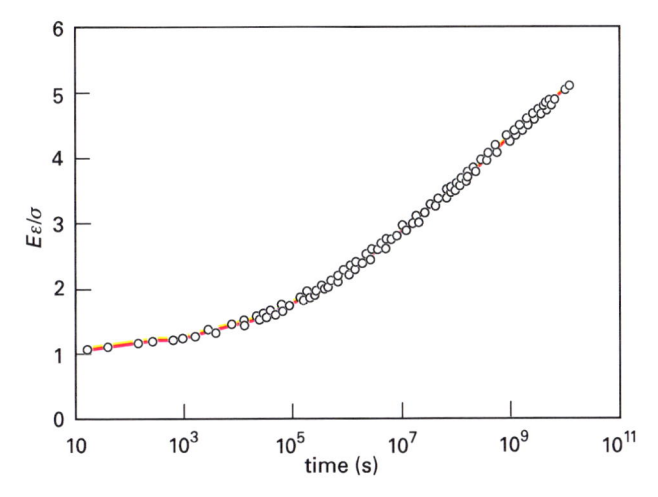

Fig. 7.7 Master curve of creep formed by moving the curves in Figure 7.6a parallel to the time axis.

the passage of much time would be required to activate the *log-linear* creep. In other words, once the creep response is activated by load it accumulates rapidly, and the rate of additional creep accumulation is fairly independent of load.

At low load levels, Figure 7.6a shows two different time regimes of creep. For low loads, Brezinski (1956) found that the initial creep deformation could be fit with a power-law equation. At high loads, the creep was log-linear for all measured time scales. If one uses plasticity to account for the nonlinearity of creep, it could be modeled as a strain-activated process. Even though the additional creep has been made linear, the plasticity still leaves the material inherently nonlinear to load.

Brezinski (1956) also found that in the primary regime (power–law relationship), creep was largely recoverable and that the secondary creep (log-linear response) was largely nonrecoverable. Although it is difficult to completely separate the recoverable and nonrecoverable components, it seems reasonable to assume that, in general, the primary power-law creep is recoverable, and the secondary log-linear creep is nonrecoverable. At some point in time, a tertiary stage of creep will occur where the creep rate begins to increase until the sample fails. The tertiary creep is typically short in duration and likely occurs because some critical level of damage to the paper has been reached.

A generic compliance curve that captures these observed behaviors at any given load level is

$$J(t) = B_0 \cdot \left[1 - \exp(b_0 t^\omega)\right] + B_1 \cdot \log(b_1 t + 1) + B_2 \cdot \exp(b_2 t^\zeta), \qquad (7.23)$$

where B_0, B_1, and B_2 are strain magnification factors; b_0, b_1, and b_2 are time magnification factors; and ω and ζ are shape factors. The first two terms represent primary and secondary creep mimicking behavior observed by Brezinski (1956), and the last term is representative of tertiary creep. The primary creep is written so that it acts as a delayed elastic response. Some possible creep curves obtained from Equation 7.23 are given in Figure 7.5.

Because we expect at least three types of creep to occur in a given paper, the variation of creep response from paper to paper will differ. Also, the measured response depends on level and duration of load. One should always characterize the paper for creep over several load levels and for sufficient duration so that it can be properly modeled. To separate out recoverable and nonrecoverable creep, creep recovery tests should be completed. This requires that one removes load after a specified time and measures the recovery of strain. Recovery may require more time than the initial creep.

7.4.2 Stress relaxation

Johanson and Kubát (1964) investigated the stress relaxation in paper (Fig. 7.8). Similar to creep, when stress is plotted as a function of logarithmic time, very little relaxation happens before a relaxation starts that is fairly linear with the logarithmic time. The fact that the relaxation modulus is distinct for each level of strain (top graph in Fig. 7.8) again demonstrates the load nonlinearity present in paper.

In creep and relaxation tests, the initial loading is assumed to be quick compared to any relaxation processes. For creep tests with a dead load weighting, this can be accomplished quite easily. Relaxation testing is typically done on a universal tester,

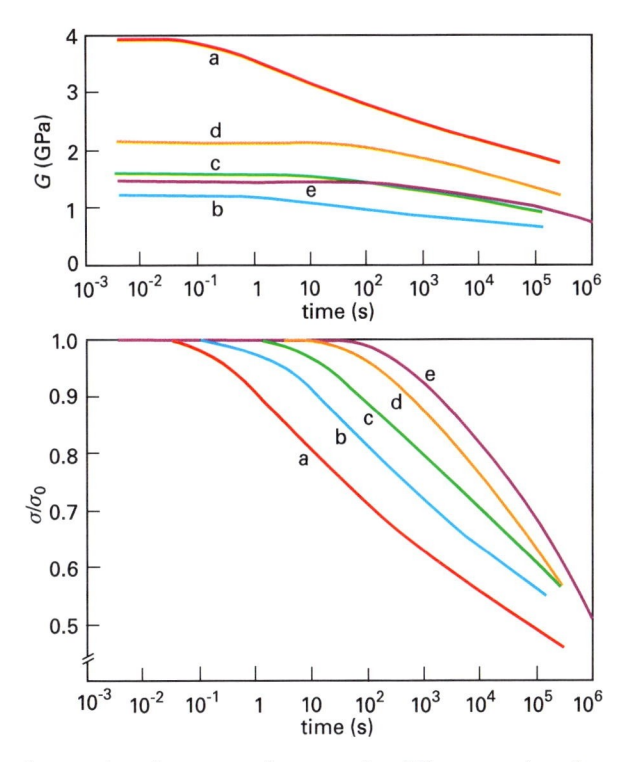

Fig. 7.8 Stress relaxation curves for paper for different strain values, showing the relaxation modulus (top graph) and stress normalized with the initial stress value (lower graph), after Johanson and Kubát (1964). Reproduced with permission from Svenska Pappers- och Cellulosaingeniörsföreningen (SPCI).

and one has the choice of loading either instantaneously or with a constant strain rate. The choice affects the presentation of results. In the first case, the relaxation modulus (Eq. 7.5) is divided by the elastic modulus. This is equivalent to dividing the stress by the ideal stress that would arise in a linearly elastic material when the strain is applied instantaneously. In the second case, the relaxation stress is divided by the stress achieved when the desired strain was first achieved. This is equivalent to using the linear elastic strain, not the total strain, to calculate the relaxation modulus (Eq. 7.5).

Consider the effect that stress relaxation has on web dynamics (Ch. 6). Under steady-state conditions the web in an open draw is under a state of constant strain. Because of relaxation, the tension in the web will be reduced compared to when no relaxation occurred. In addition, the magnitude of the stress gradients will be decreased.

7.4.3 Tensile versus compressive creep

Because paper is a fibrous network structure, it is unstable in compression. On various length scales the material will flex, buckle, and fold under compression, in essence making paper inherently weaker in compression than tension. It is also more compliant at larger stress levels and longer times. Figure 7.9 compares tensile and compressive isochrones. At small load levels the creep responses are similar, but at larger load levels compression is more complaint. Vorakunpinij (2003) showed that increased paper density prolonged the range of stress levels where tension and compression were similar.

The added deformation from structural rearrangements has the effect that in compression the creep rate does not decrease with time as much as it does in tension. Therefore, instead of the log-linear secondary creep, compressive creep is more like a power-law or even linear with time. Corrugated boxes under compression exhibit a linear secondary creep response (see Fig. 3.26) because of additional panel buckling and concentration of load into the corners of the box, as discussed in Chapter 3.4. The linearity of the secondary creep simplifies phenomenological models of corrugated board box. On the other hand, applying the techniques covered in Chapter 4.4 to creep could suggest ways to decrease not only the creep rate of a box under compression but even to get the creep rate to decay with time, in the same way as the constituting papers of the box behave.

7.4.4 Effect of the papermaking process and furnish

The basic time-dependent behavior discussed in the previous three sections is generic, but given the vast variation in paper properties that can be achieved through choice of materials and processing, the time-dependent response of paper can be modified in the manufacturing process. In this section, some effects of processing and furnish on the creep response are introduced.

Brezinski (1956) carried out creep tests on handsheets made at different levels of wet pressing. Increased wet pressing acts to increase paper density and to increase the area of bonding between fibers (Chapter 11.2.2). Figure 7.10 shows 24-hour isochrones for sheets made at different levels of wet pressing. Clearly, higher levels of wet pressing decreased creep compliance. For these sheets, the elastic modulus also

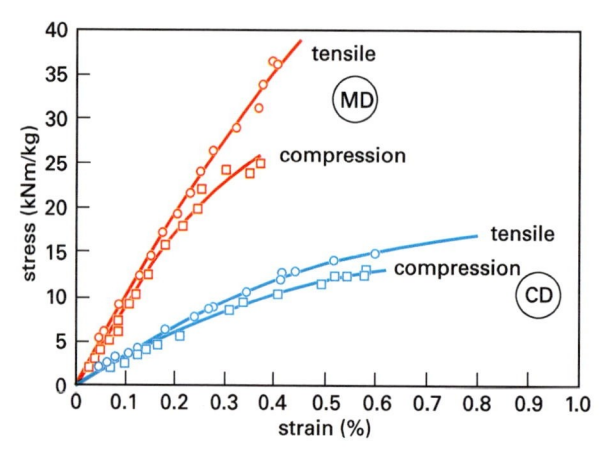

Fig. 7.9 Isochronous curves in tension and compression for both MD and CD creep, after Haraldsson et al. (1993). Reproduced with permission from Svenska Pappers- och Cellulosaingeniörsföreningen (SPCI).

7.4 Creep and relaxation properties of paper

increased with increased wet pressing. If the isochrones in Figure 7.10 are scaled by the initial modulus, they all collapse to the same curve. This suggests that a simple stress magnification can explain these effects. If we define an efficiency factor ϕ as the ratio of elastic modulus to some reference value (Seth and Page, 1981), then the normalized creep compliance $J^{norm} = \phi \cdot J(\sigma/\phi, t)$ is independent of wet pressing.

Moderate wet straining applied during drying has the effect of increasing elastic modulus, decreasing breaking strain, and slightly increasing tensile strength. Schulz (1961) showed that wet straining also decreases creep compliance. If the creep compliance is multiplied with the elastic modulus, a master creep curve can be constructed, as shown in Figure 7.11 (Coffin, 2005). Here the logarithmic time shift was linearly proportional to the degree of wet straining. Isochronous curves for various wet-strained sheets will likely not superimpose with only an efficiency factor ϕ, but the fact that a master creep curve can be formed implies that the major effect of wet straining is to shift the activation of secondary creep response to longer times.

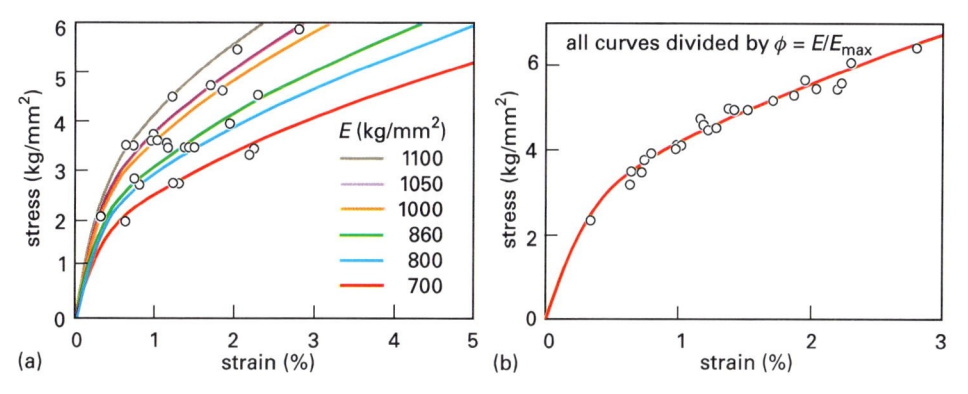

Fig. 7.10 Isochronous creep curves for handsheets made with different levels of wet pressing (a), and the same data divided scaled by the efficiency factor ϕ, based on data from Brezinski (1956). Reproduced from Coffin (2005) with permission from The Pulp and Paper Fundamental Research Society (www.ppfrs.org).

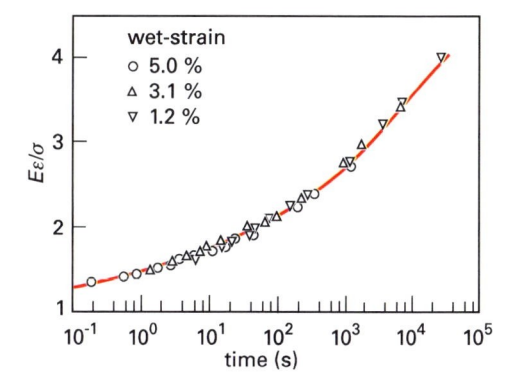

Fig. 7.11 Master curve of creep for sheets of various levels of wet straining, using data from Schulz (1961). Reproduced from Coffin (2005) with permission from The Pulp and Paper Fundamental Research Society (www.ppfrs.org).

7.5 Moisture effects

The mechanical response of paper is influenced by the environment. Both changes in temperature and moisture content will change the mechanical properties, can alter the structure, and will likely influence the distribution of stresses in a sheet of paper. The hygrothermal influences on dimensional stability are discussed in Chapter 9. Here we discuss the influence of moisture on the time-dependent properties.

7.5.1 Softening with moisture

As illustrated in Chapter 2.2, increased moisture will decrease elastic modulus. At moisture contents below 5%, the elastic modulus is fairly constant. For moderate moisture increases above this level, the modulus decreases in approximately a linear fashion. Water is considered to be a plasticizer for paper and, in essence, activates thermal relaxations that normally occur at higher temperatures. Moisture will

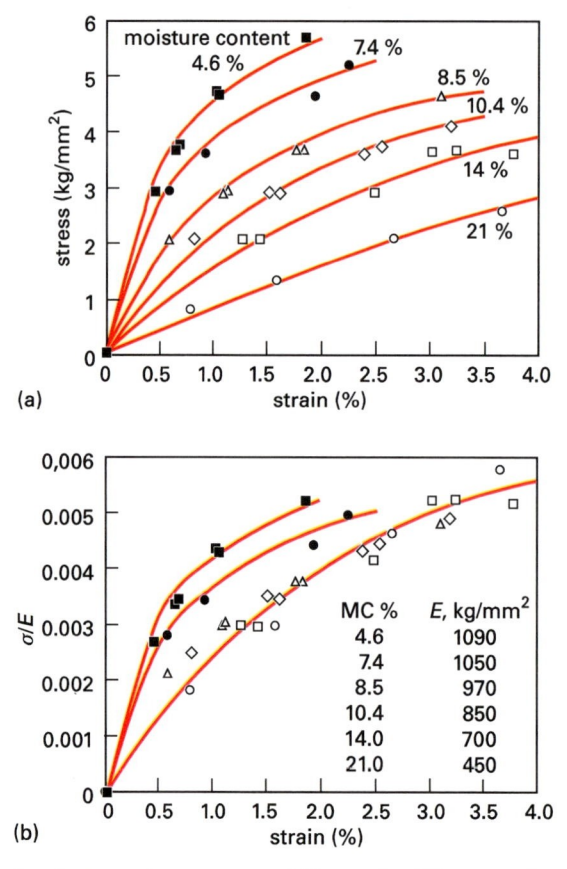

Fig. 7.12 Isochronous creep curves at 24 hours for different moisture contents (a) and the same curves with stress divided by the initial elastic modulus (b), using data from Brezinski (1956).

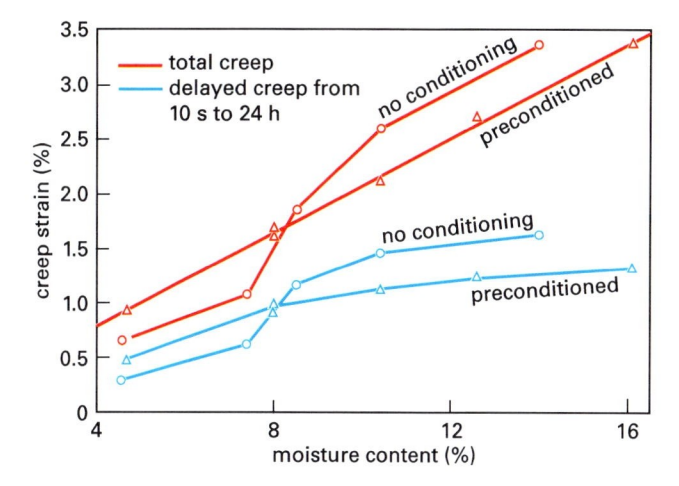

Fig. 7.13 Creep strain against moisture content, with and without preconditioning at high moisture content, using data from Brezinski (1956). Reproduced from Coffin (2005) with permission from The Pulp and Paper Fundamental Research Society (www.ppfrs.org).

also affect the time-dependent nature of paper, such as the isochronous curves in Figure 7.12a. If one scaled stress values with the elastic modulus measured at the same moisture content (Fig. 7.12b), the higher moisture content curves would converge, but it is not clear if they form one curve. Brezinski (1956) found that at high moisture contents, the logarithmic rate of change in creep compliance was independent of moisture, which would imply that secondary creep is independent of moisture content. At lower moisture contents, creep compliance is lower, but the effect can be removed by exposing the paper to high moisture before the creep test. A sample preconditioned by moisture will show a shift in the low moisture creep behavior aligning it with creep at higher moisture contents (Fig. 7.13). The main influence of the moisture conditioning is thus to lower the stiffness of paper and release the effects of drying tension applied on the paper machine.

7.5.2 Accelerated creep

Accelerated creep is one of the manifestations of the mechanosorptive effects. The term *mechanosorptive* is used to describe the coupling between mechanical and moisture effects during changing moisture contents. Accelerated creep occurs when a material is subjected to the combined action of creep load and cyclic moisture content. Figure 7.14 shows the results for two tensile creep tests. For the log-linear curve, moisture was kept constant at a high level corresponding to 80% relative humidity (RH) of the surrounding air. For the second curve, the relative humidity was at first held at 80% and then cycled between 80% and 30% every 2 hours. When the moisture cycling begins, two changes in the strain response are noted. First, there is a cyclic strain component due to the change in moisture content of the sample. The cyclic strain results from a combination of the hygroexpansion and the change in elastic modulus. The second effect is an increase in the rate of creep. It is this second effect that is called accelerated creep.

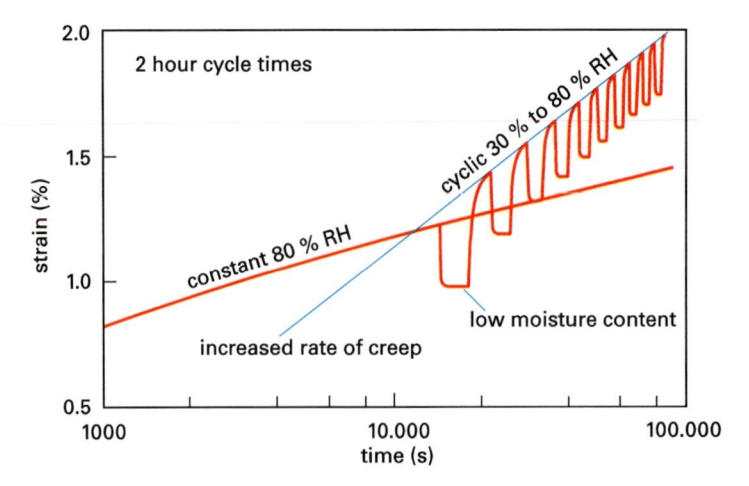

Fig. 7.14 Accelerated creep in paper.

The increased creep rate by moisture cycles leads to shorter lifetime, or the time required for the sample to fail. The box creep curve in Figure 3.37 was obtained under conditions of cyclic moisture because it is considered a worst-case scenario for box lifetime. Clearly, the increased creep rate is due to a coupling of load and moisture effects. The effect of constant moisture content on creep rate (Fig. 7.14) is not sufficient to explain accelerated creep because in that case the creep observed under the cyclic humidity should be lower than creep under the constant 80% RH. As is seen in Figure 7.14, almost no creep occurs at the low humidity.

From the viewpoint of mechanics, accelerated creep can be understood fairly easily. When relative humidity is changed, a change of moisture in the paper will happen over some time period. This transport of moisture can be expected to create some moisture gradient across the thickness of the paper. Any change in the local moisture content drives local swelling or shrinkage. Thus, a moisture gradient across the paper creates a gradient in the local hygroscopic strain, the amount of local swelling or shrinkage that the paper would achieve if no restraints were present (see Eq. 2.5). Furthermore, and perhaps more important, paper is heterogeneous, and the hygroexpansivity of the structure will not be uniform even if there is no moisture gradient.

Both moisture gradients and heterogeneity can be expected to create gradients of hygroscopic strains. The change in moisture content combined with the variation of hygroscopic strains will create stress gradients in the paper. Even though the average stress across the sheet (Eq. 7.1) is constant, some local areas will be stressed above the average. These stress gradients will persist for some time during and after a moisture change. We have also seen that paper is inherently nonlinear with respect to load (Fig. 7.6). This means that highly stressed areas will creep disproportionately more than lightly stressed areas. Therefore, as long as elevated levels of stress exist in the sheet it will produce more creep than expected from the average load.

There are two approaches that could be used to model the mechanism of accelerated creep. The first would be to include a sufficient level of modeling to account for the stress gradients and the nonlinear material behavior. This model would predict transient stress gradients, and the accelerated creep would be a natural outcome. The second approach would be to use only the average stress and moisture values and add a hygroscopic strain component that accounts for the influence of the variations. The first approach is more fundamental, and the second approach is more empirical, but either method could be used to model accelerated creep.

There are two important aspects of accelerated creep that any model should reflect:

1. Although the overall creep rates are lower at lower moisture contents, the acceleration of creep is more severe at lower moisture contents. This is because stress gradients are more severe and persist for a longer time because the material is less compliant.
2. The rate of creep accumulation is affected by the moisture cycle. The acceleration of creep with a given cycle time depends on the time needed for stress gradients to relax. If the cycle time is close to the relaxation time of stress gradients, accelerated creep will be most severe. Longer moisture cycles then reduce accelerated creep.

According to this explanation, there is nothing extraordinary about accelerated creep. In fact, tests (DeMaio and Patterson, 2005) have shown that, also in the case of accelerated creep, the effects of wet pressing can be accounted for with an efficiency factor ϕ. This implies that the same underlying mechanisms are active for both constant and cyclic moisture and that only the stress distribution is different in the case of cyclic moisture.

In Chapter 3.3.4 the practical problem of long-term stacking performance of corrugated board boxes was discussed. Accelerated creep is of prime importance to the lifetime of stacked boxes. Box lifetime is discussed in the next section.

7.6 Prediction of box lifetime

The discussion of the stacking strength of corrugated board boxes in Chapter 3 concerned primarily the short-time loading behavior. Here we consider what happens at long-time scales. The useful lifetime of a box is considered to be the time lapse that a box can safely protect its contents in the service environment. In general, during its useful life a box could be subjected to a combination of dynamic loading from transportation, dead loading from storage, and combined hygroscopic and thermal effects from changes in the environment. For this discussion, *lifetime* is the time of survival for a box subjected to a dead load in a given environment, such as the case with stacking in a warehouse.

7.6.1 Creep response of a box

Consider a box (Fig. 7.15) subjected to a compressive load P that will result in creep. Figure 7.16 shows the creep process and predominant box deformation mechanisms, described by Kellicutt and Landt (1951) and Bronkhorst (1997). When the box is first loaded, the primary deformation will be a flattening of the top flaps until the load distributes fairly uniformly around the perimeter of the box. This deformation can be large, as first noted by Malcolmson (1936). For a well-formed box, the walls will compress uniformly if the load is evenly distributed. In the secondary regime, the deformation is dominated by structural buckling of the box panels and inter-flute buckling of the liners at the corners, similarly to the behavior under short-time loading (Chapter 3.4.2). This is the longest time regime of the deformation process. As first pointed out by Little (1943) for the Box Compression Test (BCT), box perimeter and bending stiffness are important factors that would impact this regime. The tertiary phase is the initiation of failure in the box wall due to the formation of a hinge, typically along a corner.

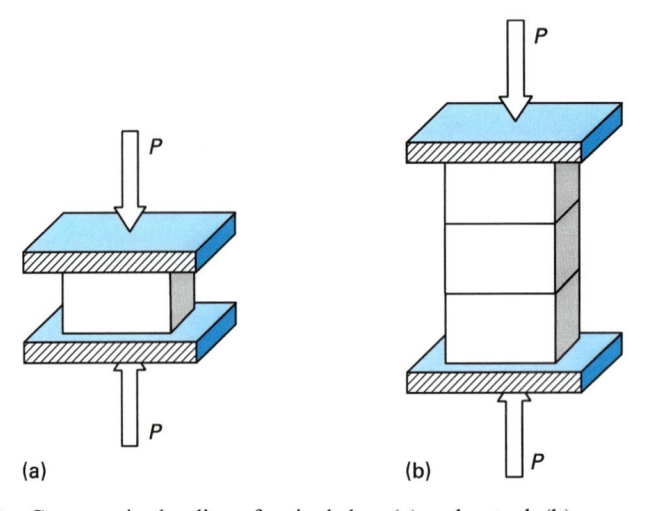

Fig. 7.15 Compressive loading of a single box (a) and a stack (b).

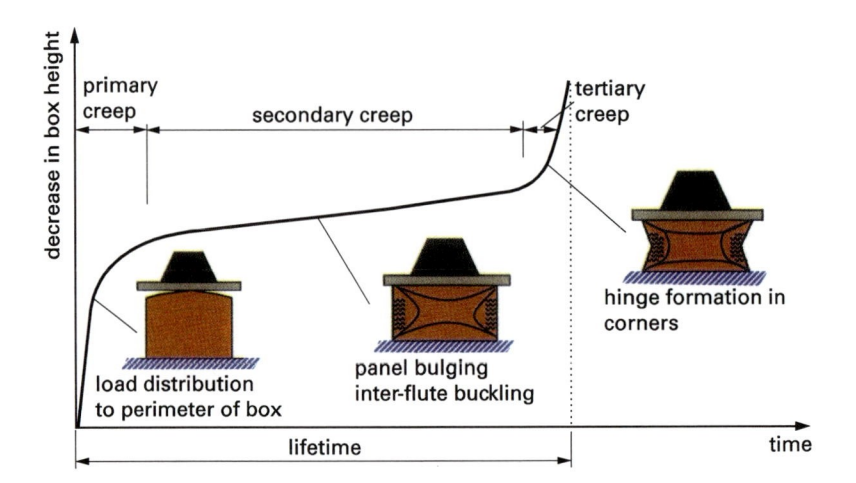

Fig. 7.16 Typical creep response of a box subjected to compression.

7.6.2 Previous equations for box lifetime

Through a series of box lifetime tests, Kellicutt and Landt (1951) found that for loads less than $0.75 \cdot \text{BCT}$, the lifetime t_{BL} could be expressed as a function of load level according to

$$t_{\text{BL}} = t_{\text{BL}_0} e^{-k_0 P/\text{BCT}}. \tag{7.24}$$

Here t_{BL_0} is the lifetime at zero load and k_0 is a load proportionality factor. Stott (1959) found similar results. In addition, he found that the sensitivity to moisture was greater for lifetime than BCT. Dagel and Brynhildsen (1959) developed a similar equation

$$t_{\text{BL}} = e^{k_1(\text{BCT}/P - 1)} - 1, \tag{7.25}$$

where k_1 is another proportionality factor. The advantage here is that t_{BL} goes to infinity when $P \to 0$ and to zero when $P \to \text{BCT}$. Equation 7.25 was developed from assumptions that the secondary creep response is linear in logarithmic time, that the rate of creep is proportional to the load, and that the failure of a box is triggered at the same displacement regardless of load level.

Observations that the lifetime of a box is dominated by the secondary creep and that the rate of secondary creep is approximately inversely proportional to time led Bronkhorst (1997) to develop a prediction of lifetime as

$$t_{\text{BL}} = \frac{\varepsilon^{\max,\text{SC}}}{\dot{\varepsilon}^{\text{SC}}}, \tag{7.26}$$

where $\varepsilon^{\max,\text{SC}}$ is the amount of secondary creep strain at box failure and $\dot{\varepsilon}^{\text{SC}}$ is the average secondary creep rate. Other researchers had developed similar relationships but allowing the secondary creep rate to be raised to some other power than negative one (Bronkhorst, 1997). Equation 7.26 refers to the fact that lifetime correlates very strongly to secondary creep rate.

In fact, testing has shown that whereas lifetime versus P gives large scatter in results, lifetime versus $\dot{\varepsilon}$ reconciles the scatter and provides good correlation (Bronkhorst, 1997). This gives the impression that the key to improving lifetime is to decrease the secondary creep rate $\dot{\varepsilon}^{\text{SC}}$. While this is an important finding, it does not help predict the lifetime of a given box. Both lifetime and secondary creep rates are measured from the same tests and only verify that the behavior illustrated in Figure 7.16 and the relationship (Eq. 7.26) hold. Seemingly identical boxes loaded with the same dead load may easily give vastly different lifetimes. While the boxes with the shortest lifetimes tend to have the largest secondary creep rates, we have no means to predict in advance which boxes give the shortest lifetime. Thus, Equation 7.26 is not pertinent for lifetime predictions.

In order to be of predictive value the equation for lifetime needs to be expressed in terms of a prescribed parameter. For stacking lifetime, the prescribed parameter is load. In essence, this is what box designers do. A box is designed to have certain strength (BCT). In a simple scenario, the stacking load is prescribed by the user, the expected conditions that the box might see are identified, and then the box designer applies a series of safety factors to determine the minimum BCT that user needs. The term *safety factor* implies that we need to increase the strength of the box to account for unknown effects. While this is clearly the conservative, safe, and ethical path to

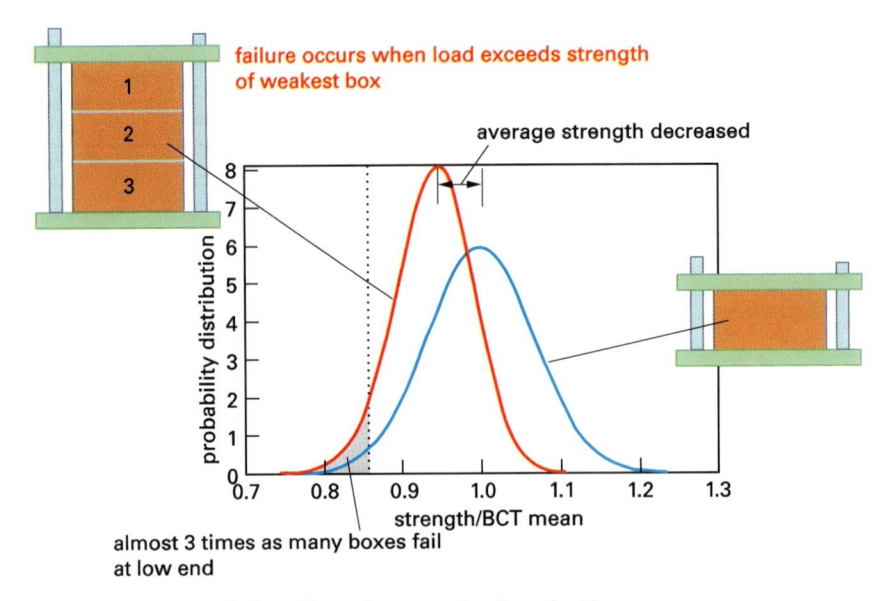

Fig. 7.17 BCT variation affects the strength of stacked boxes.

(Figure annotations: "failure occurs when load exceeds strength of weakest box"; "average strength decreased"; "almost 3 times as many boxes fail at low end"; x-axis: strength/BCT mean; y-axis: probability distribution)

take, it also implies unfounded ignorance of the mechanisms of box compression and material behavior of paper. For a given box, we certainly know why increased loading levels lead to reduced lifetime. Therefore, the idea of safety factors is misleading. These factors are inverses of load reduction factors based on the expected service environment of the box.

7.6.3 Derivation of a new equation for box lifetime

Equations 7.24 to 7.26 do not satisfactorily capture the effect of load on lifetime. Equation 7.26 relies on two measured quantities and is therefore not valid for design. The approach used by Dagel and Brynhildsen (1959) is reasonable, but Equation 7.25 does not utilize the observation that secondary creep rate is linear in logarithmic time. At failure the total compressive strain of a box is approximately

$$\varepsilon^{max} = \varepsilon_0 + \dot{\varepsilon}^{SC} \cdot t_{BL}, \tag{7.27}$$

where the primary deformation $\varepsilon_0 = \varepsilon^{max} - \varepsilon^{max,SC}$. As was discussed in Section 7.4.1, it is reasonable to assume that characteristic time constants have an exponential dependence on load level. The creep compliance would then be expected to be

$$\frac{\dot{\varepsilon}^{SC}}{P} = k_2 e^{\lambda P}, \tag{7.28}$$

where λ is the slope of the shift factor versus stress curve (cf. Eq. 7.21), and k_2 is a constant. One can also assume that ε_0 is proportional to load,

$$\varepsilon_0 = k_3 P. \tag{7.29}$$

Substituting Equations 7.28 and 7.29 into Equation 7.27 gives

$$\varepsilon^{max} = P \cdot k_3 \cdot \left[1 + \frac{k_2}{k_3} \cdot e^{\lambda \cdot P} \cdot t_{BL}\right]. \tag{7.30}$$

If the load P were chosen equal to the box compression strength (BCT) that would be measured at the equivalent environmental conditions, then the lifetime would be practically zero, and one gets $\varepsilon^{max} = k_3 \cdot BCT$. Solving Equation 7.30 for the lifetime gives the final expression

$$t_{BL} = \kappa \cdot \left[\frac{BCT}{P} - 1\right] \cdot e^{-\omega \cdot \frac{P}{BCT}}, \tag{7.31}$$

where $\kappa \equiv k_2/k_3$ and $\omega \equiv \lambda \cdot BCT$ are constants to be determined from lifetime tests, P is the applied box load, and BCT is the box compression strength measured at the equivalent environmental conditions.

7.6.4 Accounting for variability

From the literature it is clear that variability in strength must be considered in lifetime studies. Here we show what happens in Equation 7.31. Variability in lifetime could arise from all the parameters κ, ω, and BCT, but we consider only the effect of variation in BCT. We used the data of Challas et al. (1994) for 300 BCT tests and fitted a normal distribution function to the normalized BCT values (i.e., the mean set equal to 1). This gave the coefficient of variation of 0.067; we use this as a typical value of variation in BCT. The effect is illustrated in Figure 7.17. In a stack of three boxes, the chance of having a weak box is about three times greater than the chance with only one box. This shifts the whole distribution to lower values because the strength of the stack is given by the weakest box in the stack. The distribution of BCT values will also affect the variation in stacking lifetime.

We then fitted Equation 7.31 to data reported in the literature (Fig. 7.18). We found that $\omega = 20$ is reasonable for all the data, even under conditions of cyclic moisture. In contrast, the value of κ turned out to depend on box construction and moisture conditions, shifting the line in Figure 7.18 along the vertical axis. The large variability in lifetime data is apparent in the figure. We then substituted the normal distribution for BCT into Equation 7.30 in order to calculate predictions for the variability in lifetime. Figure 7.19 illustrates the result. For the load of BCT/2, the average lifetime is 45 days, but 2.5% of the boxes should fail in less than 7 days. This drastic reduction in the box lifetime arises entirely from the distribution of BCT values, with the coefficient of variation equal to 0.067. In a stack, the frequency of premature failures would be even bigger.

Through the constant λ, box lifetime is inherently tied to the creep response of the box, and the average response appears to be predictable. Damage, flaws, poor load distribution, and variable moisture content will affect lower BCT and, in turn, have a dramatic effect on lifetime. Thus, the actual stacking failures cannot be predicted from the average properties of boxes. The effect of load level to lifetime appears to directly come from the large nonlinear effect of load on creep rate, which magnifies the effect of BCT variation on box lifetimes. This explains some of the high variability in experimental data of lifetime versus load. It also suggests that the majority

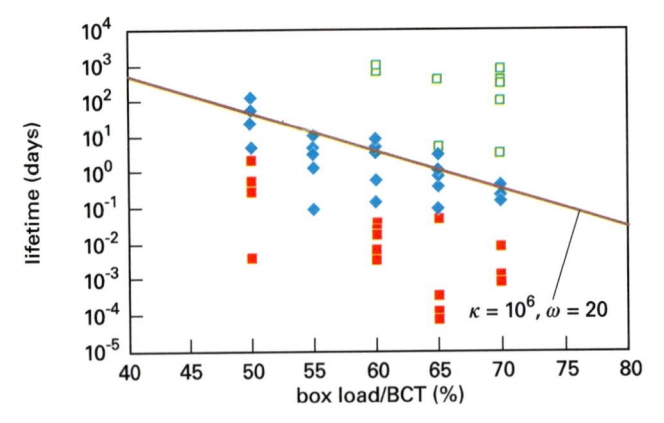

Fig. 7.18 Box lifetime versus box load from Koning and Stern (1977), each type of symbols representing one box type. The straight line is the fit of all data to Equation 7.31.

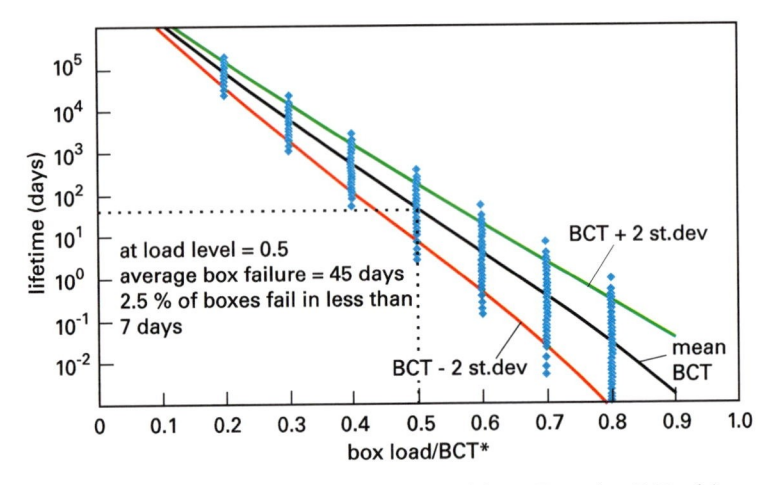

Fig. 7.19 Predicted variability in box lifetime calculated from Equation 7.31 with parameters obtained from Figure 7.18.

of stacking failures may better be attributed to the failure of a few weak boxes. This subject of variability in strength is discussed in Chapter 8.

7.7 Summary

Accounting for the time-dependent nature of the mechanical response of paper proves to be useful for many practical problems. After studying this book, the reader should better appreciate that stress relaxation and creep should be accounted for in web mechanics and box lifetime. In addition, the reader should envision that problems in fracture, bending, and even micromechanics could account for the time-dependent nature of paper. Overall, if you account for time-dependence in material behavior you must do the following:

- define the time scales involved in the problem
- separate recoverable and nonrecoverable deformation
- consider linear and nonlinear behavior
- account for differences in tension and compression
- know that time-dependent stress distributions impact responses (e.g., accelerated creep)
- consider that multiple factors may influence observed performance (e.g., variability in lifetime)

This chapter provided an introduction to the basics of time-dependent behavior. If one approaches a problem from a fundamental point, then appropriate constitutive equations must be introduced and robust testing must be completed to determine the values of model parameters. For example, the creasing and folding problem of Chapter 4, the crack tip models of Chapter 5, or the fluting problem that will be presented in Chapter 9 could all include time-dependent behavior by extending the constitutive equation.

Additionally, this chapter demonstrated that observations of creep, could be used to develop empirical equations that could prove useful, such as the prediction for box lifetime. The advantage of this approach is that a simple equation could be used to help guide box designers, similar to the McKee equation for BCT discussed in Chapter 3.

Finally, no phenomena can be considered solely on its own when it comes to understanding practical problems. The example given in Section 7.5 shows that although a reasonable prediction for lifetime can be developed, without consideration of the large variability in BCT strength, such an equation would prove of little use. Those who are able to integrate information on time-dependence with information on fracture, dimensional stability, web mechanics, variability, and micromechanics will be able to solve practical problems facing the paper and related industries.

There is still much fundamental research on creep and relaxation that is necessary. Future testing should include stages of recovery so that recoverable and nonrecoverable deformations can be separated. It would be helpful to study the time-dependence of recoverable deformation separately from the time-dependence of nonrecoverable deformation. Once this is established one could more effectively study the effects of furnish and processing on the creep response.

Very little work on the time-dependent response of paper under multi-axial states of stress has been published. In order to develop appropriate models, studies and data are required. In addition, the effect of previous loading histories on the future response of paper requires more study. It would be possible to develop an entire body of work on developing the appropriate test methods to fully characterize time-dependent stress–strain behavior.

Studies of the effect of moisture and cyclic moisture on time-dependent behavior of paper are still needed. The coupling of moisture and mechanical loads is a complex problem, and appropriate tests are needed from which results can better elucidate mechanisms and behaviors.

The reader that wishes to gain true insight into the behavior of paper cannot ignore the time-dependent nature of paper.

References

Brezinski, J. P. (1956). The creep properties of paper. Tappi J. 39(2), 116–128.

Bronkhorst, C. A. (1997). Towards a more mechanistic understanding of corrugated container creep deformation behavior. J. Pulp Pap. Sci. 23(4), J174–J181.

Challas, J., Schaepe, M., and Smith, C. N. (1994). Predicting package compression strength geometry effect. Final Report Project 3806 to the Containerboard and Kraft Paper Group of the American Forest and Paper Association.

Coffin, D. W. (2005). The creep response of paper. In: Proceedings of the Advances in Paper Science and Technology, Transactions of the 13th Fundamental Research Symposium, S. J. I'Anson, ed. (Cambridge), Vol. II, pp. 651–747

Dagel, A. V., and Brynhildsen, H. O. (1959). Tidens Inverkan pa pappladors belastingsformaga. Svensk Pappperstidning 62(3), 77–82.

DeMaio, A., and Patterson, T. (2005). Influence of fiber-fiber bonding on the tensile creep of paper. In: Proceedings of the Advances in Paper Science and Technology, Transactions of the 13th Fundamental Research Symposium, S. J. I'Anson, ed. (Cambridge), Vol. II, pp. 649–775.

Haraldsson, T., Fellers, C., and Kolseth, P. (1993). The edgewise compression creep of paperboard: New principles of evaluation. In: Proceedings of the Products of Papermaking, Transactions of the Tenth Fundamental Research Symposium, C. F. Baker, ed. (Oxford), pp. 601–637, FRC.

Johanson, F., and Kubát, L. J. (1964). Measurements of stress relaxation in paper. Svensk Papperstidning 67(20), 822–832.

Kellicutt, K. Q., and Landt, E. F. (1951). Safe stacking life of corrugated boxes. Fibre Containers 36(9), 28–38.

Ketoja, J. (2008). Rheology. Chapter 8. In: Paper Physics, 2nd Edition, K. Niskanen, ed. (Helsinki, Finland: Paperi ja Puu Oy).

Koning, J. W., and Stern, R. K. (1977). Long-term creep in corrugated fiberboard containers. Tappi J. 60(12), 128–131.

Little, J. R. (1943). A theory of box compressive resistance in relation to the structural properties of corrugated paperboard. Tappi 116(24), 275–278.

Malcolmson, J. D. (1936). The value of the compression test for corrugated boxes. Fibre Containers 21, 14–15.

Salmén, L., and Hagen, R. (2002). Viscoelastic properties. Chapter 2. In: Handbook of Physical Testing of Paper, Vol. 1, 2nd Edition, R. E. Mark, C. C. Habeger, J. Borch, and M. B. Lyne, eds. (New York, NY, USA: Marcel Dekker).

Schulz, J. M. (1961). The effect of straining during drying on the mechanical and viscoelastic behavior of paper. Tappi J. 44(10), 736–744.

Seth, R. S., and Page, D. H. (1981). The stress-strain curve of paper. In: Proceedings of The Role of Fundamental Research in Paper Making, Transactions of the Seventh Fundamental Research Symposium, J. Brander, ed. (Cambridge), pp. 421–452, FRC.

Stott, R. A. (1959). Compression and stacking strength of corrugated fibreboard Containers. APPITA 13(2), 84–88.

Vorakunpinij, A. (2003). The effect of paper structure on the deviation between tensile and compressive creep responses. Doctoral Dissertation. Institute of Paper Science and Technology, Atlanta, Georgia.

Ward, I. M., and Sweeney, J. (2004). The Mechanical Properties of Solid Polymers, 2nd Edition (Chichester, West Sussex, UK: John Wiley and Sons).

8 Statistical aspects of failure of paper products

Tetsu Uesaka

8.1 Introduction

Failure of paper or board could happen both in manufacturing and in end-use. The phenomena are typically manifested, on one hand, as web breaks on paper machines and in printing houses, and on another, in the collapses of container boxes in warehouses and logistic chains. Although these problems do not occur frequently, once they occur, the consequences are significant. Therefore, current strength specifications of paper materials all concern these less-frequently occurring problems.

One unique characteristic of these problems is that the *system* to which the paper is subjected is of prime importance. For example, the open draw sections in paper machines, reel winders, printing presses, corrugators, converting machines, trucks, rails, and warehouses. Paper is only part of the system. Therefore, failure problems concern the performance of the whole system, rather than simply the strength of paper.

Another important characteristic is that these problems are *rare* phenomena and normally exhibit enormous variability in their occurrence. This statistical nature of the problems makes it difficult and expensive to perform direct investigations of cause and effect, such as pilot trials or full-scale experiments.

This chapter first discusses the systems and statistical nature of failure of paper products using two examples: (1) web breaks on a paper machine or a printing press, and (2) collapse of container boxes, introduced in Chapters 3 and 7. These practical examples may illuminate the structure of the problem and define the related scientific problems. Based on this preparation, we then discuss the basic approaches for treating statistical and systems failure.

8.2 Practical examples

8.2.1 Web breaks in a printing press and on a paper machine

In printing houses in North America and Europe, web breaks typically happen with a frequency of a few breaks per 100 rolls. This means that one may find breaks at every 350 km, assuming that a typical paper roll (diameter about 1 m) contains 10 km of paper, and there are 3 breaks per 100 rolls. This number, however, varies considerably from one printing house to another. For example, some printing houses still run at more than 10% break rate (10 breaks per 100 rolls), whereas others are getting to below 1% break rate. In extreme cases, such as Japanese printing houses, one runs at a few breaks per 10,000 rolls (Uesaka, 2005).

The large influence of press operation was well recognized in early research. For example, Page and Bruce (1985) showed that the printing house causes much more of the difference in web breaks than the paper supplier. The same difference is seen also with web breaks on paper machines. Typical European and North American paper mills have a few breaks per day. However, on a bad day this number climbs up to more than 10 breaks per day. In contrast, one finds that in Japanese paper mills a few breaks *per month* is not unusual.

These anecdotes clearly suggest the extreme importance of the systems issues, such as the way in which the production equipment is inspected and maintained and how the paper web is handled. Good practices have been developed through years of experience, continuous improvement efforts, and consequently increased understanding of the web transport system. Problems related to tension control and system stability in the web transport system are discussed in Chapter 6.

The next question is what controls the break frequency at a given system (a specific printing press or paper machine)?

One common notion in the trade has been "the stronger the paper, the better the performance". Here *strength* can mean any one of the standard strength parameters of paper (tensile strength, tear strength, tensile energy absorption, burst strength, or breaking strain), as well as various fracture toughness parameters. However, the common observation in a number of investigations made by paper mills and printing houses has been that there are no significant differences in the strength of paper between failed rolls and nonfailed rolls, or between good months and bad months (Page and Bruce, 1985). Correlations between the strength and break frequency were found only when the number of rolls studied was extremely large, that is, between 10,000 and 50,000 (Page and Seth, 1982; Page and Bruce, 1985; Ferahi and Uesaka, 2001; Deng et al., 2007). Similar observations have been reported for paper machines. For example, Miyanishi and Shimada (1997) found that no parameter of paper strength explained web breaks on paper machines. These results again underline that the system conditions, such as how the printing press or paper machine is operated and maintained, have a dominant effect on web breaks.

Another often-cited cause for web breaks is *defects,* which are visible objects such as edge cuts, calendar cracks, slime holes, wrinkles, bursts, thin spots, sticky spots, and so forth. The most notable of these are reeling defects, such as wrinkles oriented in the cross machine direction, called crepe wrinkles, and bursts, which are frequently found in the edge rolls made from the end of a large parent roll (called the *jumbo roll*). Because these are often found in connection with an outburst of web breaks, major efforts have been made in improving the winding and reeling operations. Such incidents also prompted a series of investigations to directly observe the relationship between the presence of defects and web breaks. Moilanen and Linqdvist (1996) installed a web inspection system in the in-feed section of a commercial gravure printing press to determine if defects going into the press were related to web breaks. They observed 76 breaks in total. Of these, 73 had no relationship with the detected defects, and only 3 *might* have been caused by the defects. Ferahi and Uesaka (2001) took 50 broken samples at random from a commercial printing press and inspected them with a special optical setup to detect defects. Again, the results showed no indication of defects except one break (among the 50) associated with crepe wrinkle.

The observations discussed contradict the common belief that web breaks in printing are caused by defects. However, data from printing houses support such observations. Figure 8.1 lists the causes of breaks observed in a well-maintained commercial printing house in North America (Deng et al., 2007). Most breaks were related to the operation of the printing press, as alluded to previously. The second largest group of defects had an "unknown" cause, meaning that one could not see any obvious defects in the paper or problems in the press operation. The third largest cause of breaks was a crease in the paper. Creases are caused by a nonuniform tension profile of the paper web. It is clear that the defects papermakers are familiar with are all tailing the list of break causes. Another U.S. printing house reported similar results (Uesaka, 2005).

One can conclude that defects cause only a small minority of the breaks occurring in printing houses. Furthermore, defect-related problems can be solved effectively by implementing rigorous quality practices. The key research question is what constitutes the unknown breaks? Invisible defects are often claimed to explain unknown breaks; however, an *invisible defect* is seldom defined, and it is unclear how it would differ from the natural inhomogeneity of paper, such as the spatial distributions of mass, fiber orientation, fiber flocs, and pigment agglomerates at various length scales; fiber damages; and so forth. It may indeed be more appropriate to consider paper as a disordered material rather than a uniform continuum with defects.

From this perspective, it is natural to expect that the strength of paper varies statistically. Figure 8.2 shows the cumulative distribution function $Q(T)$ of tensile strength for two different size specimens. $Q(T)$ is the probability that the paper fails

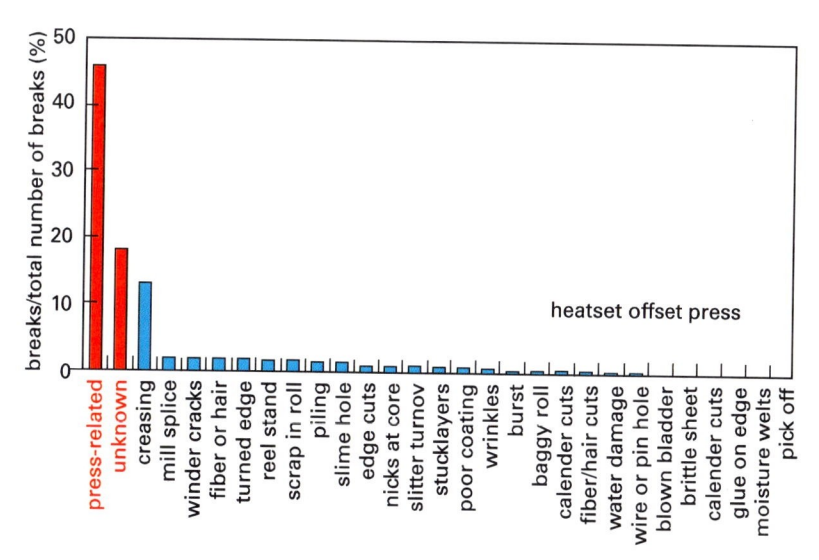

Fig. 8.1 Break cause statistics for heatset offset presses (Deng et al., 2007). Reproduced with permission from Pulp and Paper Technical Association of Canada (PAPTAC).

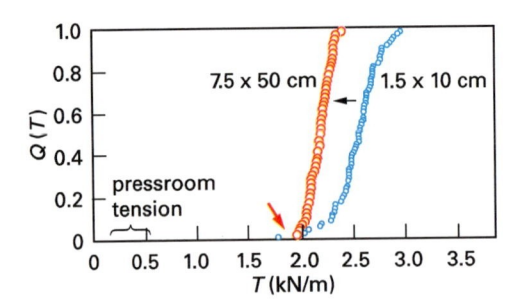

Fig. 8.2 Cumulative distribution functions of tensile strength of a newsprint for two different size specimens (1.5 cm × 10 cm and 7.5 cm × 50 cm). Tensile strength is expressed as tensile force/width.

at a tension $\leq T$. As specimen size increases, the distribution shifts to lower tensions – larger specimens are weaker – and the shape of the distribution changes. The web tensions typically seen in printing presses are also shown. They are far below the typical strength values. Therefore, in order to predict web breaks one should know the shape of the lower tail of the strength distribution at *specimen* sizes that are equivalent to the paper roll dimensions used in practice. This question is discussed in Section 8.3.

8.2.2 Stacking performance of boxes

Stacking performance is of particular importance for container boxes made of corrugated board (Chapter 3). One of the main end-use performance parameters is lifetime (discussed in Chapter 7.6). *Lifetime* is the time that the box can hold its shape without collapsing under given mechanical loading and environmental conditions (temperature and humidity). Like web breaks, lifetime has a *systems* and *statistical* nature.

Kellicutt and Landt (1951) determined the lifetime of boxes under compression by applying dead loads under different atmospheric conditions (Fig. 8.3). In the range where load is less than 75% of the static compression strength, each 8% reduction in load extended the lifetime by a factor of 8. This implies that even small variations of loading during transportation and warehousing cause enormous changes in the lifetime, which shows the importance of the whole logistic chain on the lifetime of container boxes.

Figure 8.3 also illustrates the generally large scatter in the lifetime data. Koning and Stern (1977) reported lifetime values varying from 125 hours to 2,817 hours when the box load was 50% of the maximum compressive strength, temperature 26.7°C, and relative humidity 90%. Because of the large scatter, lifetime data are often organized in terms of the secondary creep rate (Chapter 7.6); however, even such plots show very large scatter, up to 50% (Morgan, 2005; Popil and Schaepe, 2008). In the end-use it is the few "bad" boxes that cause problems when an entire stack of boxes collapses.

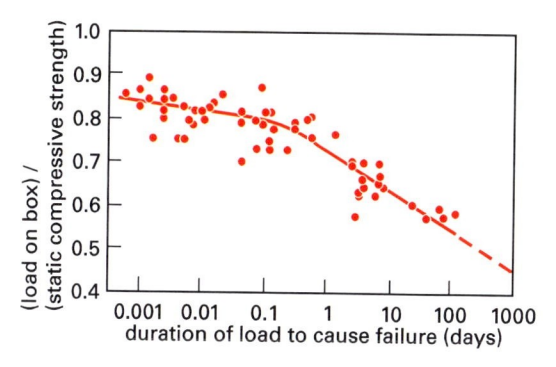

Fig. 8.3 Example of creep tests conducted at different dead loads and humidity conditions, based on data of Kellicutt and Landt (1951).

In the aforementioned studies, data were collected under nominally constant loading, temperature, and relative humidity, and for one type of box. On the other hand, in practical situations the conditions are far from constant. For example, load spikes can arise from dropping, falling, or abusive handling; and high frequency or low frequency vibrations from transportation (Brandenburg and Lee, 1988). Therefore, the mechanical conditions imposed on container boxes should be regarded as *dynamic* and random. In addition, temperature and humidity vary all the time in logistic chains, causing variable strains in the boxes (cf. Chapter 7.5.2). Therefore, the statistical nature of the end-use conditions is of extreme importance to the lifetime of container boxes.

To summarize, the problems of web breaks and box failures have the following common features:

- Systems effects, particularly in the end-use conditions, have large, sometimes dominant impacts.
- There is a large statistical variation in performance parameters (web break frequency and lifetime).
- Applied load is generally much lower than the strength values that are routinely measured.

8.3 Statistical approaches for failure in materials or systems

This section introduces statistical approaches to analyze the distributions of strength parameters and their dependence of the system size (scaling law). We first discuss time-independent failure models that may be relevant to the problems of web breaks, and then more general time-dependent models that are intended for creep failure of container boxes.

8.3.1 The chain model

The simplest but most informative model is the chain model where elements are linked in series (Fig. 8.4). The *element* can be part of the material or of a big system. If we apply a tension T, then the chain fails when its weakest element fails. This is the "weakest link" hypothesis that appears in the early work of extreme-value statistics (Fisher and Tippett, 1928; Weibull, 1939). The cumulative distribution function (CDF) of the strength values of an element is denoted $Q_0(T)$, and the CDF of the entire chain of n elements is $Q_n(T)$. We assume that the elements are statistically independent and subjected to the same tension T. Then the survival probability of the chain, $1-Q_n(T)$, is given by the probability that all elements in the chain survive,

$$1-Q_n(T) = \left[1-Q_0(T)\right]^n. \tag{8.1}$$

It is clear from this equation that as n increases (as the length of the chain increases), the survival probability of the chain decreases because $1-Q_0(T) < 1$. In other words, for large n, the strength distribution of the chain is determined by the lower tail of the element distribution $Q_0(T) \ll 1$. Under this condition, Equation 8.1 can be approximated as

$$Q_n(T) \approx 1 - \exp\{-n \cdot Q_0(T)\}. \tag{8.2}$$

Equation 8.1 represents an important scaling law called *weakest-link scaling* (WLS). For example, Equation 8.2 can be rewritten as

$$\ln\left(-\ln\left(1-Q_n(T)\right)\right) = \ln(n) + \ln\left(Q_0(T)\right). \tag{8.3}$$

When the left-hand side of the equation is plotted as a function of $\ln(T)$ or T, the system size n appears as a vertical shift $\ln(n)$ of the curve (Fig. 8.5). From Equation 8.2 one can derive various distribution functions by choosing the appropriate functional form for $Q_0(T)$. However, there is no way of *a priori* knowing the element distribution function $Q_0(T)$. One possible form for the lower tail is $Q_0(T) = (T/T_0)^k$. It leads to the well-known Weibull (1939) distribution:

$$Q_n(T) \approx 1 - \exp\left\{-n \cdot \left(\frac{T}{T_0}\right)^k\right\}. \tag{8.4}$$

Here the exponent k is called the Weibull modulus (or shape parameter), representing a spread of the distribution. In addition, if $Q_n(T)$ follows Weibull distribution (Eq. 8.4), then plotting the distribution in the manner shown in Figure 8.5 yields a straight line relationship. Therefore, this special way of plotting is called the Weibull plot. Although the Weibull distribution is, probably, most frequently used

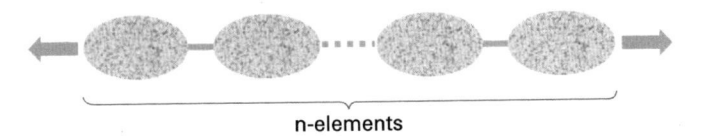

Fig. 8.4 The chain model.

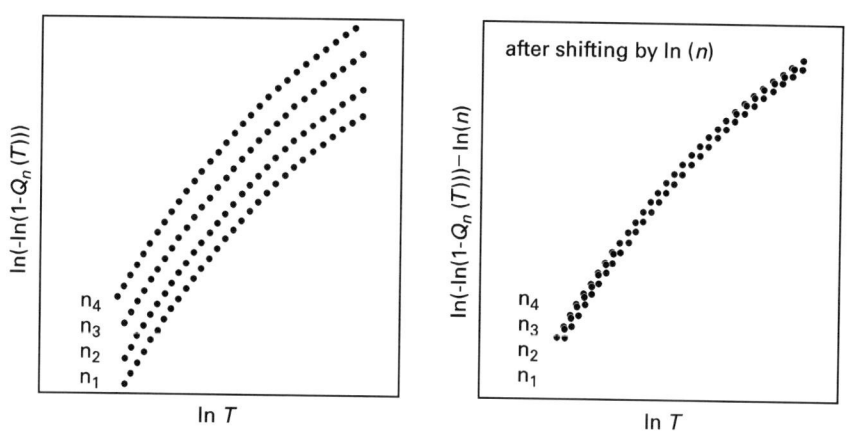

Fig. 8.5 Weakest-link scaling plot (Weibull plot).

in engineering mechanics, it is only one of the possible extreme-value distributions (Fisher and Tippett, 1928).

The most important point of the chain model is the weakest-link scaling (WLS), Equation 8.1. Although the WLS is often mistakenly called the Weibull theory, it does not require the use of the Weibull distribution. Furthermore, the WLS is not limited to one-dimensional systems, such as the chain model, and appears in general 3D systems, as is discussed later. Additionally, the tension T is to be regarded as a nominal tension that can be measured and monitored externally. One should not interpret T as the local tension in the web.

8.3.2 The bundle model

In this case the elements are placed in parallel (Fig. 8.6) so that the failure of one element does not have to lead to the failure of the bundle. Again, the failure of each element is assumed to be statistically independent. This model was first analyzed by Daniels (1945) to obtain the strength statistics of fiber bundles, and later it was investigated extensively as a model for fiber-reinforced composites and various extensions (Harlow and Phoenix, 1978, 1979, 1982; Curtin and Scher, 1991, 1992; Phoenix and Raj, 1992; Phoenix et al., 1997; Wu and Robinson, 1998; Curtin, 1999, 2000; Lu et al., 2002). In spite of its simple structure, the model shows rich characteristics similar to other more complex models.

There are two possible load-sharing mechanisms in the bundle model: the global load sharing (GLS) and the local load sharing (LLS). In the GLS one assumes that if one element fails, its load is shared equally with all the other elements, as would happen in a bundle of uncoupled fibers. In the LLS, on the other hand, one assumes that the load is shared by neighboring elements. For example, if the elements formed a two-dimensional system, then one could assume that two nearest neighbors always carry the additional load from the failed element. The LLS emulates the stress concentration at crack tips.

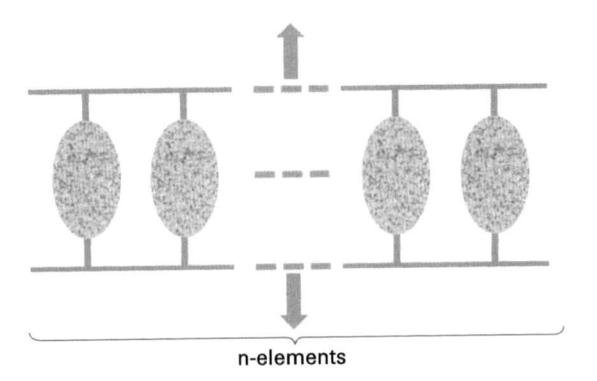

Fig. 8.6 Bundle model.

Daniels (1945) derived analytical solutions in the case of GLS, showing in particular that the coefficient of variation of strength values scales as $1/\sqrt{n}$. As the size n of the bundle increases, the distribution of strength values becomes increasingly narrow. In other words, the strength values of a large system vary little. On the other hand, the mean strength is independent of n (under certain restrictions), unlike in the chain model shown previously, so that larger system size does not mean lower strength.

The case of LLS is more interesting in relation to real structures. Wu and Leath (1999) obtained the exact recursive formulas for the strength distribution of the general bundle model under different boundary conditions. They found that if the size n of the system increases, the CDF moves first to higher tension values and then back. In other words, unlike in the chain model, the strength values of the LLS model *increase* with increasing n up to some critical size n_c. What is most important is that when $n > n_c$, the WLS equation (Eq. 8.1) emerges, and the average strength starts decreasing with increasing size. This demonstrates that the weakest-link scaling is a rather general asymptotic property of the strength distribution of large systems, regardless of the type of model or structure. Figure 8.7 shows the plot of $\ln(-\ln(1-Q_n(T)))$ against $\ln(T)$ for a model of fiber-reinforced composites (Ibnabdeljalil and Curtin, 1997). The plots for different system sizes n are vertically shifted by $\ln(n)$ to examine validity of the weakest-link scaling in the same way as in Figure 8.5. As one can see, all the data for $n = 400$ and 900 collapse on one line so that the weakest-link scaling holds. Below the size $n = 400$, the strength values are lower than what WLS predicts for the larger system sizes.

Curtin (1999) has shown that in the bundle model, the emergence of a weakest-link scaling does *not* necessarily mean that the initially weakest spot triggers the system failure. The weakest spot may indeed fail first, but because of structural inhomogeneity, local failures are created also elsewhere. With increasing load these local failures can grow, coalesce, or even stop growing, depending on the local structures. In a bundle system that is large enough for WLS to hold, the local failures form rather independent clusters of various sizes. The system fails when the largest of

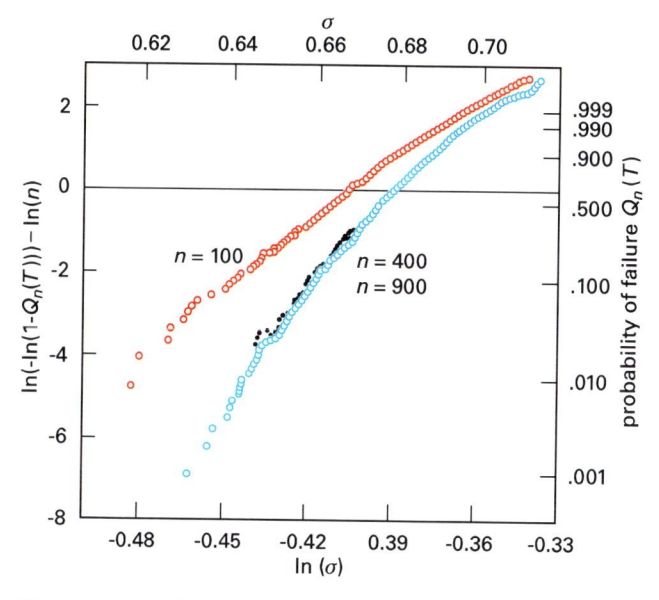

Fig. 8.7 Weibull plots of strength distributions of bundles with different sizes. Reprinted from (Ibnabdeljalil and Curtin, 1997) with permission from Elsevier.

these clusters starts growing like an avalanche. This failure process is different from the typical brittle failure of ceramic materials, for example. As is discussed later, paper seems to belong to the general class of nonbrittle, disordered materials where macroscopic failure is characterized by the previously discussed mechanism of cluster growth.

8.3.3 Time-dependent, statistical failure model

Failure phenomena are generally time-dependent. A typical example is the creep failure (Chapter 7.6) where a specimen subjected to constant load fails after a time period. In the case of fatigue failure the material fails because of cyclic or random loading even though the load itself is not large. It is also well-known that measured strength values depend on strain rate. We often observe that the higher the strain rate, the higher the strength of the paper (Fig. 2.12). In this section we introduce a phenomenological formulation of time-dependent failure by Coleman (1956, 1958) and derive some relevant formula. To the knowledge of this author, the statistics of failure are rarely discussed in the box mechanics literature. Therefore, the intention here is to provide a general framework for understanding and analyzing the time-dependent, statistical failure of container boxes.

As an example we consider the chain model in Figure 8.3. For a given loading history $T(t)$, it is useful to characterize the failure of an element by the time-to-failure or lifetime t_{BL}, not the load $T(t_{BL})$. Similar to the strength values, t_{BL} can vary from one element to the other. In the chain model the system failure is determined by the element that has the shortest t_{BL} of all the elements. In analogy to Section 8.3.1, the cumulative distribution function of lifetimes $R_n(t_{BL})$ for a chain of n elements is

$$1 - R_n(t_{BL}) = \left[1 - R_0(t_{BL})\right]^n, \tag{8.5}$$

where $R_0(t_{BL})$ is the CDF for one element. For large n, this can, again, be approximated (cf. Eq. 8.2) as

$$1 - R_n(t_{BL}) \approx \exp\{-n \cdot R_0(t_{BL})\}. \tag{8.6}$$

Then the key question is how to obtain $R_0(t_{BL})$. For a general loading history $T(t)$, the failure probability of an element depends not only on time, but also on the loading history itself. For example, intensive loading may give short lifetime and gentle and low-level loading may give long lifetime. One way to account for the effects of loading history is to introduce a "breaking function" D that is a function of the loading history $T(t)$. It plays the same role as the damage variable (Kachanov, 1986) in continuum mechanics (Chapter 5.3.5). The simplest functional form postulated by Coleman (1958) is $\dfrac{dD}{dt} = \Phi[T(t)]$ so that the rate of change of D at time t is determined by the load $T(t)$ acting at the same time (t). In other words, we presume that the higher the load, the higher the increase in D. Integration gives

$$D(t) = \int_0^t \Phi[T(s)] ds. \tag{8.7}$$

This equation shows more clearly that the internal timescale D depends on the history of T up to time t. With this preparation, it is now natural to assume that the distribution of lifetimes $R_0(t_{BL})$ for an element is a direct function of $D(t)$. Because the failure probability of the entire chain is determined by the lower tail of $R_0(t_{BL})$, we use the same functional form as for $Q_0(T)$ previously,

$$R_0(t_{BL}) \approx q \cdot \left(D(t_{BL})\right)^k, \tag{8.8}$$

where q is a constant. Combining Equations 8.6 through 8.8, we obtain the cumulative distribution function of the entire chain

$$R_n(t_{BL}) = 1 - \exp\left\{-n \cdot \left[\int_0^{t_{BL}} \kappa(T(s)) ds\right]^k\right\}, \tag{8.9}$$

where $\kappa = q^{1/k} \phi$ is a function that connects the breaking function D with the loading history. By choosing the appropriate functional form for κ, we can calculate the CDF of the system for a given loading history. Before we do that, we give the most straight-forward implication of Equation 8.9.

In the case of creep, $T(t) = T_0$ is constant for $t > 0$, Equation (8.9) gives

$$R_n(t_{BL}) = 1 - \exp\left\{-n\left[\kappa[T_0]\right]^k \cdot t_{BL}^k\right\}. \tag{8.10}$$

In other words, in the chain model creep lifetime follows the Weibull distribution. Because the exponent k is generally more than unity the probability of failure is very sensitive to the creep load T_0. The extension of Coleman's theory to the bundle model has been made by Zhang (1999).

The functional form of $\kappa[T]$ gives the dependence of lifetime on creep load. For the case of exponential form often used, the average creep lifetime is

$$\ln(\overline{t_{BL}}) = a_e - b_e \cdot T. \tag{8.11}$$

This relationship has been observed in the early studies of creep lifetime of corrugated board boxes at moderate and low loads (Kellicutt and Landt, 1951; Moody, 1966; Stott, 1959). The constants a_e and b_e can be easily determined by plotting the logarithm of average lifetime and load T. It should be noted, however, that when load T approaches zero, the lifetime in this expression approaches a finite value, implying that the box breaks without any load. This artifact means that Equation 8.11 should be regarded as a convenient fitting expression. On the other hand, if a power law is chosen for $k(T)$, the resulting lifetime

$$\ln(\overline{t_{BL}}) = a_p - b_p \cdot \ln(T) \tag{8.12}$$

goes to infinity when load goes to zero. With the constants a_p and b_p determined from experimental creep data, we can go back to Equation 8.9 to determine the CDFs for any given loading history, such as simple compression loading, cyclic loading, or random loading, all of which are routinely encountered in logistic chains. In addition, the exponent k in Equation 8.9 is related to Coefficient of Variation (COV) of the lifetime in a creep test through

$$\mathrm{cov} = \left[\frac{\Gamma\left(1+\frac{2}{k}\right)}{\Gamma\left(1+\frac{1}{k}\right)}\right]^{\frac{1}{2}}, \tag{8.13}$$

where Γ is the gamma function. Thus, k is a measure of the scatter in measured lifetime values so that the lower the number, the more the scatter there is. Again, k can be determined experimentally from the Weibull plot of Equation 8.10.

The most important point to notice is that from Equation 8.9 we can predict the box lifetime distribution in any logistic chain or warehouse if we know the distribution of static box strength and the loading history of the logistic chain.

8.4 Statistical failure of paper

In paper mills or box plants, the basic quality control parameter is currently the average of the strength properties of choice, and strength distributions are rarely measured. Accordingly, the specifications for strength properties are currently based on past experience or the empirical safety factor (the designed strength of paper [or box] divided by the maximum load that the paper [or box] is expected to experience during its use). Therefore, the next step would be to develop a rational control strategy by taking into account the statistical aspects of the performance.

In this section, we take an example of the paper web failure and introduce the applications of some of the concepts of statistical strength models discussed previously.

8.4.1 Strength distributions

Strength distributions have been measured by using standard laboratory testing devices or pilot-scale web straining devices, and they have been compared with various distribution forms, such as the Weibull distribution, double-exponential distribution, Gumbel distribution, and log-normal distribution (Sears et al., 1965; Gregersen, 1998; Korteoja et al., 1998; Uesaka and Ferahi, 1999; Wathén and Niskanen, 2006). Figure 8.8 shows a typical Weibull plot for a newsprint sample of the standard-size test specimen. The Weibull distribution (either two-parameter or three-parameter distribution) generally fits well with experimental data, but it is not necessarily the distribution function that gives the best fit (Korteoja et al., 1998). It is difficult to distinguish between the fits for different distributions because they differ little in the normal frequency range where experimental data can be obtained. Usually the lowest tail of the cumulative distribution function from experiments occurs in the range of 10^{-2} to 10^{-3}. This frequencies are much higher than in web breaks, such as the frequency of 10^{-9} (Uesaka and Ferahi, 1999). Figure 8.9 shows another Weibull plot for a larger specimen width (25 cm), obtained using an automatic, continuous web straining device. Again, the majority of points agree with the two-parameter Weibull distribution, except some data scatter in the low tail (Wathén and Niskanen, 2006).

In the experiments regarding strength distributions, awareness has increased concerning the importance of experimental conditions, including the following: (1) the size of the specimens, which we will discuss later; (2) sampling frequency and total sampling length; and (3) the uncertainty of the equipment. Regarding sampling frequency and length, strength variations originate from various sources with a large spectrum of timescales and length scales in papermaking. Significant autocorrelations of strength properties can exist at least within 1 hour of production. Therefore, random sampling throughout more than 1-hour of production is important. In printing, the typical distance between frequency breaks is 350 km. Therefore, the variations in this length scale cannot be captured with sampling procedures that rely on standard laboratory strength testing or web straining devices. Currently, the best procedure to determine strength distributions is to analyze data collected with an automatic testing device for every paper reel for an extended time period (several months or a year). Such data could be obtained at several points across

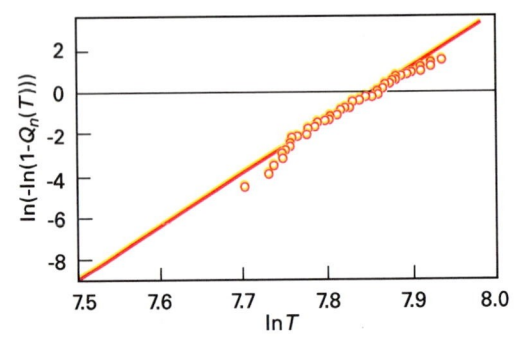

Fig. 8.8 Strength distribution of a newsprint, measured using the standard test specimens.

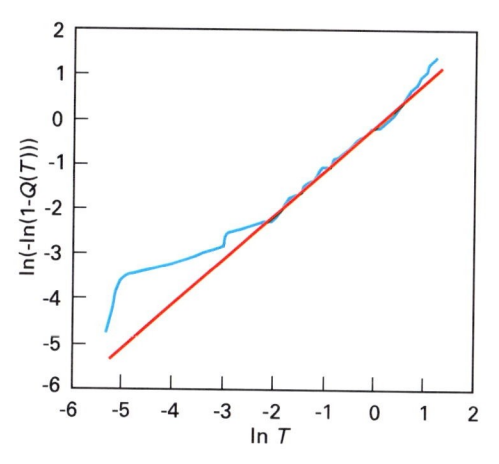

Fig. 8.9 Strength distribution of large size specimens. Redrawn from Wathén and Niskanen (2006). Reproduced with permission from Pulp and Paper Technical Association of Canada (PAPTAC).

the paper machine width. Thus, one can obtain cross-machine direction profiles of strength distributions. The small size of the specimens used poses a scaling problem, as will be discussed later, but with this practice, both the sampling frequency and sampling length are ideal, and the data are reliable as long as the testing equipment is properly adjusted and maintained.

Concerning the uncertainty of the equipment, it was observed that the determination of the shape parameters of the distribution, such as the Weibull exponent k, is much more demanding than that of the average strength. Even slight misalignments in the testing device, use of load cells of different capacities, and other equipment defects sensitively affect the distributions. This sensitivity also raises the interesting question of how the material strength couples with the uncertainty of the real boundary conditions in a printing press or paper machine.

8.4.2 Factors controlling strength distributions

The question of what constitutes the lower tail of the strength distribution has been the subject of active discussions. In the early days of research, so-called shives of mechanical pulp in the paper (small "chunks" of wood not fully disintegrated into fibers) were regarded as the culprit of web breaks and, thus, believed to constitute the lower tail of the strength distribution. Sears et al. (1965) and Adams and Westlund (1982) observed that shive-related breaks were a dominant cause of breaks in their experiments with a pilot-scale web straining device. Even in testing with small-size specimens, such as in Figure 8.2, shives were indeed observed in those samples that had the lowest breaking tension values, as indicated by the arrow in the figure. However, in real printing press environments, it is surprisingly rare to find shives in the failed samples. In the example mentioned previously (Uesaka, 2005), the direct inspection of the 50 break samples randomly collected from a printing house showed no indication of shive-related breaks.

Furthermore, during the last 30 years the shives content in mechanical pulps has decreased drastically, but a similar effect has not been seen in break rates. This discrepancy between the earlier literature and the direct investigations in printing houses may be explained if one compares the difference between the tension level that Adams and Westlund (1982) used and that of real printing presses. The former was much higher than the latter. At the very low web tensions used in printing presses, there is no guarantee that shives still trigger breaks in the same manner. Instead, it is rather natural to expect some different mechanism. For example, recall that in a disordered system the initial weakest spot does not necessarily coincide with the location of the final break. In addition, it is also highly improbable that occasional tension spikes would be sufficiently high to trigger a shives-related failure. Therefore, the phenomena observed in experiments of elevated stress levels may have no direct relevance to the phenomena that may happen at the much lower stress levels that prevail in printing presses.

Nonuniform formation (Chapter 2.5.2) has also been suspected to control the distribution of paper strength. The speculation is that low basis weight areas form *weak* spots that will be the cause of failure at low tension. However, mechanistic studies performed at a microscopic level did not support this notion (Wong et al., 1995, 1996; Korteoja et al., 1996, 1998). Low basis weight areas are sometimes shielded and reinforced by high basis weight areas so that they do not necessarily become weak spots. In addition, the complex interactions between the different length scales of disorder present in paper obscure the formation effects even more. Hristopulos and Uesaka (2004) also estimated the effect of macroscopic basis weight variation on strength distribution and concluded that the effect is insignificant. At micro-scale, a different mechanism may operate and affect the distributions. Korteoja and coworkers (1998) and Wathén and Niskanen (2006) have reported very weak relationships between strength variations and formation.

In summary, what factors control the variability of paper strength, particularly the low-strength tail, is still up for debate.

8.4.3 Strength scaling

The size effect of strength distributions is critical in predicting real web breaks in practical situations. The size effect may come from the presence of defects. *Defects* consist of (1) visible defects, that is, holes, cuts, creases, and so forth; and (2) the *natural* disorder present in paper structure (fiber or pigment agglomerates, local disturbances in fiber orientation, fiber damages, etc.). This section focuses on the natural disorder as a potential cause of the majority of web breaks, press-related breaks, and unknown breaks in Figure 8.1.

Paper structure has a wide spectrum of disordered structures, spanning from the fiber-width scale (micrometers) to the meter scale of paper machines. Although the intensity of variation tends to decrease with the length scale, it is still conceivable that one can see effects of the long-range structural correlation of web breaks. This possibility directly affects the critical length scale at which weakest-link scaling (Section 8.3.2) is expected to emerge. Accordingly, the selection of the appropriate specimen size is important in experiments. The question is whether the standard tensile specimen size (1.5 cm × 10 cm) suffices, or if it is necessary to use pilot-scale testing.

Figures 8.10 and 8.11 show Weibull plots of tensile strength distributions for newsprint samples of various lengths and widths (Figs. 8.10 and 8.11, respectively; Hristopulos and Uesaka, 2004). The curves have been shifted according to Equation 8.3, in the same way as in Figure 8.5, in order to be able to evaluate the applicability of weakest-link scaling. In the length direction, the weakest-link scaling appears to hold even at 25 mm because all the curves collapse to a single curve after the vertical shifting. This means that as the specimen length increases, the strength decreases, according to the WLS. On the other hand, in the width direction, this collapse of the curves happened only after 20 mm, suggesting that in order to ensure the WLS, the specimen width has to be at least 20 mm. It is interesting to note that, as can be read from Figure 8.11, increasing the width first shifted the strength distribution to higher tension values, and then gradually shifted the strength distribution back to the lower tensions, precisely as predicted by the bundle model (Ibnabdeljalil and Curtin, 1997).

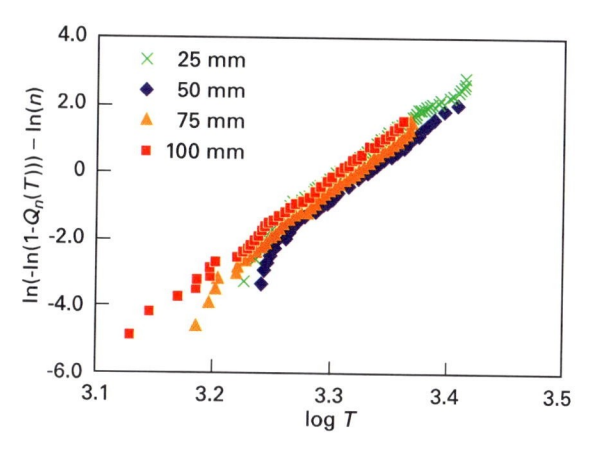

Fig. 8.10 Weibull plots for different length specimens. (Hristopulos and Uesaka, 2004). Reproduced with permission from American Physical Society (APS).

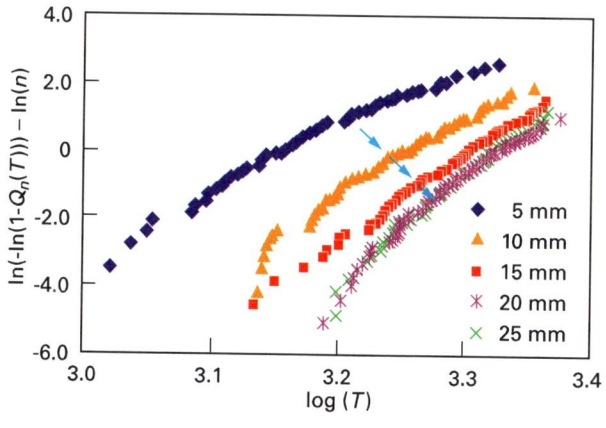

Fig. 8.11 Weibull plots for different width specimens. (Hristopulos and Uesaka, 2004). Reproduced with permission from American Physical Society (APS).

8 Statistical aspects of failure of paper products

Therefore, one cannot use the standard specimen of 15 mm width to estimate the strength distributions of paper rolls of realistic sizes by using the WLS. In the case of newsprint, the specimen width and length have to be at least 20 mm. However, if one used the standard specimens with the weakest-link scaling, then the bundle system analysis suggests that an upper estimate of web breaks is obtained (Uesaka and Ferahi, 1999). Such an estimate may be reasonable to use as a conservative evaluation of break frequencies.

8.4.4 Web break prediction

With the preparation described earlier, we attempt to predict web breaks in a printing press. We assume that the paper web can be treated as a chain of elements where each element size satisfies the WLS requirement discussed previously. This assumption is valid when the size of the reference specimen exceeds the critical length at which the WLS emerges. Another assumption is that the paper web has no visible macroscopic defects. That is, we deal with the *unknown* breaks and *press-related* breaks of Figure 8.1.

As seen in Figure 8.12, when the paper web runs through the printing system, it is subjected to tension variations in space and time. Deng et al. (2008) have reported tension variations at different positions in a variety of printing systems. The survival probability of the i-th element while running through a printing process is

$$1 - Q_0(T_{\max}(i)), \tag{8.14}$$

where $T_{\max}(i)$ is the maximum tension that the element experiences when going through the entire process. Then, the probability that all elements in the chain, say one paper roll, survive is

$$\prod_i \left(1 - Q_0(T_{\max}(i))\right). \tag{8.15}$$

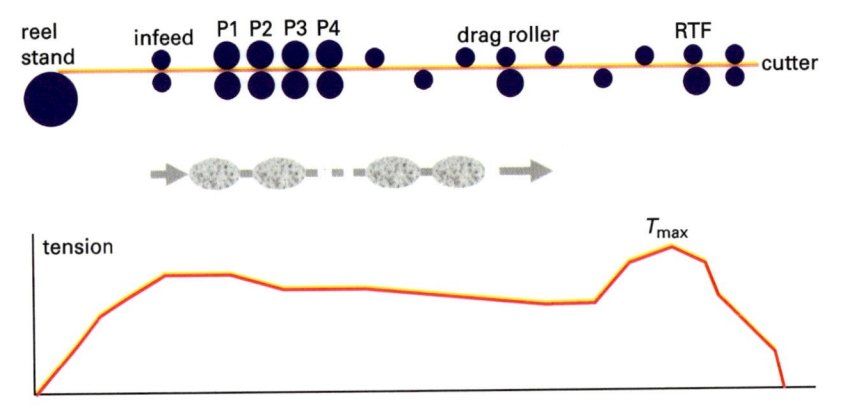

Fig. 8.12 Paper web running through a printing press system with tension variations.

Therefore, the probability that the roll fails is

$$Q_{BR} = 1 - \prod_i \left(1 - Q_0\left(T_{max}(i)\right)\right).\tag{8.16}$$

For $Q_0(T_{max}) \ll 1$ we obtain

$$Q_{BR} \approx N_{roll} \cdot \mathrm{Ex}\left\{Q_0\left(T_{max}(i)\right)\right\},\tag{8.17}$$

where $N_{roll} = A_{roll}/A_{ref}$ and A_{roll} is the total area of paper in the roll, and A_{ref} is the area of the specimens used in measuring the strength distributions. Ex{.} denotes the expectation value of the variability of the variations of T_{max}. From the measured strength distributions and known tension variations in the printing press, we can evaluate the expectation value in Equation 8.17 and, thus, estimate break frequency Q_{BR} (Uesaka, 2007; Uesaka and Ferahi, 1999).

Figure 8.13 gives a sample calculation of the number of breaks per 100 rolls in two cases that differ in the different average strength of the paper. It should be noted that in this calculation the printing press tension was assumed to be constant in time and space, which is, of course, not the case in reality. As one can see, without tension variations, there are virtually no breaks even for the low-strength paper. In Figure 8.14 break rate is calculated as a function of strain variations (time/space wise) from its average. Break rate is a very sharp function of the strain variation. These sample calculations illustrate that the printing press tension variations indeed have a great effect on web breaks. However, it should be also pointed out that these curves are sensitive to the strength uniformity of the paper that is characterized by, for example, Weibull shape parameter k. The more uniform the strength distribution, the smaller the web break frequency, as one would expect even intuitively. Field data showed that performance differences between different papers often appear also in the statistics of press-related breaks (Uesaka, 2005). Paper still matters!

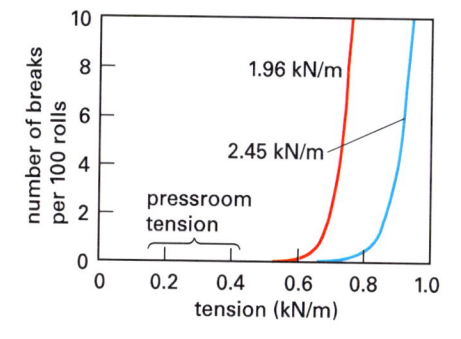

Fig. 8.13 Prediction of web breaks when tension variations are absent for the Weibull modulus $k = 19$, and two values of average tension $T_{avg} = 1.96$ kN/m and 2.45 kN/m.

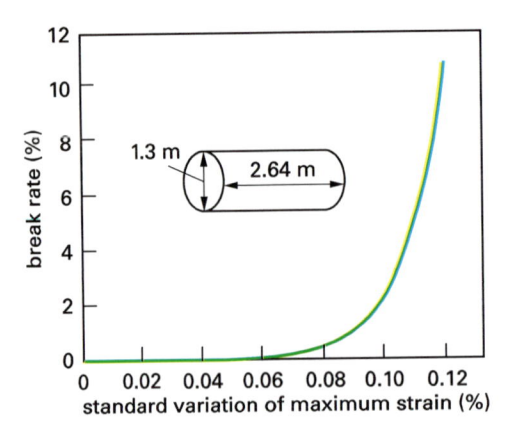

Fig. 8.14 Prediction of web breaks as a function of strain variations in a printing press. Average tension applied in the printing press 0.2 kN/m, average tensile strength of paper 2.5 kN/m, breaking strain 1.1%, and Weibull modulus $k = 15$.

8.5 Research front of statistical failure of paper

So far this chapter has discussed the statistical failure of paper on the system level. Both chain model and bundle model can represent important scaling effects; however, neither of the models can directly predict effects of structural disorder of paper or uncertainty in fiber properties. The models only address coarse-grained effects of the statistical distributions of imaginary structural elements. Lattice models may give a more detailed description of the complex fiber network structure of paper; however, the main conclusions obtained from the lattice models are surprisingly similar to those from the chain models and bundle models in terms of strength scaling. A comprehensive review of this area may be found in Alava et al. (2006). In order to answer the questions of the statistical distributions of strength and the strength scaling of fiber network structures, it is necessary to model the systems on the fiber network level.

Heyden (2000) may have been the first one to look at the size-dependence of the fiber network failure by direct simulation. A finite element method was used, by modeling the fiber as a Bernoulli beam element of circular arc and the fiber–fiber bond as a stick-slip bond with possible degradation. The dependence of average strength on sample length and sample width was investigated for 2D networks. The results showed that the network strength decreased quickly as the sample length increased from 1.2 times to 9.6 times of fiber length, whereas increasing the sample width did not show significant change in strength. These phenomena closely resemble those predicted by the chain models and bundle models.

Kulachenko and Uesaka (2010) extended the direct simulation method into a 3D network of realistic basis weight (35 g/m²). The in-plane dimensions were still limited to 3–7 times the fiber length. The fiber interaction model was assumed to be a stick-slip bond with properties similar to those measured by atomic force microscope (AFM) for wet fibers. It was found that when both the number of fibers in a given sample area and the fiber orientation distribution were kept constant for each of the simulated networks, the resulting tensile strength distribution was very uniform (the coefficient of variation was only 4%–5%), whereas the breaking strain varied considerably (the coefficient of variation was 15%–20%). The distributions also depended on the size of the specimen (width and length), as well as fiber geometries, such as fiber length and fiber curl.

With the continuous increase of computer power and improvement of algorithms, the direct simulation methods are becoming a promising tool. The target size in the in-plane dimensions may be, in the case of dry sheets, about 10 times fiber length, at which point the weakest-link scaling starts appearing. This target size with a realistic basis weight is not a far-reaching goal for today's simulation technology. The micromechanics approach is presenting a hope for understanding the statistical properties of fiber networks from the first principles.

Lastly, the statistical aspect of box lifetime has not been addressed well, although the problem has been well recognized from early days. As the demands for lightweight constituent boards further increases, the design of the statistical performance of boxes will become increasingly important.

8.6 Concluding remarks

The subject area described in this chapter is probably one of the most recognized applications of paper mechanics. However, until now the systems nature of the problem often plagued progress, posing huge challenges in sorting out, one by one, the dependent and independent factors and their interactions. In addition, the statistical nature of the problem, particularly the extreme-value statistics, presented enormous challenges for the analysis of data from trials and experiments and in understanding the basic mechanisms of the phenomena.

Although these problems have been slowly cleared out and some preliminary understandings are emerging, the area still requires new approaches to answer some outstanding questions. One of the challenges is the determination of the strength distributions in an efficient and accurate way. The counterpart is the determination of the distributions of end-use conditions, such as tension variations and dynamic loading histories in the same manner. In both cases new instrumentation is required. On the other hand, high performance computing of fiber networks and composite structures is already showing a promise in the mechanistic investigation of this problem.

References

Adams, R. J., and Westlund, K. B. (1982). Off-line testing for newsprint runnability. Paper presented at the 1982 International Printing and Graphic Arts Conference, Quebec City, Canada (CPPA Technical Section).

Alava, M. J., Nukala, P.K.V.V., and Zapperi, S. (2006). Statistical models of fracture. Advances in Physics 55, 349–476.

Brandenburg, R. K., and Lee, J.-L. (1988). Fundamentals of Packaging Dynamics, 3rd Edition (Minneapolis, MN, USA: MTS Systems Corporation).

Coleman, B. D. (1956). Time dependence of mechanical breakdown phenomena. J. Applied Physics 27, 862–866.

Coleman, B. D. (1958). Statistics and time dependence of mechanical breakdown in fibers. J. Applied Physics 29, 968–983.

Curtin, W. A. (1999). Stochastic damage evolution and failure in fiber-reinforced composites. Advances in Applied Mechanics 36, 163–253.

Curtin, W. A. (2000). Dimensionality and size effects on the strength of fiber-reinforced composites. Composite Science and Technology 60, 543–551.

Curtin, W. A., and Scher, H. (1991). Analytic model for scaling of breakdown. Physical Review Letter 67, 2457–2460.

Curtin, W. A., and Scher, H. (1992). Algebraic scaling of material strength. Physical Review B 45, 2620-2627.

Daniels, H. E. (1945). The statistical theory of the strength of bundles of threads I. Proc. R. Soc. London A 183, 404–435.

Deng, X., Ferahi, M., and Uesaka, T. (2007). Press room runnability. A comprehensive analysis of press room and mill database. Pulp and Paper Canada 108, T39–T48.

Deng, X., Parent, F., Manfred, T., Aspler, J. S., and Hamel, J. (2008). Pressroom draw variations and its impacts on web breaks, Part 1. Newspaper press trials. In: Proceedings of the 35th International Research Conference of IARIGAI, Valencia, Spain.

Ferahi, M., and Uesaka, T. (2001). Pressroom runnability. Part 2: Pressroom data analysis. Paper presented at the PAPTAC Annual Conference, Montreal, Quebec, Canada.

Fisher, R. A., and Tippett, L.H.C. (1928). Limiting forms of the frequency distribution of the largest or smallest member of a sample. Proc Cambridge Philosophical Society 14, 180–191.

Gregersen, O. W. (1998). On the assessment of effective paper web strength. Doctoral thesis. Norwegian University of Science and Technology, Trondheim.

Harlow, D. G., and Phoenix, S. L. (1978). The chain-of-bundles probability model for the strength of fibrous materials I: Analysis and conjectures. J. Composite Materials 12, 195–214.

Harlow, D. G., and Phoenix, S. L. (1979). Bounds on the probability of failure of composite materials. Int. J. Fracture 15, 321–336.

Harlow, D. G., and Phoenix, S. L. (1982). Probability distributions for the strength of fibrous materials under local load sharing I: Two-level failure and edge effects. Adv. Appl. Prob. 14, 68–94.

Heyden, S. (2000). Network modelling for the evaluation of mechanical properties of cellulose fiber fluff. Doctoral thesis. Lund University, Lund.

Hristopulos, D., and Uesaka, T. (2004). Structural disorder effects on the tensile strength distribution of heterogeneous brittle materials with emphasis on fiber networks. Physical Review B 70, 1–18.

Ibnabdeljalil, M., and Curtin, W. A. (1997). Strength and reliability of fiber-reinforced composites: Localised load-sharing and associated size effects. Int. J. Solids Struct. 34, 2549–2668.

Kachanov, L. M. (1986). Introduction to Continuum Damage Mechanics (Lancaster, UK: Martinus Nijhoff Publishers).

Kellicutt, K. Q., and Landt, E. F. (1951). Safe stacking life of corrugated boxes. Fiber Containers 36, 28–38.

Koning, J. W., and Stern, R. K. (1977). Long-term creep in corrugated fiberboard containers. Tappi 60, 128–131.

Korteoja, M., Lukkarinen, K. K., Kaski, K., Gunderson, D. E., Dahlke, J. L., and Niskanen, K. (1996). Local strain fields in paper. Tappi 79, 217–223.

Korteoja, M., Salminen, L. I., Niskanen, K. J., and Alava, M. J. (1998). Statistical variation of paper strength. J. Pulp and Paper Sci. 24, 1–7.

Kulachenko, A., and Uesaka, T. (2010). Simulation of wet fiber network deformation. Paper presented at the Progress in Paper Physics – A Seminar. Montreal, Canada (FPInnovations).

Lu, C., Danzer, R., and Fisher, F. D. (2002). Fracture statistics of brittle materials: Weibull or normal distributions. Physical Review E 65, 0671021–0671024.

Miyanishi, T., and Shimada, H. (1997). Using neural networks to diagnose web breaks on a newsprint machine. Tappi 81, 163–170.

Moilanen, P., and Linqdvist, U. (1996). Web defects inspection in the printing press – Research under the Finnish Technology Program. Tappi 79, 88–94.

Moody, R. C. (1966). How dead load, downward creep influence corrugated box design. Packaging Engineering 11, 75–81.

Morgan, D. G. (2005). Analysis of creep failure of corrugated board in changing or cyclic humidity regimes. Paper presented at the 59th Appita Annual Conference and Exhibition, Auckland, New Zealand.

Page, D. H., and Bruce, A. (1985). Pressroom runnability. Paper presented at the Newsprint and the Pressroom, Chicago, Illinois (CPPA and ANPA).

Page, D. H., and Seth, R. S. (1982). The problem of pressroom runnability. Tappi 65, 92–95.

Phoenix, S. L., Ibnabdeljalil, M., and Hui, C.-Y. (1997). Size effects in the distribution for strength of brittle matrix fibrous composites. Int. J. Solids Struct. 34, 545–568.

Phoenix, S. L., and Raj, R. (1992). Scalings in fracture probabilities for a brittle matrix fiber composite. Acta Metall. Mater. 40, 2813–2828.

Popil, R. E., and Schaepe, M. K. (2008). Dependence of corrugated container lifetime on component properties. Paper presented at the Progress in Paper Physics Seminar, Espoo, Finland, TKK (Helsinki University of Technology).

Sears, G. R., Tyler, R. F., and Denzer, C. W. (1965). Shives in newsprint: The role of shives in paper web breaks. Pulp and Paper Magazine of Canada 66, T351–T360.

Stott, R. A. (1959). Compression and stacking strength of corrugated fiberboard containers. Appita 13, 84–89.

Uesaka, T. (2005). Principal factors controlling web breaks in pressrooms – Quantitative evaluation. Appita J. 58, 425–432.

Uesaka, T. (2007). Prediction of low tail of strength distribution of paper. Proceedings of the 2007 International Paper Physics Conference, Gold Coast, Australia (Appita), pp. 115–122.

Uesaka, T., and Ferahi, M. (1999). Principal factors controlling pressroom breaks. Proceedings of TAPPI International Paper Physics Conference, San Diego, California (TAPPI Press), pp. 229–245.

Wathén, R., and Niskanen, K. (2006). Strength distributions of running webs. J. Pulp Pap. Sci. 32(3), 1–8.

Weibull, W. (1939). A statistical theory of the strength of materials, Ing. Vetenskaps Akad. Handl., Vol. 151, pp. 1–45 (Stockholm: The Royal Swedish Institute for Engineering Research).

Wong, L., Kortschot, M. T., and Dodson, C.T.J. (1995). Finite element analyses and experimental measurement of strain fields and failure in paper. Paper presented at the 1995 International Paper Physics Conference, Niagara-on-the-lake, Ontario, Canada (TAPPI).

Wong, L., Kortschot, M. T., and Dodson, C.T.J. (1996). Effect of formation on local strain fields and fracture of paper. J. Pulp Pap. Sci. 22, J213–J219.

Wu, B. Q., and Leath, P. L. (1999). Failure probabilities and tough-brittle crossover of heterogeneous materials with continuous disorder. Physical Review B 59, 4002–4010.

Wu, E. M., and Robinson, C. S. (1998). Computational micro-mechanics for probabilistic failure of fiber composites in tension. Composite Science and Technology 58, 1421–1432.

Zhang, S.-D. (1999). Scaling in the time-dependent failure of a fiber bundle with local load sharing. Physical Review E 59, 1589–1592.

Part III
Reactions to moisture and water

9 Moisture-induced deformations

Artem Kulachenko

9.1 Introduction

This chapter discusses out-of-plane deformations that moisture induces in paper. Such deformations are a challenging quality problem. We consider three phenomena known as curl, fluting, and cockling, which differ primarily in length scale and magnitude of appearance. We start with the curl problem. Instead of presenting a wordy definition of what the *curl* is, we describe a simple experiment that demonstrates the phenomenon. Take an A4 copy paper. Make sure it is flat. Cut two strips (say 10 cm by 2 cm) along the long and short edge of the sheet (Fig. 9.1). Then take a slightly wetted paper towel and rub it along the surface of each strip.

Observe what happens. Both strips will bulge out, but in different manners. In our experiment (Fig. 9.1), one strip bulged along the long dimension and the other one along the short dimension. This is what usually happens, but sometimes both of the strips twist. After about half a minute, the curvature in both strips reduces, and they return close to the original flat shape, but only for a short while. Then they continue to bend and bulge out in the opposite direction – although no external force is applied. Again, the orientation of the bulge is different in the two strips. After a couple of minutes, the shape of the strips does not change any more. You can leave these strips for a day or a week and nothing will change – they will remain curled. This is an example of the phenomenon called paper curl.

There are a number of questions that we can formulate from this experiment:

1. Why does the paper deform upon moistening?
2. Why does it deform differently in the strips taken in two orthogonal directions?
3. Why does the paper react to applied moisture by bending?
4. Why does it bend first in the direction of applied moisture?
5. Why does the strip end up having a permanent curl?
6. Why does the curl change the direction of bending during the deformation process?

This chapter explains the mechanisms behind paper curl by answering these questions one by one.

9.2 Moisture-induced deformations

9.2.1 Hygroexpansion of paper

Our first question is *why does the paper deform when exposed to moisture?* Paper is made of wood fibers that are hydrophilic, that is, they absorb moisture readily.

9 Moisture-induced deformations

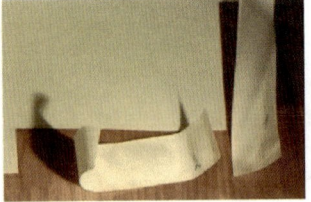

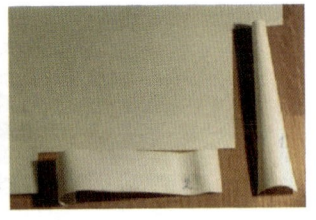

Fig. 9.1 Demonstration of paper curl: wetting the paper strips (left), reaction to the wetting (center), and final dry state (right).

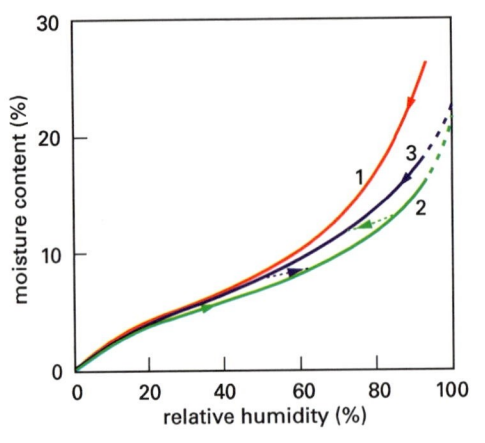

(1) desorption isotherm (never dried)

(2) Subsequent absorption

(3) desorption from a once-dried state

Fig. 9.2 Moisture content vs. relative humidity in a sulfite pulp at 20°C, after Christensen and Giertz (1965). Reproduced with permission from The Pulp and Paper Fundamental Research Society (www.ppfrs.org).

As explained in Chapter 2.5, the hydrophilic nature of wood fibers is the basis of papermaking. Figure 9.2 shows how the moisture content (MC) of chemical pulp fibers (amount of water relative to the dry weight of the fibers) changes with relative humidity (RH) at 20°C. The first line is the desorption isotherm from a never-dried state of the fibers, the second line shows subsequent absorption, and the third line shows desorption from a once-dried state. Analogous behavior, quite independent of the raw material composition, is observed with paper.

There are a few interesting things to note here. First, the moisture content curve is nonlinear, increasing rapidly both at low and high relative humidity. In normal conditions, the moisture content of paper is 5%–10%. The saturation moisture content at 100% RH ranges from 15%–30%. The second interesting aspect to notice is the hysteresis: at equal relative humidity, the moisture content during absorption is lower than during desorption. Finally, the moisture content during the initial desorption is higher than the desorption (curve 3) observed after complete drying. Discussion of the underlying mechanisms can be found in, for example, Uesaka (2001) and Walker (2006).

What happens when a fiber swells in a sheet of paper? The swelling affects the neighboring fibers that are bonded together and pushes them farther apart. The

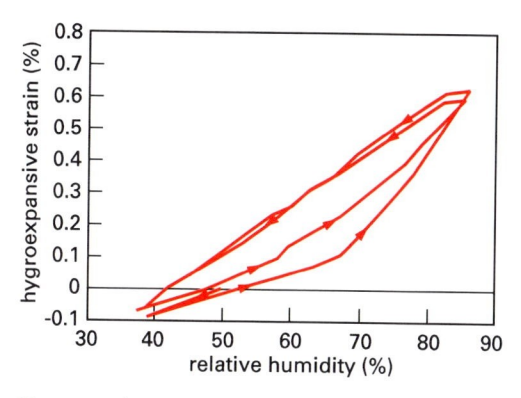

Fig. 9.3 Hygroscopic strain versus cyclic relative humidity in a freely dried laboratory sheet, data of Uesaka et al. (1992). Reproduced with permission from Pulp and Paper Technical Association of Canada (PAPTAC).

transfer of the local strains over the whole fiber network determines how much paper expands when the moisture content increases and how much it shrinks when moisture is removed (Chapter 11.6). This effect is called the *hygroexpansion* of paper. Figure 9.3 shows the hygroscopic strain that is induced in a freely dried paper sheet when the relative humidity changes. In this particular case, the hysteresis in strain originates from the moisture content hysteresis of the fibers.

The hygroscopic strain of paper is similar to the ordinary thermal expansion, with moisture being a counterpart of heat. Under normal environmental conditions, the hygroscopic strains of paper are much larger than the thermal strains (Chapter 2.2). In the present discussion, the one-dimensional constitutive law of paper includes the hygroscopic strain ϵ^{hygro} and inelastic strains $\epsilon^{inelastic}$ (such as plastic strain or creep strain):

$$\sigma = E \cdot \epsilon^{elastic} = E \cdot (\epsilon^{total} - \epsilon^{hygro} - \epsilon^{inelastic}). \tag{9.1}$$

Now we know that moisture changes cause dimensional changes in paper. The next question is *why does the paper deform differently in the strips taken in the two orthogonal directions?* We note that commercial papers are not isotropic because of the nature of the paper manufacturing process (Chapter 2.5). Can we then distinguish between the machine direction and cross-machine direction of a paper by doing the curl experiment? As Figure 9.4 demonstrates, hygroexpansion in CD is 3–5 times higher than in MD, and therefore, the curl in paper is generally stronger in CD than in MD. We can conclude that the cross-machine direction of paper likely is revealed by the stronger curl in our test strips.

We know now that paper reacts to moisture by changing dimensions, but *why does the paper react to applied moisture by bending*? Bending is always associated with a gradient of in-plane strains across the thickness of paper (Chapter 4.4). The axial strain and stress in the case of pure bending (small deflection and linear elastic response) are illustrated in Figure 9.5. The distribution of strain and stress is antisymmetric with respect to the middle plane of the sheet, and the total strain is equal to the elastic strain.

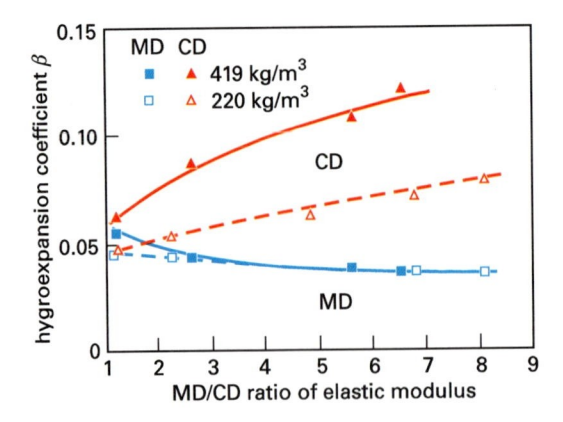

Fig. 9.4 MD (squares) and CD (triangles) hygroexpansion coefficient β (Eq. 2.5) against fiber orientation anisotropy, measured as the MD/CD ratio of elastic modulus in laboratory sheets of two densities: 419 kg/m³ (black symbols) and 220 kg/m³ (white symbols). Data of Uesaka (1994). Reproduced with kind permission from Springer New York LLC in the format Journal via Copyright Clearance Center.

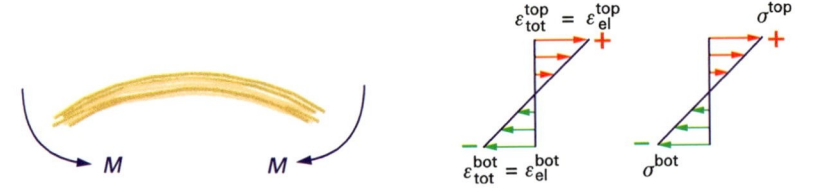

Fig. 9.5 Strain and stress distribution in the case of pure bending.

Now we have to understand where the strain gradient comes from when moisture is applied. A classical example of similar bending is the bimetallic strip. When the temperature is increased, one layer expands more than the other, creating a strain gradient, and the beam bends. Drawing on the analogy, we might speculate about the variation of elastic properties in the thickness direction of paper; however, if this was the reason for bending, then the paper in our test should always bend toward the same side, regardless of which side moisture is applied. A quick test will reveal that this is not the case. The final curl of paper is consistently toward the side where moisture was applied. Let us try to understand out why.

When we apply moisture to the top side of a paper strip, the moisture content on that side will be initially higher before moisture is transported through the thickness by diffusion. Before moisture equilibrium is reached, the top side expands more and bending takes place. Bending would not happen in the ideal case of perfectly uniform moisture content all the time, but in practice such a situation is not possible to reach with one-sided moisture application. In this respect, the hygroscopic expansion of paper is clearly different from the thermal expansion of a metal sheet. It would be hard to create curl in a metal sheet by one-sided heating because the thermal conductivity is so high.

Let us consider more closely the strain and stress profiles for nonuniform moisture distribution. Imagine that we have a linear elastic paper that is initially dry (MC = 0),

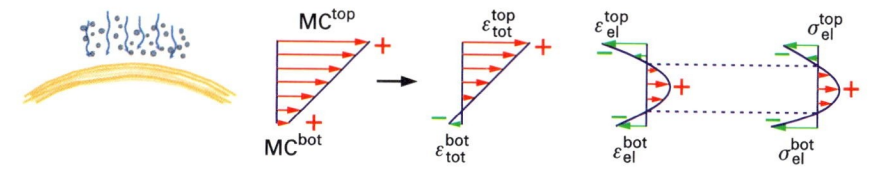

Fig. 9.6 Strain and stress distribution created by a linear distribution of moisture content (MC).

but after the one-sided moisture application it has a linear moisture distribution through the thickness. Figure 9.6 shows the total strain and elastic strain in this case. Although the deformation of paper may look similar to pure bending, there are a few differences. First, the total strain is asymmetric and can even be positive throughout the entire thickness. Second, the elastic strain is no longer the only contribution to the total strain. There is also a hygroscopic strain contribution (cf. Eq. 9.1) that can be greater than the elastic one.

One can also see from Figure 9.6 that the top side is under negative elastic strain and stress. The top side tries to expand, but the expansion is restrained by the rest of the paper. In terms of Equation 9.1, $\epsilon^{elastic} = \epsilon^{total} - \epsilon^{hygro} < 0$ even though the total strain ϵ^{total} is positive (remember we assumed linear elasticity so that $\epsilon^{inelastic} = 0$). The bottom side does not expand as much as the top because the moisture increase is smaller. The bottom layer is under compression due to the bending; only the middle part has positive stresses. Finally, the local stress distribution and local elastic strain distribution in Figure 9.6 differ from one another because the elastic modulus E depends on moisture (Fig. 2.5).

9.2.2 Effect of moisture history

Now we know that the differences in curl appearance between the MD and CD strips are caused by inherited anisotropy of paper properties, therefore, we can move on to the next question: *why does the strip end up having a permanent curl?* Sufficiently long after the moisture application, the moisture content in our test strip would return close to the original state that is determined by the ambient conditions. If the paper had been perfectly elastic, it would then have recovered the original flat shape. One might of course claim that the original moisture content is not recovered because of the hysteresis (Fig. 9.2), but the hysteresis effect is easy to remove. For example, one can use a small light bulb to dry the paper completely, and then let it return to the ambient temperature. This would remove the effects of moisture hysteresis, but the curl in paper would not be removed. The paper strips will nevertheless have the permanent curl towards the side where moisture was initially applied.

The permanence of the curl in our experiment implies that irreversible deformations are created. Such deformations can be described as plasticity or creep, or both simultaneously (Mäkelä, 2007). Creep formulations are used for time-dependent deformations of paper (Chapter 7), while plasticity is used when constitutive laws (Chapters 2–5) are applied to paper products. We use rate-independent plasticity to explain the residual curl in our test case. For alternative creep formulations, see Uesaka (2001).

It is easy to demonstrate how paper deforms permanently if bent. Just roll a paper sheet into a tube and let it go. You will see that some residual curl developed already in such a short test. The smaller the tube radius, the stronger the residual curl. However, in this experiment curl always develops in the direction of rolling – unlike our original case where the direction of curl changed during the test. Thus, we still lack the answer to the question, *why does the curl change its direction during the test*? In order to answer this question we need to look more closely at the plastic behavior of paper (Chapter 2.3). One can see that if stress is increased, plastic strain appears earlier at higher moisture contents (Fig. 2.13) and earlier in compression than in tension (Fig. 2.14).

For a quantitative demonstration of the effect of plastic strain, we make use of the stress–strain curves in Figure 9.7. They show typical tensile and compressive behaviors in the machine direction and cross-machine direction of paper. The CD compressive stress–strain curves at 10% and 90% RH are shown in detail in Figure 9.8. Assuming that unloading is completely elastic, one can see that after exposure to a tension T_{max}, plastic strain is created at 90% RH but not at 10% RH. Alternatively, unloading from a strain ϵ_{max}, the plastic strain is larger at 90% RH than at 10% RH. Applying these observations to the stress distributions in Figure 9.6 one can understand how the difference in moisture content can create a clearly larger permanent compressive strain on the top side than on the bottom side. The same happens in our experiment with the paper strips.

The effect of moisture on the inelastic deformation of paper is similar to the effect of temperature in metals or plastics, where an increase in temperature will accelerate the build-up of visco-plastic deformations. In the case of paper, one can then understand that the curl created by nonuniform moisture application must depend on the rate of moisture changes. This is controlled by the coefficient of moisture diffusion. The situation is complicated because the diffusion coefficient itself depends on the moisture content of paper (Fig. 9.9), and evaporation of moisture must in general be included in the analysis.

So the short answer to the question about the reversed direction of curl is that it is due to the *specific* moisture and stress history. Accounting for the moisture history is very important when you need to interpret the impact of moisture on dimensional changes in paper. Indeed, even the preceding moisture and stress history of the paper we used (Fig. 9.1) may influence the magnitude of curl observed, but it cannot explain the direction that is always toward the side of moisture application.

The stress history in paper is created in the manufacturing process (Chapter 2.5). In particular, the paper web is under MD tension during the drying process. Using simple terminology, this creates *internal stresses* in the paper. These stresses can be released when paper is moistened the first time because moisture softens paper. The release of dried-in strains is manifested through the irreversible shrinkage after the first moistening–drying cycle of a paper that was initially dried under restraint (Fig. 9.10). In a freely dried sheet, the irreversible shrinkage is insignificant (Fig. 9.3). Subsequent moisture changes happen without considerable irreversible changes in hygroexpansion.

The release of eventual dried-in strains will influence curl if the dried-in strains are nonuniform in the thickness direction of paper. Indeed, one sometimes adjusts the temperatures of the last drying cylinders on the paper machine so that the paper

9.2 Moisture-induced deformations

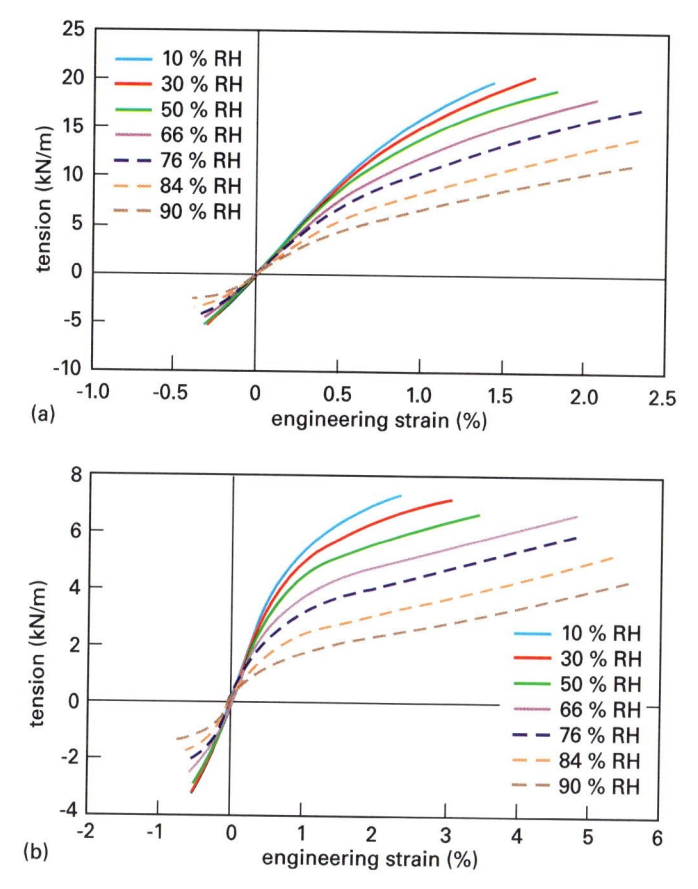

Fig. 9.7 Tension-strain curves in MD (a) and CD (b) at different relative humidities (RH) for a paper made of chemical pulp; data courtesy of Petri Mäkelä, Innventia.

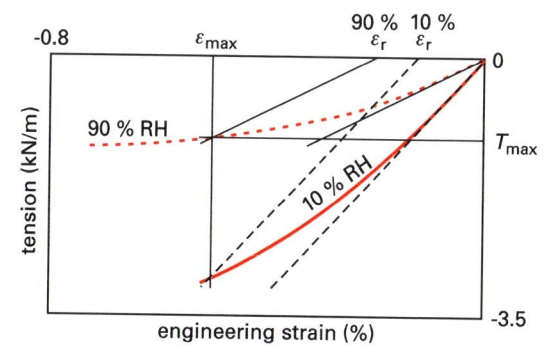

Fig. 9.8 Compressive tension-strain curves in CD of the paper in Figure 9.7 at 10% RH and 90% RH.

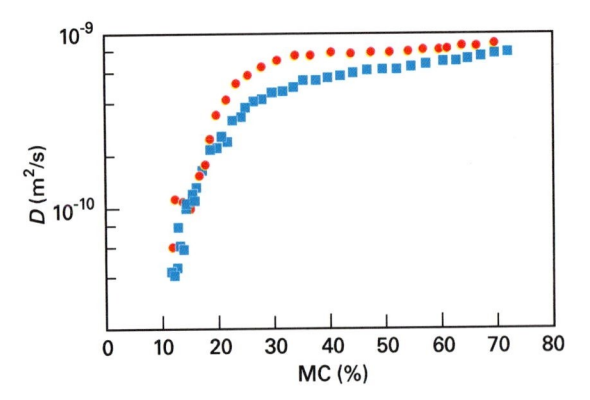

Fig. 9.9 Diffusion coefficient against moisture content at room temperature for paper made of an unrefined (circles) and refined chemical pulp (squares) (Topgaard and Söderman, 2002). Reproduced with kind permission from Springer Science+Business Media B.V.

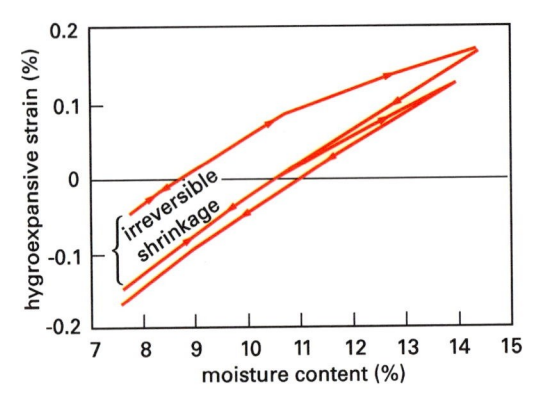

Fig. 9.10 Irreversible shrinkage during moisture cycling in a sheet dried under constraints (Uesaka et al., 1992). Reproduced with permission from Pulp and Paper Technical Association of Canada (PAPTAC).

produced is flat when it comes out from the paper machine. The "hidden" curl will then re-emerge if the moisture content of paper increases too much.

Let us summarize the mechanisms that take place in our experiment in Figure 9.1:

1. Water applied is absorbed by the top side of the paper strips.
2. The fibers in the paper react to water by swelling, and the paper expands. The expansion is nonuniform throughout the thickness because the moisture distribution is nonuniform.
3. The nonuniform hygroexpansion creates a strain gradient that makes the top side longer. As a result, the paper bends downwards.
4. The nonuniform hygroexpansion also causes considerable negative stresses on the top side of the sheet, which leads to compressive plastic strains.
5. The applied water evaporates away and the paper returns to the original moisture content (or close).

6. In the end, the top side has larger negative residual strains, giving the paper a permanent curl toward the top surface.

The discussion has shown that the hygroexpansion of paper and the deformations induced by it can be very complex. In printing, for example, ink and water are applied in a fraction of a second and under pressure (Chapter 10). Analyzing such conditions can be a formidable task, not the least because the application of liquid water causes new relaxation and expansion phenomena in addition to those discussed here (Ketoja et al., 2001). For realistic measurements, one should be able to detect the hydroexpansion caused by liquid water in realistic dynamic conditions. Because the moisture history is so important, different experimental setups may produce very different results. The remainder of this chapter discusses two examples of the paper deformations caused by *liquid* water.

9.3 Fluting

9.3.1 Tension wrinkling

Fluting is a waviness you may find in the inked areas of a glossy print (Fig. 9.11) that is oriented in the machine direction of the paper. It often appears with two-sided, heavy ink coverage. Fluting occurs only in web-fed, heatset, offset printing (see Oittinen and Saarelma, 2009, for information on different printing processes). In this printing process printing color and water are applied by rollers on the surface of a moving web that is then dried within seconds. It covers most of the higher quality printing market. Fluting usually appears with a wavelength of 1–2 cm.

When trying to understand the fluting problem, the most difficult part is to recreate the conditions that paper experiences in the heatset printing press. As we already know, deformation of paper is history-dependent. Making wrong assumptions

Fig. 9.11 Fluting in a magazine. Reprinted from Kulachenko et al. (2007) with permission from Elsevier.

about the loading history leads to misleading conclusions. Despite various views on this phenomenon, some observations are shared by everyone who deals with fluting. The most important observation is that tension *and* heat are required to cause fluting. In the web-fed heatset offset printing process, the paper web is under tension. Ink and water are applied onto the paper web in the printing stations and then dried intensively with heat in the drying section. So our first question is, *what is the role of the tension in creating fluting?*

A thin material, such as paper, may wrinkle due to tension alone (Fig. 9.12). This can easily be demonstrated by pulling a piece of paper, cloth, kitchen wrap, and so forth. There are basically two reasons for the tension wrinkling. The first reason is the boundary conditions. Usually you cannot pull paper without laterally restraining at the ends. These restraints alone can cause negative normal stresses in paper. The second reason is the nonuniformity of stress across the paper width. In a paper web, the CD tension profile is often nonuniform, which can also cause negative stresses. Because of these two factors, the free contraction of paper is prevented, and paper escapes the constraints through buckling.

The wavelength of wrinkles created by web tension is relatively large and depends on the web width and length of the draw. When the distance between boundaries is long, it is simply impossible to create the 1–2 cm wavelength observed in fluting merely by applying tension. When heat is applied in addition to tension, the wrinkles are sharper and narrower than with tension alone (Hung, 1986). On the other hand, heat alone creates out-of-plane distortions that do not exhibit the wavy fluting appearance (Strong, 1984). The question is, then, what changes with the heat application? Why does it create sharper waves?

Heat removes water from the web, and the web shrinks. The rate of water removal is very high on a heatset machine – the web must be dried in less than 1 second. In the open draw of the dryer, shrinkage increases web tension, and it has been demonstrated that the wrinkling wavelength decreases with tension (Coffin, 2003). The web tension in the heatset dryer is usually 250–500 N/m, while Hung (1986) showed that a thin paper web may show fluting after high-intensity drying even at a tension of 100 N/m or less; in practice, the web tension cannot be decreased to such low levels that the resulting fluting would be acceptable. In other words, the tension increase caused by moisture loss cannot solely explain why the wrinkling wavelength decreases with heating. What other aspects associated with moisture loss could then influence fluting?

As noted previously, moisture changes may cause considerable expansion or shrinkage of paper, particularly in CD. In the printed product the areas that are

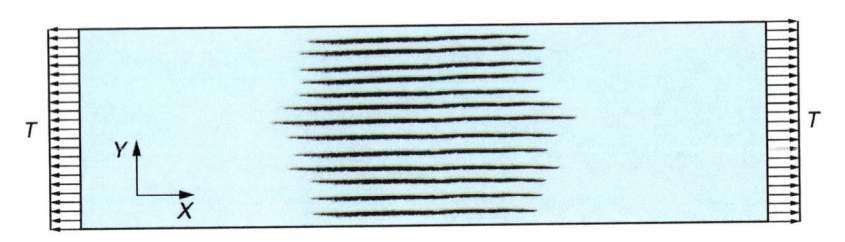

Fig. 9.12 Tension wrinkles. Reprinted from Kulachenko et al. (2007) with permission from Elsevier.

covered with ink have experienced larger moisture changes and received more water than nonprinted areas. The ink layer also retards the evaporation of moisture. This means that there is a difference in the hygroexpansion between image and non-image areas. As a result, compressive forces are created and lead to buckling (Fig. 9.13).

The effect of the difference in hygroexpansion is easily demonstrated with an inkjet printer. A wavy image is formed if a full-color image is printed that does not fill the entire sheet, but it is also created if one prints a high-quality image over the entire sheet (Fig. 9.14). To produce this image, a lot of water is applied on a sheet that moves slowly. Fluting forms at the boundary of the printed and nonprinted surface. The wavelength will depend on how fast one prints. The faster the printing speed, the higher the fluting wavelength. When the waviness is once created, a specific deformation history is created that favors making the existing buckles stronger rather than triggering new buckling at a different wavelength. In the end of printing, the wavelength is exactly the same as in the beginning. This is again a demonstration of the importance of the loading history.

Many authors have used the difference in hygroexpansion as the explanation of fluting (for a review of the literature, see Kulachenko et al., 2007), but the supporting

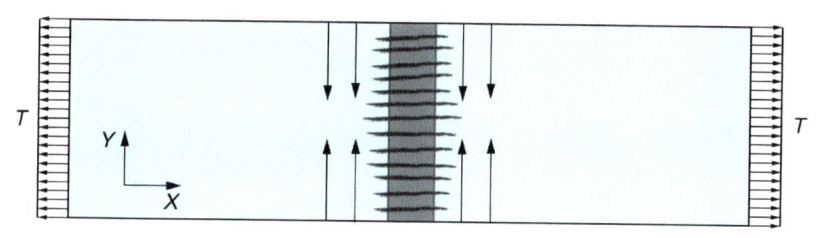

Fig. 9.13 Tension wrinkles caused by differential shrinkage. Reprinted from Kulachenko et al. (2007) with permission from Elsevier.

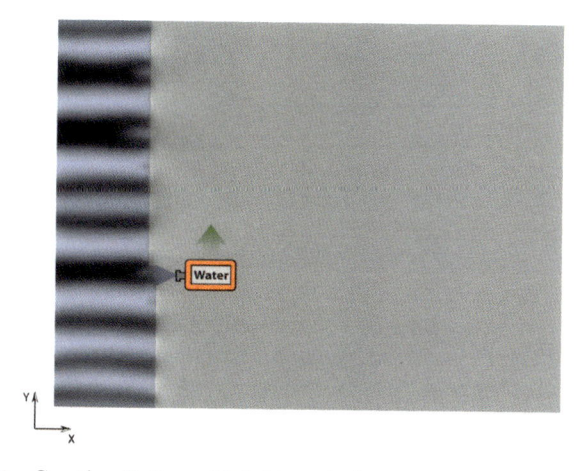

Fig. 9.14 Creating fluting with inkjet printing.

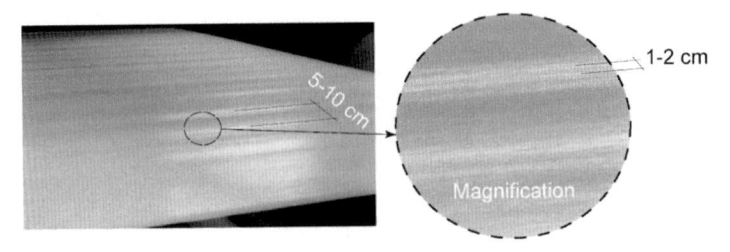

Fig. 9.15 Printing trial showing both fluting (1–2 cm wavelength) and tension wrinkles (5–10 cm wavelength). Reprinted from Kulachenko et al. (2007) with permission from Elsevier.

experimental observations are inconsistent. Furthermore, printing trials have demonstrated that fluting can occur even when the entire web is covered by ink or when no ink is used, as in Figure 9.15. In this case one can see waviness with a wavelength of 5–10 cm that is typical of tension wrinkles. However, on a closer look one can see another superimposed waviness that has a significantly shorter wavelength, 1–2 cm, close to that reported as fluting. The question is now what causes these small waves? Even a second question arises from Figure 9.15, namely, if the waviness appears everywhere in this case, why is it visible only in the printed areas of a final product?

9.3.2 Effect of small scale strain variations

In the discussion of fluting, it has been assumed that paper has homogeneous properties. In reality, the basis weight, density, thickness, porosity, and fiber orientation vary at all in-plane length scales starting from the microscopic to macroscopic. This is reflected in the physical properties of the paper as a material. Thus, paper naturally has strain variations, even under a uniform applied load. Additionally, the moisture changes associated with printing are not uniform because of variations in porosity, and, because of variations in fiber orientation, hygroscopic strains are not even. This implies local compressive strains that may cause local buckling. However, the local buckling cannot happen without triggering disturbances. We can demonstrate this with the help of numerical analysis.

Let us consider a sheet of paper as a continuum material with orthotropic properties. The sheet is subject to a uniform MD tension without any CD constraints (Fig. 9.16). This means that the sheet may contract freely in CD. If the sheet has perfectly uniform properties it does not buckle because there is no source of compressive forces. Next, we introduce spatial strain variations by letting moisture decrease uniformly except in randomly distributed spots of diameter 0.6 cm, where moisture is maintained constant. The decrease of moisture causes negative hygroexpansion and creates compressive forces that act on the "wet" spots.

We are interested in capturing the buckling instability and post-buckling shapes and amplitudes. Figure 9.17 shows the result of such a simulation for a 1% decrease in the moisture content (Kulachenko et al., 2007). Despite the random distribution of spots, the local strain variations cause rather regular wrinkling. The wavelength of the wrinkles is 1–2 cm as observed in fluting and independent of the spot size (varied

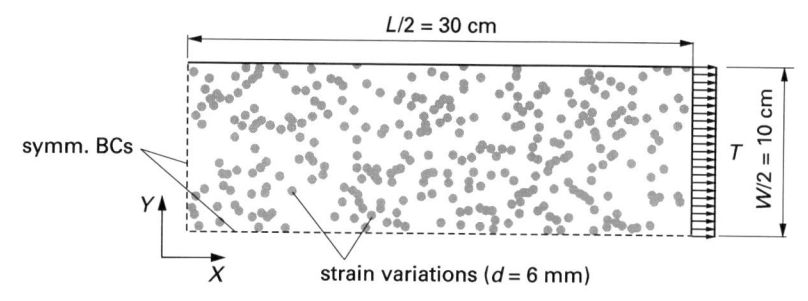

Fig. 9.16 Random pattern of "wet" spots in a model web with symmetric boundary conditions. Reprinted from Kulachenko et al. (2007) with permission from Elsevier.

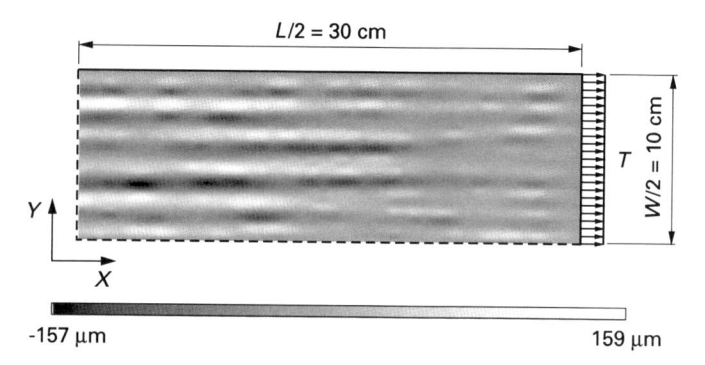

Fig. 9.17 Out-of-plane deformations caused by the "wet" spots in Figure 9.16. Reprinted from Kulachenko et al. (2007) with permission from Elsevier.

from 0.3–1.2 cm). This suggests that tension wrinkles triggered by a combination of nonhomogenous in-plane strain field may be the mechanism behind fluting.

The result shown in Figure 9.17 came from the 1% moisture difference between the spots and the rest of the paper. The instability itself was triggered already when the difference was about 0.5%. One may naturally wonder how realistic a 0.5% moisture difference might be. When convection drying is used in heatset printing, the total moisture loss is 3%–6%, and convection drying is known to cause clearly stronger fluting than infrared drying (Hung, 1986). The difference of the two drying methods is demonstrated in Figure 9.18. With fast infrared drying the wrinkled pattern was irregular, and the most visible wavelength was 3–4 cm (Fig. 9.18a). In contrast, drying with hot air (convection) gave a regular waviness with the wavelengths of 1–2 cm across the whole width. In the experiment, this pattern appeared almost immediately as the heated air was applied (Kulachenko et al., 2007). Presumably, the difference comes from the fact that infrared drying generally removes moisture more uniformly than convection because the latter interacts more strongly with the paper structure and its nonuniformities.

Upon completion of drying, both paper samples in Figure 9.18 had roughly equal moisture contents but very different out-of-plane deformation patterns. These residual deformations could not be reproduced by the simulation because the paper was modeled as a linearly elastic material – once uniform moisture content is restored, out-of-plane deformations disappear. As in the case of curl, it is important to capture the entire loading history because paper undergoes irreversible inelastic changes that depend on the loading history.

We have now shown that during printing fluting can appear in both image and non-image areas. So why is it visible mostly only in the printed areas of a final product? This is not just an optical effect (Kulachenko et al., 2007). One possible explanation could be that ink may prevent paper from returning to its original form once the moisture content has evened out. However, the stiffness of the ink layer is relatively low compared to paper stiffness.

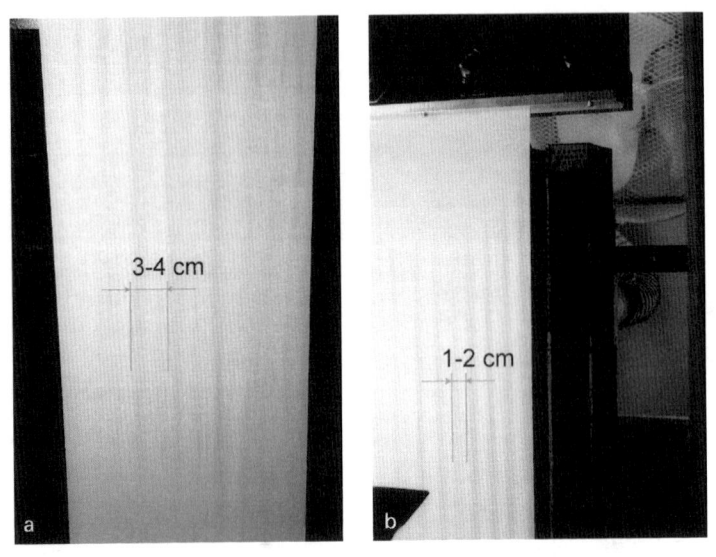

Fig. 9.18 Tension wrinkles created by fast IR drying (a) and convection drying (b). The 70 × 22 cm² paper sheets were subjected to 90% RH for 1 hour before tension and drying were applied. Reprinted from Kulachenko et al. (2007) with permission from Elsevier.

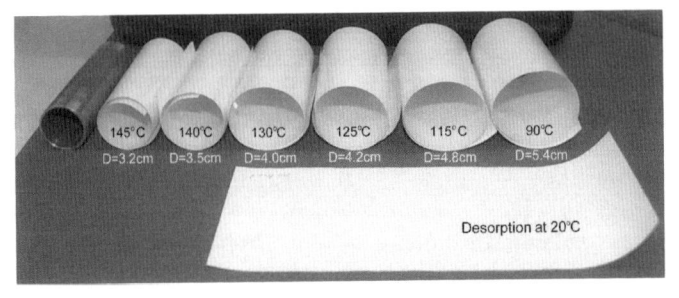

Fig. 9.19 Comparison of residual curvature in paper samples that were rolled into the metal tube and then dried in a convection oven to different temperatures. The permanent roll diameter D of the samples after reconditioning is shown. The reference sheet was dried at room temperature by desorption in vacuum. Reprinted from Kulachenko et al. (2007) with permission from Elsevier.

One important difference between printed and unprinted areas is that the printed areas heat up more and stay hot longer than unprinted areas (MacPhee et al., 2000). The surface temperature of paper in a commercial web-fed heatset offset dryer may range from 110–200°C. The effect of the temperature was demonstrated with a simple experiment (Fig. 9.19). Paper strips were rolled inside a metal tube of 3 cm diameter. A reference strip was vacuum-dried at room temperature, while the rest of the samples were dried in a convection oven so that air passed through the tube and removed water. As the temperature in the oven gradually increased, tubes were taken out at different temperatures between 90°C and 145°C. When the first tube was taken out, the paper was already fully dried. It was verified separately that drying time had no influence, only the highest temperature reached. Figure 9.19 clearly shows how higher temperature increases the residual curvature. This means that at higher temperatures, more irreversible deformation is created under the same strain state. This gives us an explanation for the stronger fluting observed in the printed image areas.

9.3.3 Fluting vs. cockling

Cockling is a phenomenon manifested in irregular, small-scale, out-of-plane distortions of paper. Some authors use the words *cockling* and *fluting* interchangeably simply because it is sometimes difficult to discriminate these two phenomena. For example, Figure 9.20 shows an unprinted A4 paper sample with what would be characterized as cockling, but it also contains some irregular waviness similar to fluting. It is commonly believed that some sort of small scale in-plane strain variation is responsible for cockling. The incompatibility in the strain field creates negative stresses that lead to local buckling. The main point of debate has been the reason for this variation, which could be fiber orientation, moisture variations during drying, and the two-sidedness of paper.

Small-scale strain variation can create both cockling and fluting. Tension is normally required to produce fluting, while cockling may occur also in a copy machine

Fig. 9.20 Example of cockled paper. Reprinted from Kulachenko et al. (2007) with permission from Elsevier.

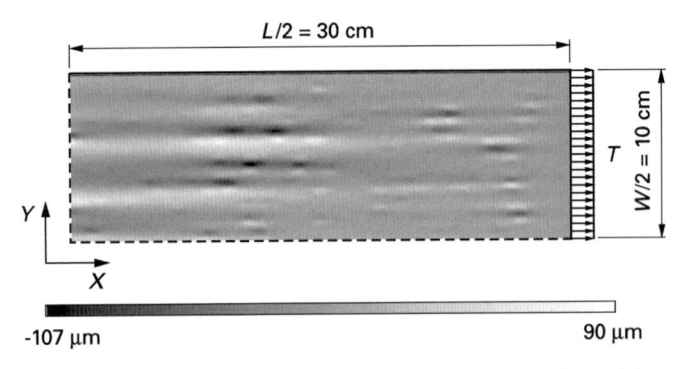

Fig. 9.21 Out-of-plane deformations analogous to Figure 9.17 but with a reduced coverage of the "wet" spots. Reprinted from Kulachenko et al. (2007) with permission from Elsevier.

and during papermaking. In a copy machine there is no tension, but in a paper machine the web is under tension. Figure 9.20 is an example of cockling developed during papermaking. So why do we see cockling and not regular fluting from the paper machine, even though paper is under tension? A clue can be found in the drying experiment of Figure 9.18, where the distortion pattern at the end of drying depended on the drying method. In the paper machine, paper is dried mainly with contact drying (in contrast to offset printing where convection drying is used). Contact drying creates fewer moisture variations than the convection drying (Hashemi and Douglas, 2003). If we return to our numerical experiment (Fig. 9.16) and reduce the coverage of the wet spots by one-half, we see cockling. The result (Fig. 9.21) is a less pronounced deformation pattern where the individual cockles are more distinct. The same trend is observed if the spot coverage approaches 100%, when the spots cover almost the entire paper area. Thus, the intensity and spatial density of strain variations seem to govern the transition between cockling and fluting when tension is applied.

9.4 Summary

In this chapter, we studied the effect of moisture by investigating the physics behind three different problems associated with moisture change – curl, fluting, and cockling. Let us summarize the important considerations for studying these phenomena.

Start with the modeling. We can solve the moisture transport and constitutive equations separately. The moisture history should be determined first, using diffusion equations or more advanced models where moisture and temperature are decoupled and the paper is considered as a multiphase material (see, e.g., Hashemi and Douglas, 1999). When the moisture history is determined, we can calculate the hygroscopic strains in the constitutive equations. Assuming that we have all the governing equations and know the boundary conditions, we still need to specify the following material properties:

- Relationship between moisture and hygroscopic strain.
- Elastic and inelastic properties as a function of moisture and position in the sheet.
- Orientation of principal material axes if the paper is anisotropic.

We also need to account for previous moisture/load histories, which can appear through irreversible shrinkage, for example.

When it comes to experimental studies, it is imperative to be fully aware of the moisture histories and discriminate between the structural curl due to a gradient in material properties, dried-in curl due to previous histories, and curl due to the actual moisture gradient. If a particular application is in question, it is important to reproduce the conditions of how the moisture is applied or removed as closely as possible to be able to analyze the system. As we demonstrated, the dynamics, boundary conditions, and the time-scale of the phenomenon have a big impact on the stress-state and deformations. You can attain the same deformation pattern having different deformation history.

References

Christensen, P. K., and Giertz, H. W. (1965). The cellulose-water relationship. In: Consolidation of the Paper Web, F. Bolam, ed. (Cambridge, UK: The British Paper and Board Makers' Association).

Coffin, D. (2003). A buckling analysis corresponding to the fluting of lightweight coated webs. In: Proceedings of the International Paper Physics Conference (Victoria, B.C., Canada), pp. 31–36.

Hashemi, S. J., and Douglas, W.J.M. (1999). Through drying of paper from mechanical and chemical pulp blends: Transport phenomena behavior. Drying Technology 17, 2183–2217.

Hashemi, S. J., and Douglas, W.J.M. (2003). Moisture nonuniformity in drying paper: Measurement and relation to process parameters. Drying Technology 21, 329–347.

Hung, J. Y. (1986). Paper Fluting – Results of TEC Studies. Phase II (DePere, WI, USA: TEC Systems).

Ketoja, J. A., Kananen, J., Niskanen, K. J., and Tattari, H. (2001). Sorption and web expansion mechanisms. In: The Science of Papermaking, C. F. Baker, ed. (Oxford, UK: The Pulp and Paper Fundamental Research Society), pp. 1357–1370.

Kulachenko, A., Gradin, P., and Uesaka, T. (2007). Basic mechanisms of fluting formation and retention in paper. Mechanics of Materials 39, 643–663.

MacPhee, J., Bellini, V., Blom, B. E., Cieri, A. D., Pinzone, V., and Potter, R. S. (2000). The effects of certain variables on fluting in heatset web offset printing. Technical report, Web Offset Association (Alexandria, VA, USA).

Mäkelä, P. (2007). The effect of moisture ratio and drying restraint on the stress relaxation of paper. In Proceedings of the International Paper Physics Conference, Appita, Gold Cost, Australia, pp. 169–177.

Oittinen, P., and Saarelma, H. (2009). Print Media – Principles, Processes and Quality, 2nd Edition (Helsinki, Finland: Paperi ja Puu Oy).

Strong, W. (1984). Graphic Arts Bulletin 11.

Topgaard, D., and Söderman, O. (2002). Changes of cellulose fiber wall structure during drying investigated using NMR self-diffusion and relaxation experiments. Cellulose 9, 139–147.

Uesaka, T. (1994). General formula for hygroexpansion of paper. Journal of Materials Science 29, 2373–2377.

Uesaka, T. (2001). Dimensional stability and environmental effects on paper properties. In: Handbook of Physical Testing of Paper, R. E. Mark, C. C. Habeger, J. Borch, and B. Lyne, eds. (New York, NY, USA: Marcel Dekker), pp. 115–171.

Uesaka, T., Moss, C., and Nanri, Y. (1992). The characterization of hygroexpansivity of paper. J. Pulp Pap. Sci. 18, J11–J16.

Walker, J. (2006). Primary Wood Processing (Dordrecht: Springer), pp. 69–94.

10 Mechanics in printing nip for paper and board

Tetsu Uesaka

10.1 Introduction

In impact printing processes, such as offset, gravure, letterpress, and flexo printing, paper goes through a printing nip where ink is applied under compression (Fig. 10.1). Although the process is simple and straightforward, there are a number of important questions related to the printing nip, especially regarding how the ink is transferred to paper and what kinds of forces are applied on paper surface. Many of the questions are directly associated with the print quality, which may include the following:

- Print density: the mean density of the printed image
- Print uniformity: the spatial uniformity of ink density at microscopic and macroscopic length scales
- Ink demand: the amount of ink required to achieve a certain print density
- Linting: the problem caused by fibers, ray cells, and other fiber debris when they are pulled out of the paper surface; deposited onto the printing blankets, plates, or rollers; and then transferred back to the paper with the ink at some other spots, disturbing the printed image
- Picking: the problem of small unprinted spots where a portion of coating layer and/or vessels are picked up from paper surface during printing
- Delamination of paper in the printing nip, illustrated in Figure 5.9

Another important, but not well-recognized issue is the impact of the nip on web mechanics, especially breaks and fan-out. Web breaks often occur in an open draw section, but there are a number of observations showing that the way the printing press is operated and the materials of the printing nip, especially the offset blanket, have a large impact on web breaks (Uesaka, 2005). *Fan-out* refers to the transverse widening of the paper in printing that can lead to the misregistration of different colors, thereby destroying printed color images. Fan-out is usually considered to be a result of the hygroexpansion of paper when it picks up water during printing; however, fan-out can also be produced by the printing nip without any water.

Printing nip has been a "black box" for the industry and researchers, and it is only recently that serious mechanistic studies have started to disclose what actually happens. In this chapter we focus on those recent mechanistic studies. Readers who are interested in a broader background of printing and associated problems are recommended to consult an appropriate text book (e.g., Kipphan, 2001).

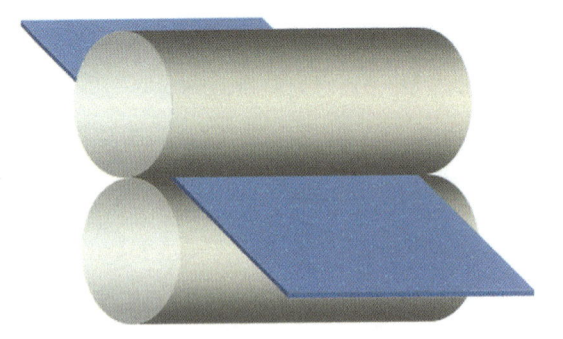

Fig. 10.1 Rolling contact in the nip of a web transport system.

10.2 Nip mechanics in offset printing of paper

Though often regarded as a simple compression of paper, the mechanics of the printing nip is actually a complex problem of a rolling contact. For example (Fig. 10.1), the rotating rolls transfer the paper with friction through the nip. During its passage through the nip, the paper is squeezed successively from the front end to the trailing end. This means that instead of simple compression, each volume element of the web receives a compression pulse and experiences shear stresses that change their directions. Because paper is an inelastic material, the loading history of shear-compression-shear can make its behavior different from what it would be in the simple compression case. It is, therefore, important to describe accurately the stress–strain behavior of the whole printing nip, not just the paper.

The outer surface of an offset printing roll, called the *blanket,* is made of a rubbery material and is normally treated as a hyper-elastic material (Fung, 1965) in solid mechanics. This means that the stress components are unique functions of the strain components, but the stress–strain relationships may be nonlinear. Another important feature of hyper-elastic material models is that they use a special strain measure that can properly represent very large deformations encountered in the use of rubbery materials.

For paper, the nonlinear stress–strain behavior, illustrated in Figures 2.15 and 2.16, is often treated as elastic–plastic behavior. For isotropic materials, the corresponding constitutive relations are well established (Fung, 1965; Belytschko et al., 2000; Fung and Tong, 2001). However, in the case of orthotropic materials, such as paper, the yield conditions and hardening rules must be chosen carefully so that the constitutive equations satisfy basic frame-independence and material symmetry. The outcome is an increased number of material constants and material functions that must be determined experimentally. This sometimes makes the problems intractable both experimentally and numerically. The search for appropriate elastic–plastic constitutive equations for paper is still in progress (Mäkelä and Östlund, 2003).

The analysis of nip mechanics relies on the conservation laws of mass, momentum, and energy; nonlinear constitutive relations; and boundary conditions. The boundary conditions involve contact problems where prescribed displacement changes to prescribed traction, creating a geometrical nonlinearity. Therefore, so far all the analyses of nip mechanics have required numerical methods, particularly nonlinear

10.2 Nip mechanics in offset printing of paper

finite element methods (Belytschko et al., 2000). Figure 10.2 shows the finite element model used for the rolling contact problem of an offset printing nip proposed by Wiberg (1999). In the model, the paper contacts with the roll under a nip of width 2a. The offset blanket consisted of three layers of hyper-elastic (Mooney, 1940; Ogden, 1986) and linear elastic orthotropic materials, and the paper was modeled with an elastic–plastic Karafillis-Boyce model.

The coordinate axes of the model of Wiberg (1999) are shown in Figure 10.3. Figure 10.4 shows the distribution of cross-machine direction (CD) strain of the paper in the nip. The CD strain is generally positive, corresponding to paper widening, and peaks near the edge of the paper. A similar distribution is seen in the machine-direction (MD) strain (Fig. 10.5). The positive strain values arise from the extrusion of the paper due to the compression. Note that the strain levels are 0.1%–0.2% which is comparable to longitudinal strains (or draw) typically applied in open draw sections of printing presses. In the center of the cylinder, the paper is subject to about 0.1% strain, and at the edges the strain increases up to 0.4%, twice the typical yield strain at which permanent elongation is created in paper.

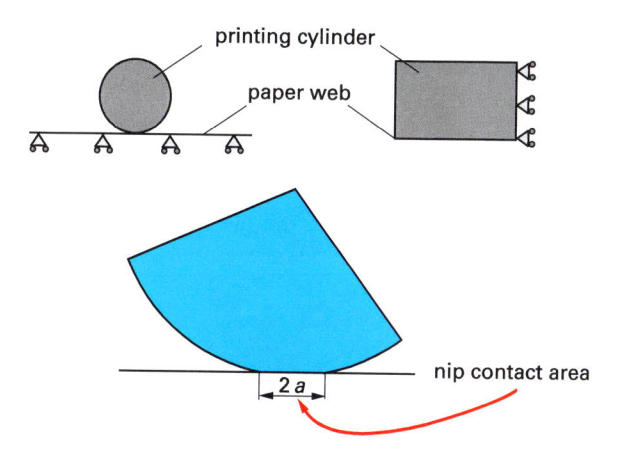

Fig. 10.2 Finite element model formulation for the rolling contact in offset printing nip (Wiberg, 1999). Reproduced with permission from the author.

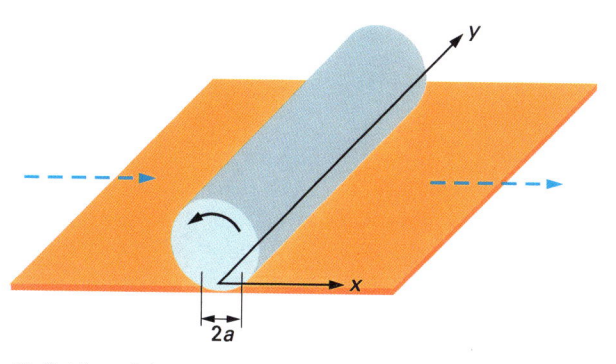

Fig. 10.3 Definition of the coordinates for printing nip problems, with 2a denoting the nip width.

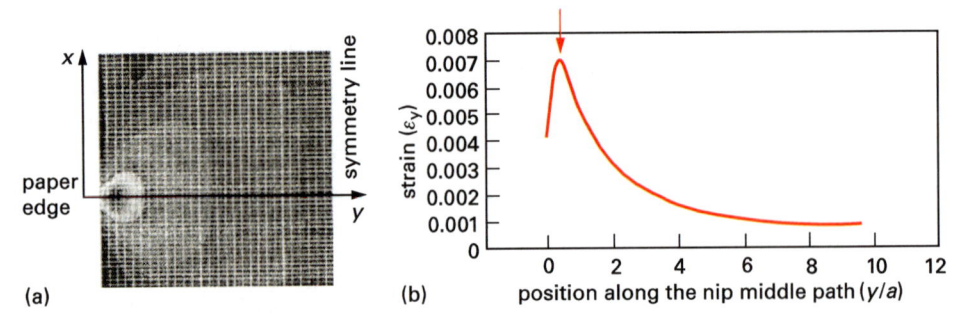

Fig. 10.4 CD strain in the paper plane (a) and against the distance from the paper edge along the nip center line (b). The CD strain has its maximum near the paper edge (Wiberg, 1999). Reproduced with permission from the author.

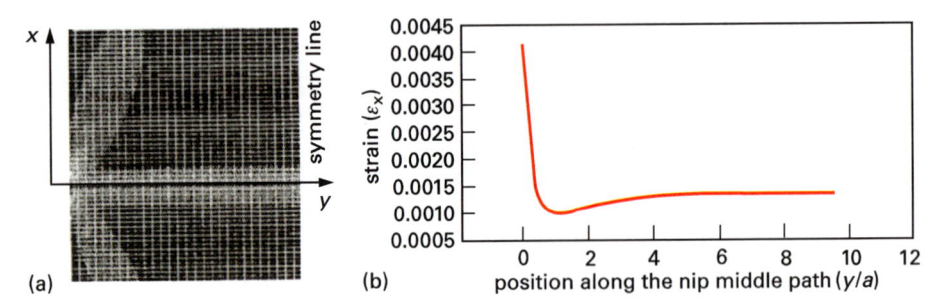

Fig. 10.5 MD strain in the paper plane (a) and against the distance from the paper edge along the nip center line (b). The MD strain has its maximum at the paper edge (Wiberg, 1999). Reproduced with permission from the author.

The implication is significant. In a normal web-fed printing press, the paper web is drawn from one section to the next (Chapter 6). Figure 10.6 illustrates the effect of adding a 200 N/m tension to the web, ignoring the transport velocity jump (or strain jump) created in the nip. When the strain jump due to the draw is included, the strain variation in the nip could be as shown in Figure 10.7. In other words, the paper web experiences a very high strain spike in the nip, in addition to the strain due to the draw, especially at the web edges. The total strain could be very significant in comparison with the breaking strain of mechanical pulp–based printing papers ($\approx$ 1%). The extrusion effect in the nip is known to depend on the blanket layer composition (Wiberg, 1999), which in turn could affect, for example, web breaks in a printing press. This conforms well to the observations that the choice of offset blankets has a large influence on web breaks (Uesaka, 2005).

Figure 10.8 shows the variation of CD strain along the traveling direction of the paper web. In the nip the paper web is widened by the compression, but this widening is partially recovered after the web leaves the nip. Nevertheless, part of the deformation remains and is observed as a fan-out that is of mechanical origin rather

10.2 Nip mechanics in offset printing of paper

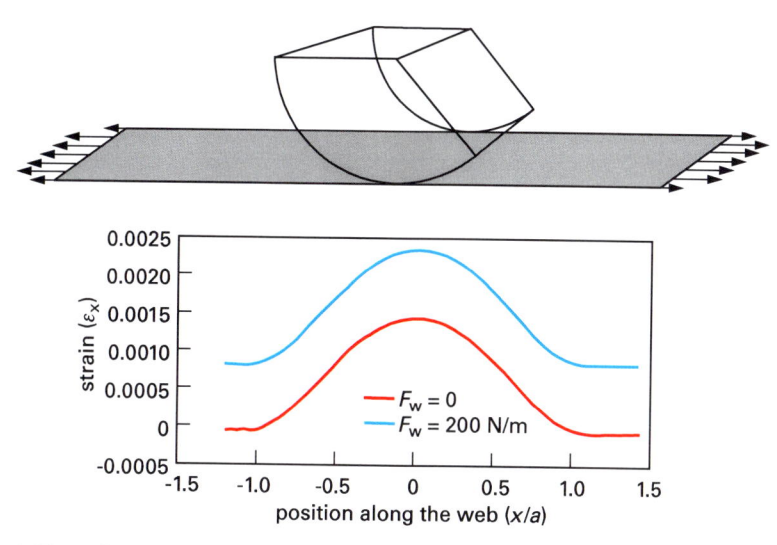

Fig. 10.6 MD strain along the traveling direction of the web with and without a superimposed web tension (Wiberg, 1999). Reproduced with permission from the author.

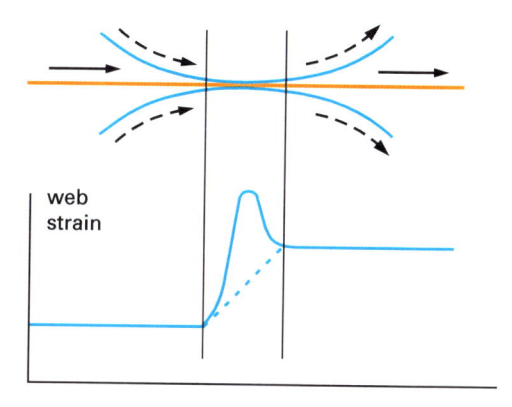

Fig. 10.7 Schematic of the MD strain of a web passing through a printing nip when draw is applied.

than hygroexpansion. In a real printing nip, hygroexpansion is superimposed on this mechanical extrusion strain. Considering that the typical CD hygroexpansion of paper is 0.15% when 1% moisture is added, then the mechanical extrusion strain of 0.1% is far from negligible. This was confirmed by a controlled printing press trial (Gomer and Lindholm, 1991). However, fan-out (often called *misregistration*) is not a major concern in today's printing presses because automatic registration systems can compensate for it. The only situation where fan-out arises as a problem is at a roll change during the printing operation when a new roll shows different fan-out behavior than an expiring roll.

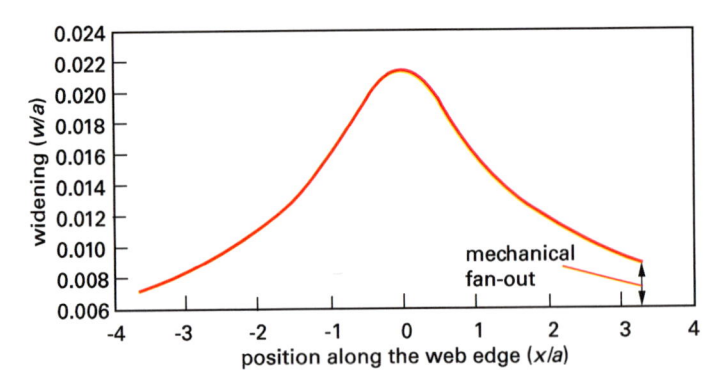

Fig. 10.8 CD at the web edge against the traveling direction of the web. Note the difference in the widening between inlet and outlet (Wiberg, 1999). Reproduced with permission from the author.

10.3 Nip mechanics in flexo post printing of corrugated board

Flexo printing is the major printing method for corrugated board packaging. One can either print on the linerboard before the corrugated board is manufactured (this process is called pre-printing), or print on the corrugated board (called post-printing). The "striping" of corrugated board (Fig. 10.9) in post-printing is one of the outstanding print quality issues. This print quality defect is obviously related to the corrugated structure, but the appearance and severity are greatly affected by the details of nip mechanics. (Zhang and Aspler, 1995; Netz, 1997; Hallberg et al., 2005). One unique feature of flexo post-printing is that the corrugated board is squeezed very lightly ("kissing" contact) to avoid damaging the corrugated board structure. Therefore, all the materials can be assumed to be elastic, even though displacements may be large, involving contact problems analogous to the case of offset printing.

Figure 10.10 shows a finite element model constructed to investigate the pressure variations in flexo post-printing (Holmvall and Uesaka, 2007, 2008a, 2008b). The model contained a portion of the corrugated board between two flute tips with symmetric boundary condition. The linerboard and fluting board were assumed to be linear elastic orthotropic materials, and the two layers of the composite printing plate were represented with models by Mooney-Rivlin and Hill-Ogden (Mooney, 1940; Hill, 1978; Rivlin, 1984).

The predicted pressure variations on the top linerboard are shown in Figure 10.11 for different values of the elastic modulus of the photo-polymer layer (an outer surface of the printing plate). The printing pressure varied considerably in the width direction between the flutes. Making the printing plate cover softer (more flexible) alleviated this variation, but it also reduced the overall printing pressure. Further parametric studies showed that if one compensated the overall reduction of printing pressure by increasing the print compression, the nonuniformity of pressure returned. Therefore, making the plate softer is not a real solution to the nonuniformity of pressure. This was also experienced in a converting plant. The model calculations

Fig. 10.9 Striping in flexo post-printing of corrugated board (Holmvall, 2010). Courtesy of Martin Holmvall.

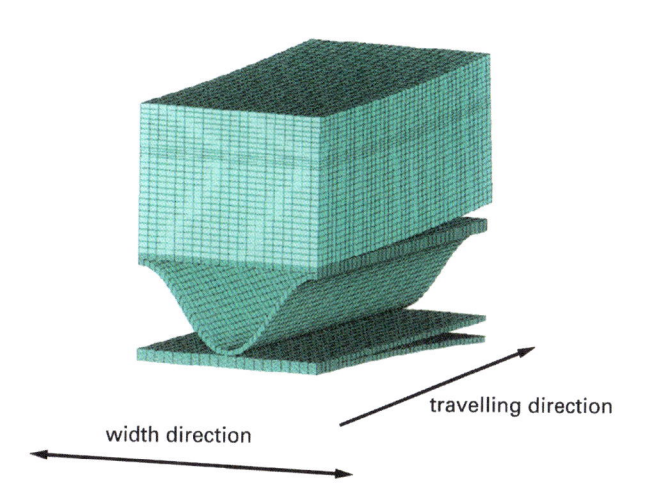

Fig. 10.10 Finite element model for the contact mechanics of a corrugated board in a flexo post-printing nip (Holmvall et al., 2007).

also made it possible to design a better printing plate that included using a hyperelastic material with a yield stress, for example, elastic foam, to achieve uniform pressure distribution (Holmvall and Uesaka, 2007).

In converting plants the striping problem is often considered to be identical with the washboarding of corrugated board. *Washboarding* refers to a geometric imperfection where the linerboard sags between flute peaks. This arises if too much adhesive is used in the corrugating process or if the bending stiffness of the linerboard

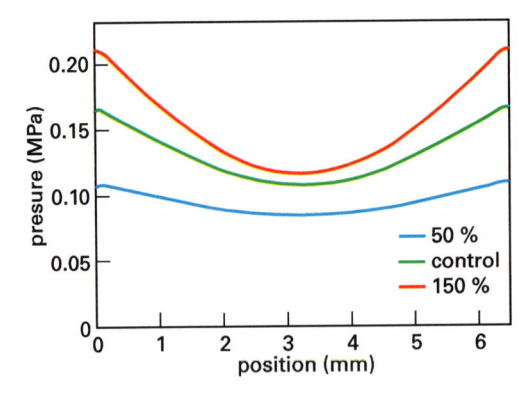

Fig. 10.11 Effect of the elastic modulus of the photopolymer cover of a flexo printing plate on the printing pressure in the model of Figure 10.10, with the flute tips located at the figure edges (Holmvall et al., 2007).

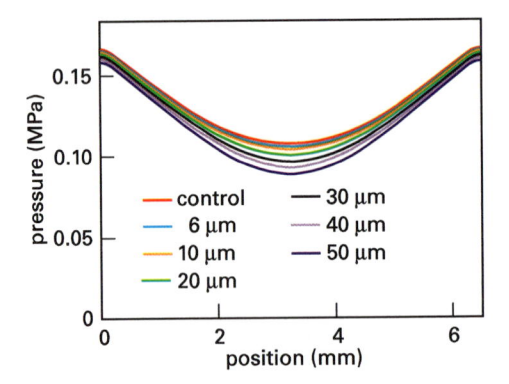

Fig. 10.12 Effect of different degrees of the washboarding of a corrugated board on the printing pressure in the flexo post-printing model of Figure 10.10 (Holmvall et al., 2007).

is low. Figure 10.12 illustrates the effect of washboarding on the printing pressure variation. One considers that a 50 μm deflection in washboarding is very high and rarely seen in reality. Still, it gives only a very modest pressure variation, aside from a slight change in the average pressure. This means that washboarding cannot make a large contribution to striping. Also, the effects of other types of physical and geometric imperfections of the corrugated board have been systematically quantified (Holmvall and Uesaka, 2008a, 2008b).

10.4 Micro-fluidics of ink in printing nip

At the microscopic level, the printing process has two targets. First, the ink should be deposited only on the very surface of the substrate, without penetration deep into the paper structure, so that the maximum print density and best color fidelity can be achieved with a minimum amount of ink. Second, the ink should be placed precisely in the intended positions to enhance image sharpness and definition. It is, therefore, important to understand how ink interacts with paper and how it is transferred into the paper surface.

Understanding the ink–paper interactions from the first principles is one of the most difficult challenges in paper mechanics and physics. First, paper is nonuniform and inhomogeneous with surface pores of various sizes and geometries; hence, one should understand the different length scales of interaction between the ink and paper. Second, one must consider the coupled problem of simultaneous deformations of solids (the paper and printing cylinders) and fluids (the ink and air). Thirdly, the ink and air form a multiphase flow problem. The fluid interfaces evolve through the entire printing process because of the ink film compression, penetration, retraction, and splitting. Fourthly, the ink itself is a suspension of different fluids, pigments, and additives (i.e., a complex fluid), and as such, it may be a non-Newtonian fluid.

Recently, there have been a few serious attempts to tackle this complex problem. The analysis starts from the multiphase flow of ink and air because tracking the motion of interfaces is the essence of the ink–paper interaction problem. A diffused interface model, called the *phase-field model,* has been used (Dubé et al., 2008, 2009; Holmvall, 2010; Homvall et al., 2010). The interface is treated as a smooth transition between the ink and air phases rather than a discontinuity. Thus, one can avoid the singularity problem of sharp interface models (Antanovskii, 1995; Jacqmin, 1999, 2000). The equilibrium distribution of the two phases is determined by equilibrium thermodynamics, whereas their dynamics are governed by the continuity and Navier-Stokes equations, together with the Cahn-Hilliard convection-diffusion equation (Cahn and Hilliard, 1959). The problem is then to solve these equations for a continuous binary fluid with appropriate boundary conditions.

The numerical method used was a finite difference projection method on a staggered grid. The boundaries between the fluids (ink and air) and the solids (printing cylinders and paper structure) must be analyzed carefully because the solid mechanics problem and the fluid mechanics problem are coupled. The complete two-way coupling has not yet been achieved, and the solids are treated as rigid phases with immersed moving boundaries (Homvall et al., 2010).

Dubé and coworkers (2008) performed the simulation of droplet "printing" onto a model network of capillaries. Here an ink droplet is pressed onto the model surface and then retracted, like in a printing process. The capillary systems investigated included a single vertical capillary, a series of vertical capillaries with varying diameters and numbers, and a combination of vertical and horizontal capillaries (Fig. 10.13). It was shown that at a given porosity, the retention of ink, so-called ink-holdout, is controlled by small capillaries whose diameter is well below 1 µm (Fig. 10.14). When the horizontal capillaries were included, they further increased fluid penetration and retention. In uncoated papers, the pores in fiber surfaces, not between fibers, have the correct size. In coated papers, such pores are present between and within the coating pigment particles. It is well known from the design practice of inkjet papers that porous coatings that create nano-scale pores on the top surface are very effective in enhancing color fidelity.

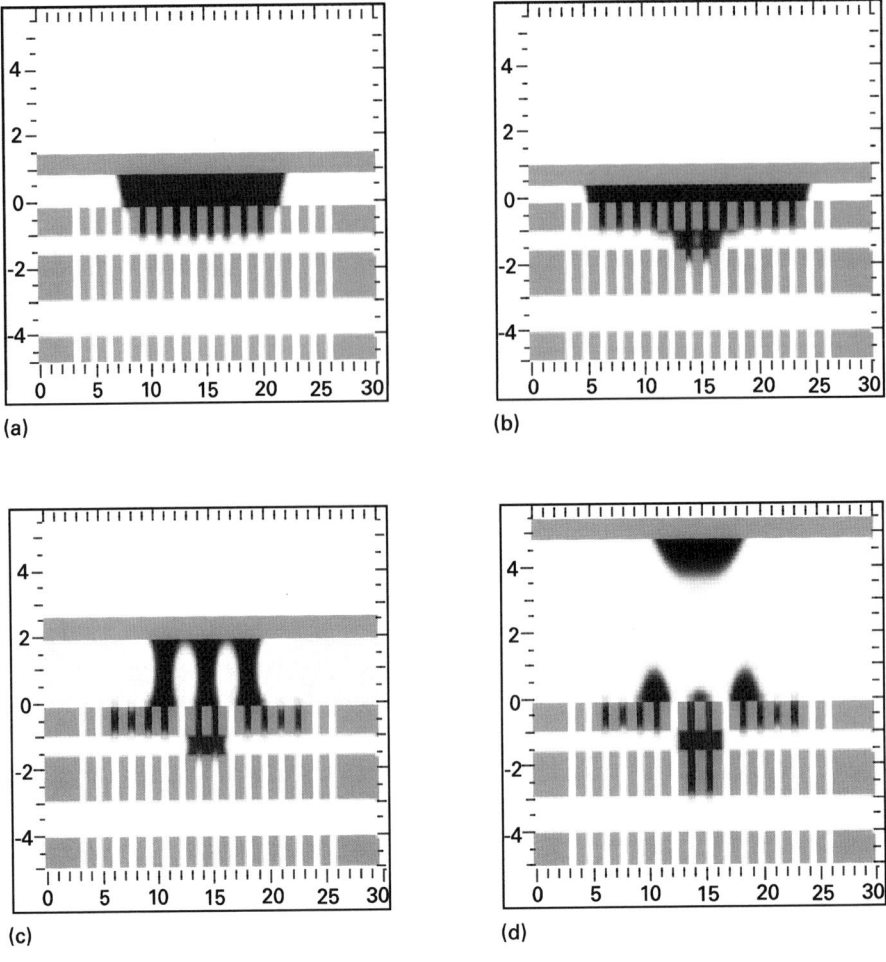

Fig. 10.13 Numerical simulation of an ink dot pressed onto a model network of capillaries. The initial diameter of the dot is 10 µm, and the vertical pore radius is 0.5 µm (Dubé et al., 2005). Reproduced with permission from Pulp and Paper Technical Association of Canada (PAPTAC).

When the typical pore size is much larger than a few micrometers, the fluid droplet tends to split in the retraction stage, and little is retained in the pores. The same phenomenon was observed in the simulations with fiber networks, where the model structures were created using a fiber deposition and consolidation model (Drolet and Uesaka, 2005). Figure 10.15 shows the entire process of fluid transfer and splitting on a fiber network. As both Figures 10.13 and 10.15 demonstrate, the fluid–paper interactions in the printing nip are limited on the very top surface of paper to depths less than a few micrometers. There is little penetration into the paper structure, which is understandable considering that the relevant timescale is only milliseconds. In this timescale, the ink behaves more like an elastic gel rather than a fluid. Penetration could, therefore, happen only after the nip passage, over much longer timescales, and with a different mechanism.

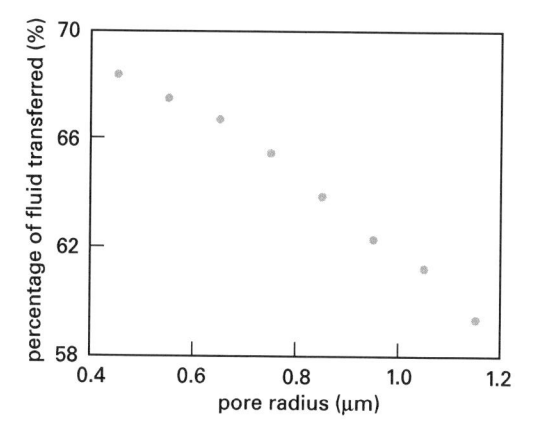

Fig. 10.14 Amount of fluid transferred against pore radius at a fixed pore volume fraction of 0.29 in the model capillary system of Figure 10.13 (Dubé et al., 2005). Reproduced with permission from Pulp and Paper Technical Association of Canada (PAPTAC).

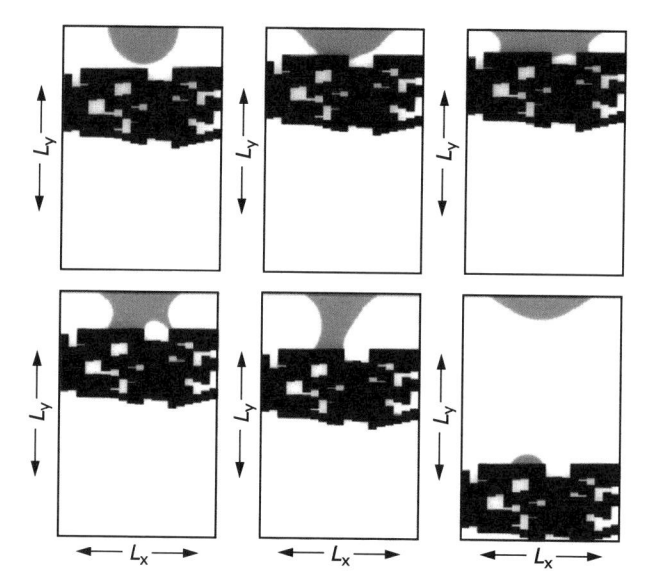

Fig. 10.15 Numerical simulation of an ink dot pressed onto a model fiber network. The initial diameter of the dot is 22.3 μm. Reprinted from (Holmvall et al., 2010) with permission from Elsevier.

10.5 Concluding remarks

As noted earlier, nip mechanics is a challenging research area. The relevant stress–strain properties are often nonlinear, and the structures often undergo large deformations that are geometrically nonlinear. Also, the coupling of fluid mechanics and solid mechanics, and multiphase flow should be solved. Therefore, the numerical solutions require state-of-the-art techniques, which are under development.

Although serious attempts using mechanics are relatively recent, important insights have been accumulated. For example, in an offset printing nip, the paper web is subjected to significant strains that can easily contribute to web breaks. These strains depend on the properties of the offset blanket. Sensitivity to the structure of printing cylinders was also demonstrated for flexo post-printing. These studies suggest that there is ample opportunity to design printing materials and printing systems using nip mechanics.

Regarding the ink transfer problem, preliminary studies based on first principles have provided insight into the ink–paper interaction that is different from the explanations based on ad hoc models, such as the Lucas-Washburn theory or Stefan's law. It is the length scale (ink thickness, pore diameter, etc.) and time scale (passage time through the nip) that make the traditional approaches difficult. Approaching the reality by solving the mutual fluid–solid coupling and the complex fluid problem of the ink is still difficult. One of the most important areas where current understanding is lacking concerns the interaction of the stochastic paper structure (including the coating layer) with the printing plate and ink. This lack of understanding means that we lack the rational surface design criteria and surface measurements. Novel mathematical and numerical apparatus are currently being prepared for use in the near future (Kassner et al., 2005).

References

Antanovskii, L. K. (1995). A phase field model of capillarity. Physics of Fluids 7, 747–753.

Belytschko, T., Liu, W. K., and Moran, B. (2000). Nonlinear Finite Elements for Continua and Structures (New York, NY, USA: Wiley).

Cahn, J. W., and Hilliard, J. E. (1959). Free energy of a nonuniform system. III. Nucleation in a two-component incompressible fluid. J. Chemical Physics 31, 688–699.

Drolet, F., and Uesaka, T. (2005). A stochastic structure model for predicting sheet consolidation and print uniformity. In: Advances in Paper Science and Technology, C.F. Baker, ed. (Lancashire, UK: The Pulp and Paper Fundamental Research Society), pp. 1139–1154.

Dubé, M., Drolet, F., Daneault, C., and Mangin, P. J. (2008). Hydrodynamics of fluid transfer. J. Pulp Pap. Sci. 34, 174–181.

Dubé, M., Drolet, F., Daneault, C., and Mangin, P. J. (2009). Penetration of microscopic drops into paper structure. In: Proceedings of the Papermaking Research Symposium, Kuopio, Finland.

Fung, Y. C. (1965). Foundations of Solid Mechanics (Englewood Cliffs, NJ, USA: Prentice-Hall).

Fung, Y. C., and Tong, P. (2001). Classical and Computational Solid Mechanics (Singapore: World Scientific Publishing Company).

Gomer, M., and Lindholm, G. (1991). Hygroexpansion of newsprint as a result of water absorption in a printing press. In: Proceedings of the 43rd TAGA Annual Conference, Rochester, New York, pp. 268–282.

Hallberg, E., Rättö, P., Lestelius, M., Thuvander, F., and Odeberg, G. A. (2005). Flexo print of corrugated board: Mechanical aspects of the plate and plate mounting materials. TAGA Journal 2, 16–28.

Hill, R. (1978). Advances in Applied Mechanics, Vol. 18 (New York, NY, USA: Academic Press).

Holmvall, M. (2010). Nip Mechanics, Hydrodynamics and Print Quality in Flexo Post-Printing. Doctoral thesis. Mid Sweden University, Sundsvall.

References

Holmvall, M., and Uesaka, T. (2007). Nip mechanics of flexo post-printing of corrugated board. J. Composite Materials 41, 2129–2145.

Holmvall, M., and Uesaka, T. (2008a). Print uniformity of corrugated board in flexo printing: Effects of corrugated board and halftone dot deformations. Packaging Technology and Science 21, 385–394.

Holmvall, M., and Uesaka, T. (2008b). Striping of corrugated board in full-tone flexo post-printing. Appita Journal 61, 35–40.

Homvall, M., Lindström, S., and Uesaka, T. (2010). Simulation of two-phase flow with moving immersed boundaries. Int J. for Numerical Methods in Fluids. DOI: 10.1002/fld. 2484.

Jacqmin, D. (1999). Calculation of two-phase Navier-Stokes flows using phase-filed modeling. J. Comput Phys 155, 96–127.

Jacqmin, D. (2000). Contact-line dynamics of a diffused fluid interface. J. Fluid Mechanics 402, 57–88.

Kassner, M. E., Nemat-Nasser, S., Suo, Z., Bao, G., Barbour, J. C., Brinson, L. C., Espinosa, H., Gao, H., Granick, S., Gumbush, P., et al. (2005). New directions in mechanics. Mechanics of Materials 37, 231–259.

Kipphan, H., ed. (2001). Handbook of Print Media: Technologies and Production Methods (Berlin: Springer).

Mäkelä, P., and Östlund, S. (2003). Orthotropic elastic-plastic material model for paper materials. Int J. Solids Struct 40, 5599–5602.

Mooney, M. (1940). A theory of large elastic deformation. J. Applied Physics 11, 582–592.

Netz, E. (1997). Washboarding and print quality of corrugated board. Packaging Technology and Science 11, 145–167.

Ogden, R. W. (1986). Recent advances in phenomenological theory of rubber elasticity. Rubber Chemistry and Technology 59, 361–383.

Rivlin, R. S. (1984). Forty years of non-linear continuum mechanics. In: Proceedings of 9th International Congress of Rheology, Acapulco, Mexico, pp. 1–29.

Uesaka, T. (2005). Principal factors controlling web breaks in pressrooms – Quantitative evaluation. Appita J. 58, 425–432.

Wiberg, A. (1999). Rolling contact of a paper web between layered cylinders with implications to offset printing. Licentiate thesis. Royal Institute of Technology, Stockholm, Sweden.

Zhang, Y. H., and Aspler, J. S. (1995). Factors that affect flexographic printability of linerboard. Tappi 78, 23–33.

Part IV
Material properties

11 Micromechanics

Kaarlo Niskanen

11.1 Introduction

In the development of new paper materials, one must know how different fiber types and other components influence the properties of paper. Often this kind of information is obtained experimentally. One prepares laboratory sheets of alternative compositions and determines which is best. Such a purely experimental approach can require a lot of effort and is further troubled by the fact that paper properties depend not only on the composition but also on the papermaking process (see Chapter 2.5.2). Paper prepared in the laboratory is different from paper prepared on a paper machine, even if the two have the same composition. Having a physical understanding of the underlying mechanisms can improve efficiency in product development because then one can use laboratory sheets to model "real" papers.

In this chapter we consider the mechanisms through which the papermaking process and fiber properties influence the in-plane mechanical properties of paper. We do this by presenting model equations for the paper properties. Even though the model equations illustrate the principal effects, they are still only simplified approximations. More realistic representations could be readily obtained from numerical fiber network simulations of desired detail.

The most serious problem in the microscopic analysis of paper properties is the lack of real, measured data on the fiber properties. Measurement methods and experimental data are available primarily for the geometric fiber dimensions. Only some scattered data are available for the mechanical properties of fibers and interfiber bonds. Those data are of questionable value because the mechanical properties of fibers and bonds in a paper sheet are influenced by the papermaking process, as is discussed later. The effect of the papermaking process on the fiber properties is a very important special feature of paper. One may be completely mislead if one assumed that paper can be described as a network of ideal fibers that have some prescribed inert mechanical properties.

It is also instructive to consider how the problem of microscopic characterization is solved, for example, in solid state physics. There, one interprets macroscopic observations against a given theory to derive apparent values of microscopic parameters. The same is possible in the case of paper if a self-consistent and quantitative microscopic model is available, which is what we aim to achieve in this chapter. A valid fiber network description must naturally be consistent with experimental observations, including the relationships between the different properties of paper. These relationships, as in solid state physics, can then be used to infer microscopic "in-situ" fiber properties that would otherwise be difficult to determine.

We begin this chapter by considering the statistical geometry of the network structure, followed by a discussion of the elastic modulus. The mechanisms of stress transfer and fiber activation that are presented there also form the basis for the qualitative presentation of the stress–strain properties, microscopic fracture process, and hygroexpansion of the fiber network.

11.2 Fiber network structure

The structure of paper is obviously significant when one considers the mechanical properties. Fibers make up paper only if the fibers bond to each other. More bonds per fiber means, in qualitative terms, stronger paper. Thus, before we can focus on the mechanical properties of the fiber network, we have to determine how to describe the network structure. Of particular importance is the length of fiber segments between neighboring bonds on a fiber.

11.2.1 Two-dimensional network

In the ideal sense, the structure of paper is a planar random network of slender fibers. The length of the fibers varies and is typically between 1–3 mm, much higher than the typical thickness of a paper sheet, which is 0.1 mm. The length of the fibers is also much larger than their width and thickness, which range from 10–50 μm. Our discussion of paper structure concentrates on this ideal type of planar network structure, which consists of fibers that are, for most of their length, aligned parallel to the plane of the paper sheet, while making some bends up and down to conform to the shapes of the neighboring fibers.

The amount of fibers per unit area of paper is characterized by *basis weight*, the mass per unit area. Typical basis weights of printing papers and office papers are 40–100 g/m^2, while paperboards are heavier, with basis weights extending to 400 g/m^2 and higher. Fibers are not arranged in distinct layers, but a good rule of thumb is that if one would make a point-wise measurement through the thickness of a typical office paper, then approximately 10 fibers would be detected at any given point because the basis weights of fibers are 5–10 g/m^2.

As described in Chapter 2, papermaking fibers are processed mechanically to increase their conformability. At the very least this causes local ruptures to the fiber cell wall, but it may even break the fibers in pieces. The mechanical treatment is especially strong in mechanical pulps where only a fraction of the pulp material consists of whole fibers that furthermore have many clear defects in them. Most of the material in a typical mechanical pulp is in the form of partly disintegrated or cut sections of fibers, broken pieces of the fiber wall lamella, and fibril material broken off from fiber walls. The small fiber fragments and fibrils are called *fines* in the papermaking context. The simple picture of paper as a fiber network, as portrayed later, is therefore insufficient in the description of mechanical pulps. Instead, the simple random fiber network approximation applies to papers made of chemical pulp (Fig. 11.1).

11.2 Fiber network structure

Fig. 11.1 SEM image of the surface of a paper made of chemical pulp, approximately 1 mm² in size.

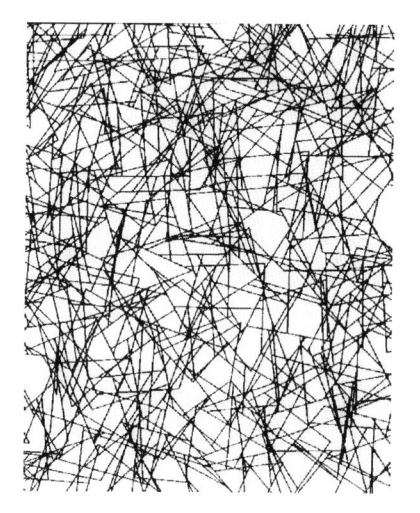

Fig. 11.2 Two-dimensional approximation of paper structure. Not shown are the "dangling" free ends of fibers that extend beyond the last inter-fiber contacts. Courtesy of Jan Aström.

A particularly simple approximation of the fiber network structure of paper is one where the thickness and width of fibers is ignored. The result is a two-dimensional random fiber network of the kind shown in Figure 11.2. The crossing points or "bonds" between fibers divide the fibers into segments. These fiber segments and the bonds between them obviously control the mechanical properties of the paper. The length distribution of such fiber segments is exponential (Kallmes and Corte, 1960a, 1960b). The mean segment length in this ideal two-dimensional case is inversely proportional to the total fiber length per unit sheet area. External loads are transferred from fiber to fiber through the bonds between fibers. As the number of bonds per fiber increases, fiber segments become shorter, and the network becomes stiffer and stronger. This is discussed in more detail in Section 11.3.2.

The applicability of the statistical geometry model of Kallmes and Corte (1960a, 1960b) is limited to paper sheets of very low basis weight. At basis weights above a few g/m^2, the limited conformability of real fibers prevents inter-fiber bonding at every point where the planar projections of two fibers cross. As a result, open space between fibers forms also in the thickness direction of the sheet (Niskanen et al., 1997). In the basis weight range of real paper, above 40 g/m^2, the length of fiber segments is controlled by the paper density and not by the basis weight.

11.2.2 Densification mechanisms

Figure 11.3 illustrates paper structure with a cross-sectional image of a paper sheet made of mechanical pulp. The roundish or rectangular loops are cross-sections of unbroken fibers. The open space inside them is the *lumen*. In thin-walled fibers the lumen collapses easily in the papermaking process, especially if high pressure is applied when pressing water out of the wet paper web (cf. Chapter 2.5.2). The irregularly shaped strands are various pieces of the fiber cell wall, which are created in the mechanical processing of the fibers. They can be large fragments of the fiber wall, or fines.

In mechanical pulps the dimensions of fiber fragments and fines particles vary. They are relatively large and rigid so that they can extend to the open spaces between fibers (Fig. 11.3). This is in contrast to chemical pulps, where fiber fragments are less common and fines are small and slender and, therefore, easily collapse along the fiber walls. Thus, chemical pulps give a paper structure that depends primarily on the whole fiber dimensions. With mechanical pulps the paper structure depends strongly on the shape and size distributions of the fiber fragments and fines, not only on those of the fibers.

The density of paper is created in the papermaking process when the wet mat of fibers, fiber fragments, and fines is pressed in the thickness direction to remove water. The compression of the mat deforms the particles variably to an extent that depends on their conformability. For example, stiff fibers (typically of high fiber wall thickness) resist compression and lead to a low-density, relatively open paper structure. Later in the drying stage, surface tension forces of water still remaining in the paper may pull flexible fines particles to contact with fiber surfaces. Depending on the fines properties, these contraction forces may even cause densification of the network "backbone" formed by fibers (Sirviö et al., 2003).

The conformation of fibers may happen in many ways, such as the shearing and twisting of fibers and collapse of the open fiber lumens. In contrast, the pure axial

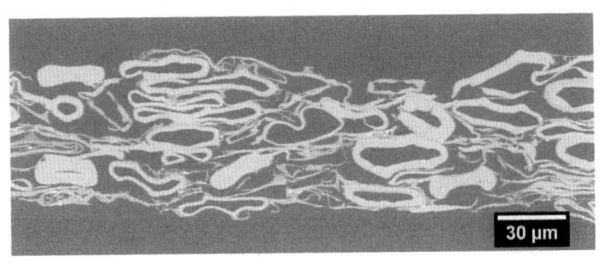

Fig. 11.3 Cross-section image of a paper made of mechanical pulp, courtesy of Gary Chinga, PFI.

bending of fibers is relatively rare and insufficient to account for paper densification in pressing (He et al., 2003). This is presumably due to the fact that fiber width is of similar magnitude as a typical distance between fiber-to-fiber contact points and that fibers seldom cross at right angles. In summary, it is misleading to think of the paper densification process as one where narrow fibers are being bent between far separated contact points. It is thus also misleading to think that the measurement of the axial bending flexibility of wet fibers would suffice to quantify the complexity of fiber conformation during the densification of paper. Unfortunately, a representative measurement method for fiber conformability has not been presented.

11.2.3 Statistical geometry of real fiber networks

Next, we consider the statistical properties of the network structure of paper. The network structure that we are interested in primarily consists of fibers and is determined by the mean fiber dimensions. This is particularly appropriate for papers made of chemical pulp. In the case of mechanical pulps, the sizes of particles (fibers, fiber fragments, fines) vary much more, and the geometric paper structure depends on the particle size distribution and not solely on the mean values of particle dimensions. However, even in the case of mechanical pulp, the mechanical properties of paper depend primarily on the connected network formed by long fibers. They have many bonds between them and, therefore, can give paper stiffness and strength. When compared with long undamaged fibers, short fibers and fiber fragments have fewer bonds and, therefore, contribute less to the in-plane mechanical properties of paper, as is explained in depth in Section 11.3.2. Fiber fragments and small-size fines contribute primarily through their effects on inter-fiber bonding.

The fiber network geometry has two important aspects: first, the distribution and mean value of fiber segment lengths; and second, the distribution and mean value of inter-fiber bond areas. The variability in fiber segments and the variability in inter-fiber bonds have different origins. Fiber segment lengths are governed by the random fiber network, in analogy to the two-dimensional case in Figure 11.2. The same is true for the number of bonds but *not* for the areas of individual bonds. The latter should depend on the surface structure of fibers and the bonding properties of eventual fines and fiber fragments at the inter-fiber bonding sites. Thus, the characterization of inter-fiber bonding is essentially an empirical task of detecting the frequency of different factors that influence the area (and mechanical properties!) of individual inter-fiber bonds. Our discussion is focused in the general statistical geometry of the fiber segments and not in the areas of inter-fiber bonds.

We consider first the network structure in the thickness direction of paper, and then proceed to the in-plane structure. In the end, we come up with an expression for the mean in-plane distance from one fiber to the next and argue that this mean distance is equal to the mean value of "*free* segment lengths" l_{free}. Here the term *free* refers to the part of the total fiber segment that is not bonded with other fibers (Fig. 11.4)

In order to define the network geometry, assume that a scanning line is drawn in the thickness direction of a cross-section of paper, such as Figure 11.3. The line crosses fiber cross-sections and open spaces between them. Obviously, the "heights" h_i of the inter-fiber spaces give some measure for the porous paper structure. Notice that there are no distinguishable convex "pores" in paper (aside from the lumen

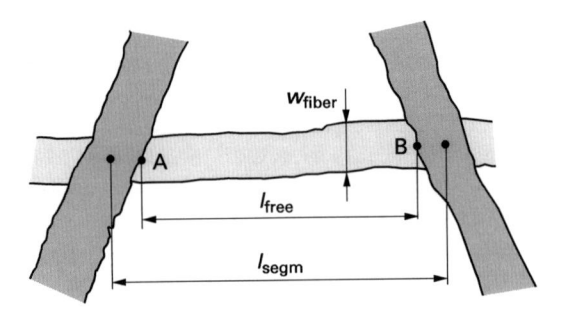

Fig. 11.4 Definition of the free segment length l_{free} or the distance between points A and B. Here it is assumed that the crossing fibers are bonded to each other over the entire overlap area. The total length of the fiber segment is l_{segm} and the width of the fibers is w_{fiber}.

inside uncollapsed fibers). The porous structure is hyperbolic. Every point of the inter-fiber pore space can be connected to any other point in the pore space with a line that never crosses a fiber. Therefore, we use the term *average pore space height* for the average value $\langle h \rangle$. (Throughout this chapter, the brackets $\langle \; \rangle$ denote average values that are calculated over the statistical sheet structure.)

The average number of fibers on a typical scanning line gives the average number n_{cov} of fibers that there are on any point of the sheet area. This number n_{cov} is called the *coverage* of the paper. It might be perceived as the number of fiber layers in the paper, but we prefer not to use that terminology because the fibers in paper are not arranged in well-defined layers. A typical value in paper is $n_{\text{cov}} = 10$, for example, in ordinary office printing paper.

Next, we define the pore volume fraction of the paper Φ, often called *porosity*, given by the total volume of pores divided by the paper volume. For this purpose we determine the total height of pores along a typical scanning line, $\sum h_i$. The average value of this is $(n_{\text{cov}} - 1) \cdot \langle h \rangle$ because there are $(n_{\text{cov}} - 1)$ inter-fiber spaces (some of them not empty). When we divide this by the average thickness of the paper, we obtain the pore volume fraction of the paper. Now comes an important point. The average "top-to-bottom" thickness of the paper would be $(n_{\text{cov}} - 1) \cdot \langle h \rangle + n_{\text{cov}} \cdot d_{\text{fiber}}$, where d_{fiber} is the mean thickness of fibers. If the top-to-bottom thickness was used to calculate the pore volume fraction, then the result would depend on the coverage of the paper. Pore volume fraction would be systematically lower at low coverages, even if the local network structure inside the paper would look the same as at high coverages. Therefore, we use the paper thickness minus d_{fiber} as the reference thickness and define the pore volume fraction as

$$\Phi = \frac{\sum h_i}{\sum h_i + (n_{\text{cov}} - 1) \cdot d_{\text{fiber}}} = \frac{\langle h \rangle}{\langle h \rangle + d_{\text{fiber}}}. \tag{11.1}$$

It follows that the mean pore space height is

$$\langle h \rangle = \frac{\Phi}{1 - \Phi} \cdot d_{\text{fiber}}. \tag{11.2}$$

A more detailed analysis (Niskanen et al., 2002) shows that the pore space heights h_i obey an exponential distribution consistent with the assumption that fiber positions are completely random and independent of each other. Minor deviations from the exponential distribution occur at both the low and high ends of the distribution. In the low end, deviations arise from the "exclusion domain" along fiber sides, where other fibers cannot enter because of limited conformability. This makes invalid the assumption of no correlations (complete randomness). In the high end, paper thickness puts an upper limit on the height of the inter-fiber pore space.

Since it applies in the thickness direction of the paper sheet, the assumption of uncorrelated random fiber positions should be reasonably good also along any in-plane line. Even though fiber flocculation and hydrodynamic disturbances induce inter-fiber correlations in the real papermaking process, they should have only small effects on the size distribution of pore space – otherwise those correlations would show up also in the thickness direction of paper. It follows that Equation 11.2 should apply also to the "horizontal" or in-plane pore space dimensions. If the in-plane orientation of fibers is ignored (i.e., that fibers do not cross the image plane at right angles), the mean pore space width would be given by

$$\langle x \rangle = \frac{\Phi}{1-\Phi} \cdot w_{\text{fiber}}. \tag{11.3}$$

Here w_{fiber} is the mean width of fibers. In fact, the mean width should be larger because in the cross-sectional image the apparent width of fibers is larger than w_{fiber}. However, Figure 11.5 demonstrates that measured pore space widths are clearly smaller than those calculated from Equation 11.3.

The plausible reason for the disagreement is shown in Figure 11.6. As discussed by He et al. (2003), fibers are usually not "lying flat" parallel to the sheet plane. Instead, fibers are often twisted or tilted out of the plane by some angles θ. Then the in-plane cross-section of a fiber can easily be smaller than the fiber width, if the fiber width is larger than the fiber thickness (Fig. 11.6). This is the case for thin-walled chemical pulp fibers and fiber wall fragments of mechanical pulp. In the thickness direction the effect of the fiber twist is smaller, which renders Equation 11.2 more precise.

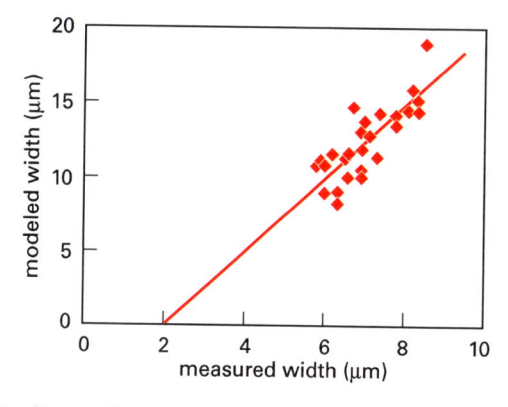

Fig. 11.5 Comparison of modeled (11.3) and measured pore width values in laboratory sheets made of different chemical pulps, Courtesy of Kari Niemi, KCL.

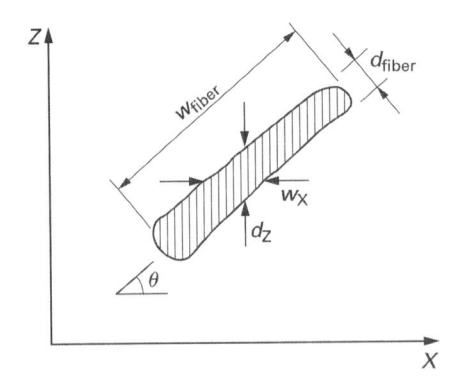

Fig. 11.6 Illustration how the twist of a fiber by angle θ decreases the in-plane cross-section $w_x(\theta)$ and increases the thickness-directional cross-section $d_z(\theta)$.

Equations 11.2 and 11.3 should thus be replaced with

$$\langle h \rangle = \frac{\Phi}{1-\Phi} \cdot \langle d_z(\theta) \rangle. \tag{11.2'}$$

and

$$\langle x \rangle = \frac{\Phi}{1-\Phi} \cdot \langle w_x(\theta) \rangle. \tag{11.3'}$$

Here $\langle d_z(\theta) \rangle$ and $\langle w_x(\theta) \rangle$ are the mean values of fiber cross-sections in the thickness and in-plane directions of the cross-sectional image (Fig. 11.6), calculated over the distributions of in-plane and out-of-plane orientations of the fiber axis. These mean cross-sections depend on the sheet pore volume fraction because the twisting of fibers is driven by the densification of paper structure. Another effect arises from the in-plane orientation distribution of fiber axes. For any fiber that crosses the image plane at an oblique in-plane angle, the in-plane cross-section w_x is larger than in Figure 11.6. Without a realistic calculation, the net effect of the twist angle and in-plane orientation angle cannot be known, but Figure 11.5 suggests that $\langle w_x(\theta) \rangle$ can be smaller than the fiber width w_{fiber}.

Previously, we argued that the mean width of pore space is equal to the mean length of free fiber segments. According to Equation 11.3', the mean length of free segments is thus linearly proportional to the in-plane projection of fiber cross-section and depends on the pore volume fraction Φ of paper. It is often more practical to use the density of paper ρ (mass per unit area and thickness of a sheet) instead of Φ. The two are related through

$$\Phi = 1 - \frac{\rho}{\rho_{fiber}}. \tag{11.4}$$

Here ρ_{fiber} is the mean mass of a fiber divided by the mean volume taken up by the fiber. This apparent fiber density ρ_{fiber} depends on the degree of fiber lumen collapse.

For fully collapsed lumens, ρ_{fiber} is equal to the fiber wall density, usually assumed to be 1550 kg/m³.

A remark on the measurement of the density of paper is in order. The apparent thickness of paper is measured with flat plates placed in contact with the sample surfaces. Given the irregularity of paper structure, the resulting apparent density gives a lower bound value for the density of paper. A higher value is obtained if the thickness of a stack of paper samples is measured. In this case the artifact created by plate surfaces is smaller, and a smaller piling thickness is obtained. A truly representative effective thickness value and hence effective density would be obtained from cross-sectional images in the manner described in the derivation of Equation 11.1. Equation 11.4 relates porosity with the effective density.

If we then use Equation 11.3 for simplicity, we obtain the following expressions for the mean of the total fiber segment length l_{segm} and the length of the free fiber segment length l_{free} (defined in Fig. 11.4),

$$\langle l_{segm} \rangle = \frac{\rho_{fiber}}{\rho} \cdot \langle w_x(\theta) \rangle \propto \frac{\rho_{fiber}}{\rho} \cdot w_{fiber}, \tag{11.5}$$

$$\langle l_{free} \rangle = \left(\frac{\rho_{fiber}}{\rho} - 1\right) \cdot \langle w_x(\theta) \rangle \propto \left(\frac{\rho_{fiber}}{\rho} - 1\right) \cdot w_{fiber}. \tag{11.6}$$

The latter version of both equations ignores the effect of fiber twist. Figure 11.6 suggests that w_{fiber} may be 50% smaller than $\langle w_x(\theta) \rangle$. In spite of the underlying cross simplifications, these equations are useful in understanding paper structure. More accurate information can be obtained directly from cross-sectional images such as Figure 11.3, or from simulations of the network structure (Niskanen et al., 1997; Drolet and Uesaka, 2005).

The in-plane mechanical properties of paper are controlled by the structure of the fiber network and, more precisely, by the backbone formed by intact fibers or long fiber sections. As the cross-section in Figure 11.3 shows, the paper structure may contain also thin fiber wall fragments and fines particles. These influence the geometric network structure and, for example, the optical and permeability properties of paper; however, they have at most a minor contribution to the stiffness of the fibrous backbone. Therefore, when considering the in-plane mechanical properties of paper, the nonfiber particles are to be excluded from the fiber segment length evaluation, and thus, the fiber width w_{fiber} in Equations 11.5 and 11.6 excludes the width of thin fiber wall fragments and fines particles. The situation would be different if one were interested in, for example, the air permeability of paper.

11.2.4 Key structural factors when engineering the mechanical properties of paper

The previous discussion identified the key factors to consider when altering the pore geometry of paper. The size distribution of the pore space depends on the density (or pore volume fraction) of paper and on the mean cross-sectional dimensions of intact fibers and long fiber fragments. The relevant cross-sectional dimensions are those prevailing in the dry paper, including the effect of eventual lumen collapse. Lumen collapse is prevalent when fiber wall thickness is small.

In the papermaking process, the pore volume fraction of paper is generally reduced by the pressure pulses applied. The density may decrease by the machine-direction draws (Baum et al., 1984) and increase by the cross-machine-direction shrinkage that the paper web experiences. The densification is resisted by the stiffness of the wet fibers, and eased by the conformability of them. The precise meaning of the stiffness or conformability is unknown, but it is not the simple flexural flexibility of wet fibers. He et al. (2003)

Fines material and fiber fragments that are highly swollen in the wet state should increase the densification of paper structure during drying (Sirviö et al., 2003), but they may also resist the densification by wet pressing. Lignin-rich fines fractions probably act in the opposite manner. In mixtures of different fibers, the pore fraction, density, or thickness of paper seldom vary in linear relation to the mixing ratio. The expected nonlinearity with different mixtures would be relatively easy to estimate using numerical network simulations.

11.3 Elastic modulus

As shown in the preceding chapters, elastic modulus is the central mechanical property of paper. The following discussion of the fiber network mechanics demonstrates how the papermaking process influences the elastic properties of individual fibers. In this respect, paper is clearly different from fiber reinforced composites and other materials where inert fibers are bonded together to form a network. It is not possible to understand how the papermaking process affects the elastic properties of paper without understanding how the fiber properties change.

11.3.1 The effect of paper density

Observations show that the elastic modulus of paper is strongly influenced by the density of paper. In Figure 11.7, the different values of density were achieved by varying the pressure applied on wet chemical pulp sheets in laboratory. The density of chemical pulp can also be increased by increasing the conformability of fibers using the mechanical treatment called refining. At a given density, the elastic modulus of paper is higher when more refining is used. The behavior shown in these figures is typical. For more examples, see Chapter 5 in Paper Physics (Niskanen, 2008).

With mechanical pulps, the effect of wet pressure on elastic modulus is similar, though often weaker than with chemical pulps. In contrast, the effect of refining is very different because refining usually increases fiber cell wall damage and fiber fragmentation in mechanical pulps. Therefore, paper density increases in part because the portion of small fiber fragments and fines increases, and these smaller particles can fit into the inter-fiber pore spaces. Thus, the connection between fiber network structure and paper density is different in chemical pulps and mechanical pulps, which is reflected in the elastic modulus of paper at a given density.

One should notice that if the thickness of paper were increased without changing any fiber properties, then the elastic modulus E (with the units GPa) should be exactly proportional to density ρ, or E/ρ = constant, because the only thing that

11.3 Elastic modulus

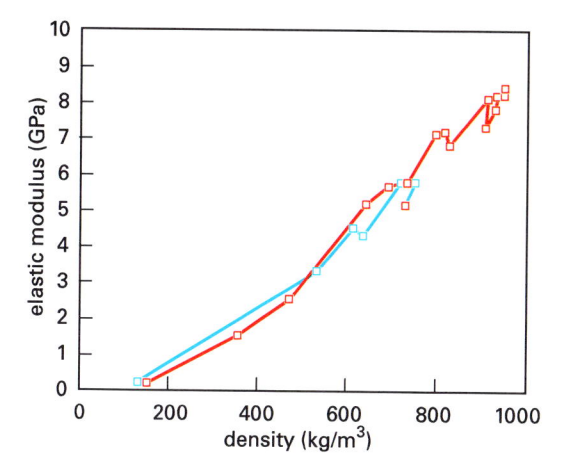

Fig. 11.7 Elastic modulus against density using data from Alexander and Marton (1968). The density of laboratory sheets was increased by increasing the wet pressure. Shown are data for three different chemical pulps, each connected with a line.

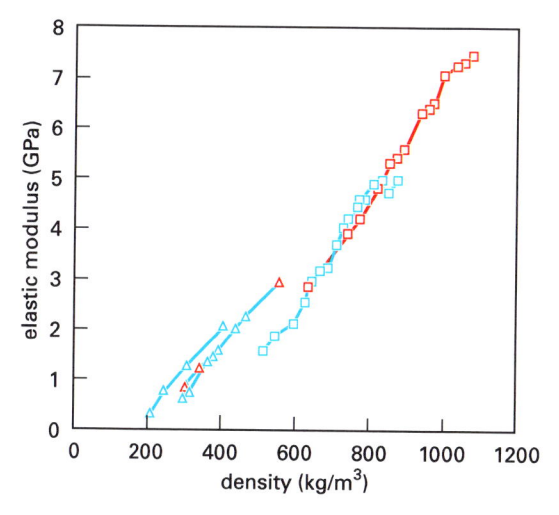

Fig. 11.8 Elastic modulus against density using data from Alexander and Marton (1968). The density of laboratory sheets was increased by increasing the refining. Shown are two different chemical pulps (squares and triangles) and several wet pressures, each refining series being connected with a line.

changed was paper thickness. The exactly same fibers would carry external load as before. The behavior seen in Figures 11.7 and 11.8 cannot be explained by this thickness effect alone.

We will now focus on the elastic modulus of paper made of chemical pulp fibers. The main purpose is to clarify the underlying microscopic mechanisms, not to argue that the models as such are complete or precise. It would, however, be straightforward to obtain more detailed numerical predictions using fiber network simulations.

11.3.2 The shear-lag mechanism

A common approach to model the elastic modulus of paper is to draw analogy from the shear-lag model of short fiber composites, see Chapter 12.3.4. One considers a characteristic "test fiber" that is aligned parallel to the externally applied stress. The rest of the fiber network is treated as the homogeneous matrix medium of a composite. The axial tensile stress of the test fiber is then constant in the middle of the fiber and goes to zero at fiber ends, over a critical length l_{crit} (see Fig. 12.4). When applied to paper with an isotropic in-plane fiber orientation distribution (Page, et al., 1979), the shear-lag model gives

$$E = \frac{1}{3} \cdot E_{fiber} \cdot \frac{\rho}{\rho_{fiber}} \cdot \left[1 - \frac{l_{crit}}{l_{fiber}}\right] = \frac{1}{3} \cdot E_{fiber} \cdot \frac{\rho}{\rho_{fiber}} \cdot \left[1 - \frac{w_{fiber}}{l_{fiber} \cdot RBA} \sqrt{E_{fiber}/2G_{bond}}\right]. \quad (11.7)$$

Here E_{fiber} is the elastic modulus, w_{fiber} the width, and l_{fiber} the length of fibers; and G_{bond} is the shear modulus of inter-fiber bonds. Isotropic orientation distribution gives rise to the pre-factor 1/3. RBA stands for a *relative bonded area*, the fraction of total fiber surface area that is bonded to other fiber surfaces. Directly measured values for RBA are seldom available when one uses Equation 11.7, in spite of opposite claims that are common in the paper technology literature (cf. Chapter 5 in Niskanen, 2008). Instead, RBA can be estimated from cross-sectional images of the type in Figure 11.3 and connected to the pore size distribution (Niskanen et al., 2002), leading to

$$RBA = \frac{a}{\langle h \rangle} \cdot \left[1 - \frac{1}{n_{cov}}\left(1 - e^{n_{cov}}\right)\right] \approx a \cdot \frac{\rho}{\rho_{fiber}} \propto \frac{w_{fiber}}{\langle l_{segm}\rangle}. \quad (11.8)$$

Here a is the measurement resolution (pixel size), and the coverage $n_{cov} = 10$ in typical paper sheets. The last equality follows from Equation 11.5. It then follows that, in paper, the critical length l_{crit} and elastic modulus can be represented as

$$l_{crit} = \lambda \cdot \frac{\rho_{fiber}}{\rho} \cdot w_{fiber} \quad (11.9)$$

and

$$E = \frac{1}{3} \cdot E_{fiber} \cdot \left(\frac{\rho}{\rho_{fiber}} - \lambda \cdot \frac{w_{fiber}}{l_{fiber}}\right). \quad (11.10)$$

The parameter λ depends on the relation of bond and fiber stiffness, so that stiff bonds give small λ. In geometric terms, λ can be interpreted as the number of fiber segments that, in effect, carry no load.

The shear-lag equation (Eq. 11.10) seems to be valid for chemical pulp sheets of low basis weight but not for chemical pulp sheets of ordinary basis weights. This is demonstrated in Figures 11.9 and 11.10 where the density dependence of elastic modulus is plotted for pulps of different fiber length. In both cases, fiber length was reduced by manually cutting fibers. If the shear-lag mechanism was correct, then the apparent value of λ obtained by fitting Equation 11.10 to the data should not be influenced by fiber length. In the two figures, this is the case if basis weight does not exceed 20 g/m², but not if basis weight is 60 g/m², a value typical of ordinary paper.

The fits made to the shear-lag model in Figures 11.9 and 11.10 imply that the critical length is $l_{crit} / l_{fiber} \approx 0.55$ for the shorter fibers in both cases. In other words, less than half of the fiber length would carry load effectively in both cases. While this is quite plausible at the low basis weights, with only a few inter-fiber bonds per fiber (the sheets are close to the percolation threshold), it is not plausible at ordinary basis weights with lots of bonds per fiber. For example, in Figure 11.5 the free segment length is 6–8 µm, while fiber width in that case is 40 µm. The same order should apply in Figure 11.10,

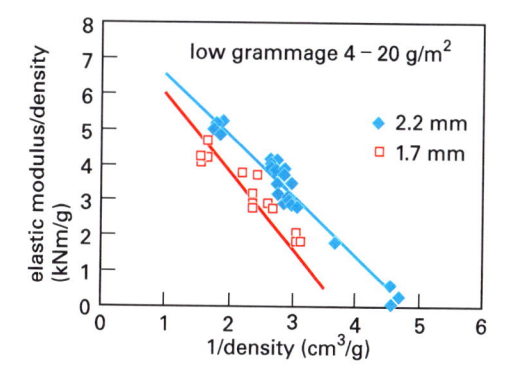

Fig. 11.9 Elastic modulus over density, E/ρ against inverse density $1/\rho$ using data of Hollmark et al. (1978) for thin laboratory sheets (basis weights 4 to 20 g/m²) made of chemical pulp fibers of two mean lengths, 2.2 mm and 1.7 mm. The apparent density increased with basis weight. The lines are calculated from (11.10) keeping constant all parameters except fiber length.

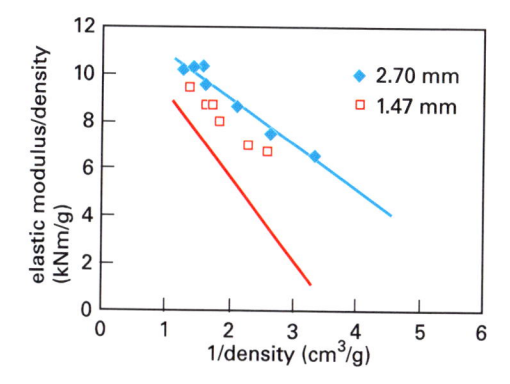

Fig. 11.10 Elastic modulus over density, E/ρ against inverse density $1/\rho$ using data of Seth (1990) for ordinary 60 g/m² laboratory sheets made of chemical pulp fibers of two mean fiber lengths, 2.7 mm and 1.47 mm. The apparent density was increased by increasing the pressure applied on the sheets. The lines are calculated from (11.10) keeping constant all parameters except fiber length.

implying that the shorter fibers should have on the average $1.47/0.05 \approx 30$ bonds. If indeed l_{crit}/l_{fiber} were 0.55 in this case, then the inter-fiber bonds would have 5–10 times smaller shear modulus G_{bond} than the bonds in the low-basis weight sheets. That is not possible.

We conclude that the observed density dependence of the elastic modulus in papers made of chemical pulp cannot, in general, arise from the shear-lag mechanism; something else is needed to explain the experimental observations. The shear-lag mechanism can suffice at unusually low basis weights and at low densities because, in these cases, the number of bonds per fiber is low and therefore the effectiveness of stress transfer through the bonds makes a large contribution to the elastic modulus of paper. The case of low density or low bonding degree may also apply in papers made of mechanical pulp.

11.3.3 The activation mechanism

At least with chemical pulps, the most plausible mechanism that causes the increase of the elastic modulus with increasing density is the "activation" of fibers and fiber segments during the drying of paper (Giertz, 1964). Activation means that the axial elastic modulus of a typical fiber or a typical fiber segment is higher if it has been held under tension during drying. The existence and significance of this mechanism has been demonstrated in many experiments of drying single fibers (Jentzen, 1964), measuring fibers cut loose from a paper sheet (Wuu et al., 1991), and testing the effect of drying tension on paper sheets (Htun and de Ruvo, 1978).

Given the helical fibril structure of wood fibers (see Chapter 12), it is easy to accept that tension applied during drying can increase the axial alignment of the fibrils, which in turn leads to higher axial modulus. One can also follow the drying of single fibers under a microscope to see how they twist and bend if no tension is applied. In a paper sheet, the tendency for such deformations leads to reduced elastic modulus unless drying shrinkage is prevented.

During drying, paper shrinks unless in-plane tension is applied, as explained in Chapter 9. The shrinkage is driven above all by the shrinkage of the fiber wall in the transverse directions. In the axial direction, the fiber wall shrinks much less. If paper shrinkage during drying is not prevented, the transverse fiber shrinkage and axial fiber twisting and bending leads to a "slack" paper structure in the dry state because the fiber segments are deformed. When stretched, such a network has low elastic modulus. External elongation primarily straightens fiber segments but does not create axial tension in them. On the other hand, if paper shrinkage is prevented, a tightened fiber network results because fiber shrinkage is prevented. Any external elongation of the dry paper immediately creates axial tension in fiber segments. This mechanism is the activation of fiber segments (Giertz, 1964).

The drying shrinkage tendency of paper increases with increasing density and with the mechanical refining of fibers, which, especially in chemical pulps, opens up the wet fiber wall structure and increases the transverse fiber shrinkage during drying. With mechanical pulps, a larger part of the increase in the shrinkage tendency of paper arises from swollen fines material that collapses during drying and pulls fibers closer together. The collapse of fines may influence the activation of fibers in a manner that differs from the transverse shrinkage of fibers.

The shear-lag model discussed above ignores the effect of fiber activation. It, therefore, cannot account for the effects that drying tension, or the lack thereof, has on the elastic modulus of paper. On the other hand, if the drying shrinkage tendency of wet paper is small, then the shear-lag mechanism may alone suffice to explain the elastic modulus of paper. This can be the case with mechanical pulps or with stiff chemical pulp fibers that give low density to paper.

11.3.4 Elastic modulus of activated fiber network

In this section we show how to analyze quantitatively fiber activation and, thereby, show what effects different fibers and papermaking conditions have on the elastic modulus of paper. The approach includes the shear-lag effects. An earlier version of the model was presented by Sirviö et al. (2008).

We focus on the axial elastic stiffness of fiber segments. This is a practical simplification. In reality, fibers are three-dimensional objects that may carry stresses also in the transverse directions, but the complete description of this would be far beyond our needs. We thus consider a system of spring-like fiber segments. The axial elastic moduli of the segments vary because of different degrees of activation.

Consider the typical fiber segment in Figure 11.11. During drying, fibers shrink mainly in the transverse direction. Therefore, the free length of the fiber segment changes in drying because of two factors: first, the crossing fibers shrink in the transverse direction, say by δ_{fiber} (<0); and second, the whole sheet is strained or allowed to shrink, say by δ (>0 or <0, respectively). If δ_{slack} (>0) is the minimum stretch needed to straighten the segment, then activation happens if $l_{free} \cdot \delta_{slack} < l_{segm} \cdot \delta - w_{fiber} \cdot \delta_{fiber}$, or

$$l_{free} \leq l_{act} = \frac{\delta_{fiber} - \delta}{\delta - \delta_{slack}} \cdot w_{fiber} \equiv \alpha_{act} \, w_{fiber}. \tag{11.11}$$

Here the activation parameter α_{act} compares the transverse shrinkage of fibers to the fiber "slack" that remains after accounting for the macroscopic drying strain of the paper.

As discussed previously, the probability distribution of free segment lengths l_{free} can be assumed to be exponential with its mean value given by Equation 11.6.

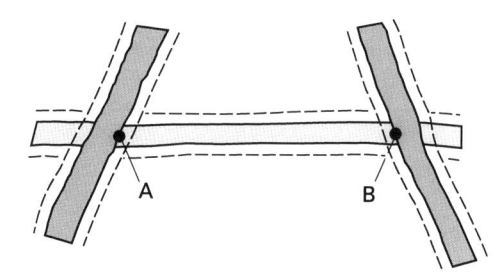

Fig. 11.11 Schematic mechanism of fiber segment activation when paper shrinkage is prevented. As in Figure 11.4, the points A and B define the fiber segment l_{free}. It is strained when the crossing fibers shrink in the transverse direction. The initial wet-state width of the fibers is show by the dashed lines.

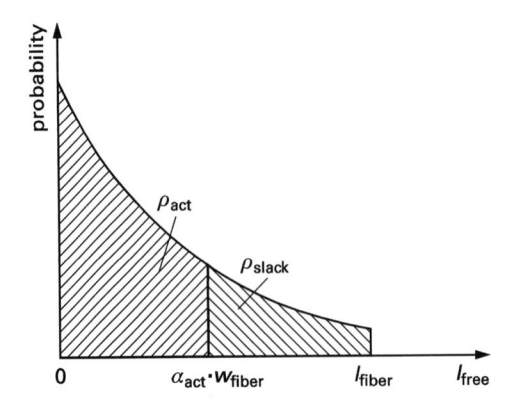

Fig. 11.12 The length distribution of free fiber segments is exponential, with a cut-off at fiber length. The segments shorter than $\alpha_{act} \cdot w_{fiber}$ are active in drying and the rest remain inactive or slack.

Therefore, it is easy to calculate the fraction of fiber segments that are active (Fig. 11.12):

$$p_{act} = \left[1 - \exp\left(-\alpha_{act} \cdot \frac{\rho}{\rho_{fiber} - \rho}\right)\right] \cdot \left[1 - \exp\left(-\frac{l_{fiber}}{w_{fiber}} \cdot \frac{\rho}{\rho_{fiber} - \rho}\right)\right]^{-1}$$

$$\approx \left[1 - \exp\left(-\alpha_{act} \cdot \frac{\rho}{\rho_{fiber} - \rho}\right)\right]. \qquad (11.12)$$

This equation captures the essentials of fiber activation. Fiber segments are all active if the activation parameter $\alpha_{act} \geq l_{fiber}/w_{fiber}$ and all inactive if $\alpha_{act} = 0$. The approximation applies in the case of typical paper grades made of chemical pulp, where paper density $\rho = 500\text{--}700$ kg/m^3 or one-third to one-half of fiber (wall) density $\rho_{fiber} = 1550$ kg/m^3, and fiber length is much higher than fiber width, $w_{fiber} \gg l_{fiber}$.

Up to now we have ignored the shear lag that always occurs at fiber ends. Given the nonuniform structure of paper, it is even possible that a single fiber may contain several disconnected sections that are active and carry load, but separated by sections that are slack and carry little or no load. The critical length l_{crit} introduced previously should describe even the stress variation for each of these loaded sections. Thus, at least λ active fiber segments are needed in a section before load in the middle of the section can reach the full level (cf. Eq. 11.10). Then the fraction of fiber segments that are unloaded is (Fig. 11.13)

$$p_{unloaded} = \left[1 - p_{act}^{\lambda}\right] \cdot \left[1 - p_{act} \frac{l_{fiber}}{w_{fiber}} \cdot \frac{\rho}{\rho_{fiber}}\right]^{-1} \approx 1 - \left[1 - \exp\left(-\alpha_{act} \cdot \frac{\rho}{\rho_{fiber} - \rho}\right)\right]^{\lambda} \qquad (11.13)$$

The final approximation again applies in the case of typical paper grades made of chemical pulp.

Thus, we know also the fraction of fiber segments $p_{loaded} = 1 - p_{unloaded}$ that become active during drying and therefore carry most of the load when the dry paper is

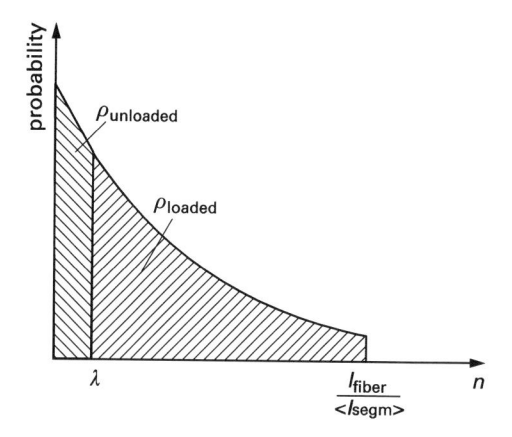

Fig. 11.13 The probability of having exactly n active fiber segments in a section is proportional to $p_{act}{}^n$ and has a cut-off at $n = l_{fiber}/<l_{segm}>$, that is, at the number of segments per fiber. If $n \leq \lambda$, then the chain of active segments is still not fully loaded.

stretched. It is reasonable to assume that in dry paper all active fiber segments carry equal load. At the very least, one may argue that this is the case in the end of drying because highest local stresses relax fast during drying. Finally, we replace the real continuous range of elastic moduli in the fiber segments with just two values, E_{act} for loaded and hence active segments, and $E_{unloaded}$ for unloaded segments. Then the elastic modulus of paper can be approximated by

$$E = \frac{1}{3} \cdot \frac{\rho}{\rho_{fiber}} \cdot \frac{E_{act}}{1 + p_{unloaded} \cdot (E_{act}/E_{unloaded} - 1)}. \tag{11.14}$$

Here we have ignored the fact that active fiber segments are shorter than inactive segments. Figure 11.14 shows the prediction obtained for increasing density through wet pressure. This behavior can be compared with the experimental data. For comparison with Figures 11.7 and 11.8 it is useful to observe that 1000 kg/m³ correspond to $\rho/\rho_{fiber} = 0.6$ if fiber density $\rho_{fiber} = 1550$ kg/m³.

At the lowest densities, perhaps below $\rho/\rho_{fiber} = 0.2$, the model is not realistic because it omits the fact that elastic modulus has a percolation threshold so that modulus goes to zero at some $\rho > 0$ (cf. Fig. 11.9). It is also clear that fiber segment activation cannot apply at high densities where the free segment length is similar to or smaller than fiber thickness (and therefore also smaller than fiber width). Under such circumstances one cannot ignore the transverse and shear elastic moduli of fibers. Page and Schulgasser (1989) have shown that one must then treat paper as a high-density laminate that consists of fiber wall segments and not fiber sections. The difference between the two arises from the nonzero microfibril angle of the fiber wall (cf. Chapter 12.2.1) and shows up in the anisotropy of stiffness tensor of paper.

The fiber parameters in Equations 11.11 through 11.14 are usually not readily available, and direct comparison with experiments is, therefore, not possible. However, if we use the approximation in Equation 11.13, then the elastic modulus of

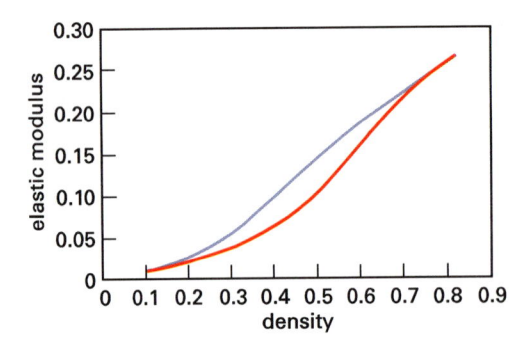

Fig. 11.14 Elastic modulus E/E_{act} against density ρ/ρ_{fiber} calculated from (11.11)–(11.13); with two values of $\alpha_{act} = 2$ and 4 (blue and red curve, respectively), $\lambda = 2$, $E_{unloaded}/E_{act} = 0.1$, and $w_{fiber}/l_{fiber} = 0.02$.

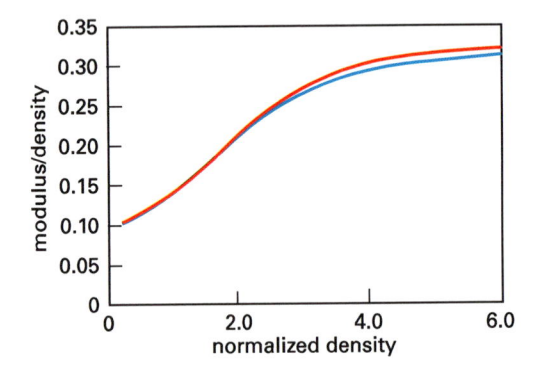

Fig. 11.15 The curves of Figure 11.14 represented so that $(E/E_{act}) \cdot (\rho_{fiber}/\rho)$ is plotted against the normalized density factor $\alpha_{act} \rho/(\rho_{fiber}-\rho)$.

paper depends on the specific combination $\alpha_{act} \rho/(\rho_{fiber} - \rho)$ of the activation parameter α_{act} and paper density ρ. Figure 11.15 shows how the curves in Figures 11.14 collapse almost completely on a single curve, if modulus on the y-axis is divided by density, and density on the x-axis is replaced with $\alpha_{act} \rho/(\rho_{fiber} - \rho)$. Equation 11.14 shows that the shape of this curve depends on the modulus ratio $E_{unloaded}/E_{act} = 0.1$ of inactive and active fiber segments, meaning that this aspect of the fiber properties cannot be simply scaled away. This may cause complications in the interpretation of experiments.

Consider now Figure 11.16, which shows the effect of wet pressing (i.e., changing only paper density) on the elastic stiffness E/ρ for different chemical pulp fibers. If we do the same transformation as we did from Figures 11.14 to Figure 11.15, then all the data can be made to collapse reasonably well on a single curve. The value of the activation parameter α_{act} was selected so that the collapse is as good as possible. We could then use these values of α_{act} to compare the activation potential of the different

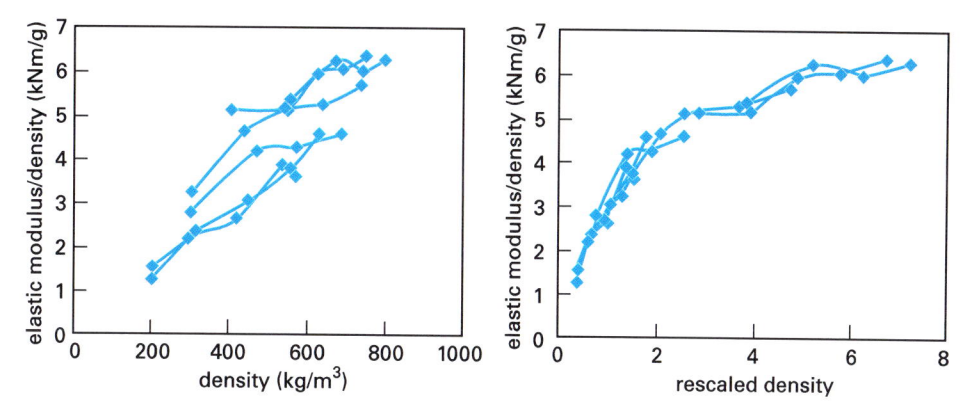

Fig. 11.16 Elastic modulus E/ρ against density on left and against rescaled density $\alpha_{act}\rho/(\rho_{fiber}-\rho)$, with $\rho_{fiber} = 1550$ kg/m³, on right, based on data of Luner et al. (1961). The density of the laboratory sheets made of different chemical pulps (each connected with a line) was changed by the pressure applied on wet sheets. The activation factor α_{act} was chosen differently for each pulp to maximize the data collapse.

pulps. The activation potential is dependent on the fiber shrinkage (controlled by swelling) and slackness (controlled by morphological factors) as prescribed in Equation 11.11. Even if in this particular case the modulus ratio $E_{unloaded}/E_{act}$ does not seem to vary between the pulps, the opposite can happen in other cases.

Machine-made ordinary paper grades have anisotropic elastic modulus and, thus, differ from the isotropic laboratory sheets that are used in research and displayed in the previous figures. The modulus of paper is usually two to four times larger in the running direction of the machine than in the transverse, cross-machine direction (see Chapter 2.5.2). The anisotropy is caused by two factors: fiber orientation and drying stresses. Because of the in-plane orientation distribution of fibers, the orientation factor in Equation 11.14 would be more than one-third in the machine direction and less than one-third in the cross-machine direction.

During drying on the paper machine, machine direction tension must be applied to prevent web fluttering. The transverse drying tension is zero at the web edges because the edges are free to contract. Adhesion to dryer cylinders accumulates into a nonzero drying tension in the middle of the web. The anisotropy in drying tensions leads to an orientation dependence of fiber activation so that fibers aligned in the machine direction have higher elastic modulus than fibers aligned in the cross-machine direction (Wuu et al., 1991). This effect has been shown to account for the shape of experimentally observed variations of elastic modulus across paper webs (Niskanen, 1990).

11.3.5 Key factors when engineering the elastic modulus of paper

The previous discussion identified the key factors influencing the elastic modulus of paper. In the case of typical chemical pulps and paper densities, Equations 11.11

through 11.14 give a qualitative account for the effect of paper structure and activation potential of fibers. In the case of relatively low densities or low activation potential of fibers, as may happen with mechanical pulps, the situation is not as clear because systematic experimental data is less abundant. However, it is probable that in low densities the shear-lag model, and hence the shear stiffness of inter-fiber bonds (cf. Eq. 11.7) can have an important effect on the elastic modulus of paper. Figures 11.9 and 11.16 suggest means of resolving what the dominant mechanism is. At very high densities, yet another mechanism arises as the elastic modulus becomes dominated by the fiber wall elastic properties, as discussed by Page and Schulgasser (1989).

The alternatives to alter the paper structure – more accurately the fiber network structure – and thereby the elastic modulus of paper were discussed in Section 11.2. In situations where fiber segment activation is the dominant factor, its influence is governed by the activation factor α_{act} (Eq. 11.11). Activation, and hence the elastic modulus, increases if paper is strained during drying ($\delta > 0$) and decreases if it is allowed to shrink ($\delta < 0$). The same happens if the drying shrinkage of paper can be maintained while the transverse drying shrinkage of fibers (quantified by $\delta_{fiber} < 0$) is increased. Fiber shrinkage is the opposite of fiber swelling, which can be increased by mechanical refining of the pulp. Fiber curl and other deviations from straightness reduce activation. This effect is purely geometric and represented by δ_{slack} (>0) in Equation 11.11. Also, fiber length (relative to fiber width) has a positive effect on the elastic modulus of paper (cf. Eqs. 11.12–11.13), but this effect is relatively small, as illustrated by Figure 11.10.

If wet fiber swelling or paper density is relatively low, then fiber segment activation is ineffective. The elastic modulus still increases with paper density but now according to the shear-lag mechanism (Eq. 11.7). In this case the important pulp properties are fiber length and inter-fiber bonding. Aside from the changes in geometry of fiber network (Section 11.2), the stiffness of individual inter-fiber bonds is enhanced by fines material and fiber fragments that are highly swollen in the wet state. Lignin-rich fines fractions seem to act in the opposite manner (Sirviö et al., 2003).

In the shear-lag case, fines and fiber fragments influence the elastic modulus of paper primarily through the shear stiffness of inter-fiber bonds. Equation 11.7 shows how the elastic modulus contribution of short fibers or fiber fragments decays with decreasing length. In cases dominated by fiber activation, the effect of fiber length is weaker. When considering mixtures of different fibers, always present in reality, the effects on the elastic modulus of paper can be reasonably well estimated by averaging the equations presented previously. The effects on paper structure are more difficult to estimate and often are possible only with numerical network simulations.

11.4 Stress–strain behavior, creep, and bond opening

This section considers what happens in the fiber network when paper is strained. Typical stress–strain curves of paper were illustrated in Chapter 2, and creep curves were shown in Chapter 7. One could argue that paper behavior is a representation of fiber behavior, and therefore, one should study single fiber properties in order to understand the behavior of paper. However, it is clear that the properties one might measure for single fibers cannot fully reflect what happens in paper. We believe that it would be more productive to focus on extracting fiber properties from laboratory sheet properties than on testing single fibers.

From the preceding discussion of the elastic modulus of paper we can infer that in cases where fiber segment activation is relevant, the stress–strain behavior of a fiber segment will depend on whether or not the segment is active. In addition to activation, the contribution to the macroscopic stress carried by paper depends on whether or not the segment is loaded. This latter effect is dominant in cases that are governed by the shear-lag mechanism, such as in some mechanical pulps.

Also, the creep response of fiber segments varies. The externally applied stress is taken up predominantly by the fiber segments that are loaded already at infinitesimal strains, such as in the end of paper drying and in the measurement of elastic modulus. As these fiber segments creep longer, the geometry of the fiber network changes. In cases where fiber segment activation is important, the result is that even inactive segments become loaded. Thus, the stress on the initially loaded segments decreases and the macroscopic creep rate decreases. If the activation mechanism is insignificant, only fiber ends are unloaded. Even though also in this case creep changes the network geometry, it is improbable that the number of load carrying segments would increase. In fact, it is possible that bonds at the fiber ends open and the critical length l_{crit} increases.

From a modeling perspective, both the stress–strain and creep behavior of paper are difficult because finite (noninfinitesimal) strains alter the network geometry. This creates spatial correlations in the distribution of local stresses and strains. For example, the stress acting on a fiber segment of low stiffness is "shielded" by neighboring segments, and as a result, local strains in the neighborhood are more uniform than the segment moduli alone would imply. The conclusion is that one cannot calculate the stress–strain curve or creep curve of paper by simply calculating an average of the corresponding curves for single fiber segments. Fiber network simulations would be needed for that purpose.

Our discussion of the stress–strain and creep behavior of paper is consequently limited to qualitative observations of the behavior of fiber segments. We start by considering what happens if the simple fiber segment in Figure 11.17a is strained. At small strains, the whole segment (including the bonded ends) elongates elastically. Then at some point, the local stress at the "bond corners" A and B exceeds some limit value, and the bond starts to open there. The opening process continues if the elongation is further increased and may eventually result in the situation shown in Figures 11.17b. The onset of the gradual bond opening process must depend on the activation of the fiber segment. In an inactive segment, any "slack" in the free segment part (between points A and B) must first be stretched out before bond opening can begin.

The gradual opening of inter-fiber bonds in paper straining has been detected in light scattering measurements (Sanborn, 1962). From the increase in the light scattering coefficient, it was inferred that the loss of inter-fiber bonding area was linearly proportional to the plastic strain of paper. Giertz and Rødland (1979) have also reported direct measurements where the strain in bonded fiber segments was generally larger than in the free segments. In summary, the plastic strain of paper follows, at least in part, from the gradual opening of inter-fiber bonds. Also, free fiber segments can naturally contribute to the plastic strain of paper. This is especially true

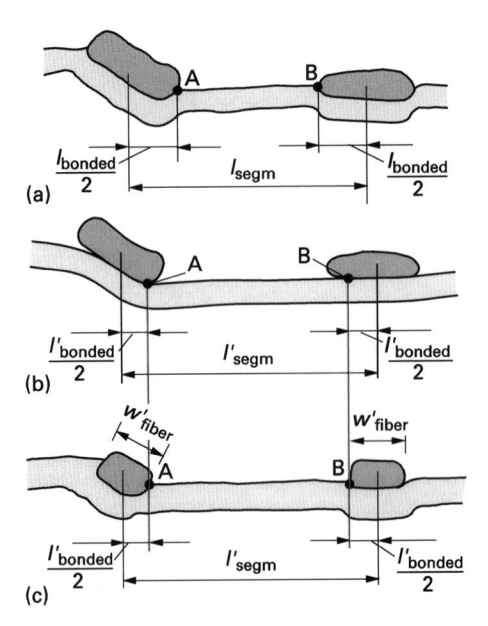

Fig. 11.17 Schematic drawing of the bond opening process. When the segment in (a) is strained, gradual bond opening leads to the situation in (b). The bonded segment length has decreased to l'_{bonded} and the free segment length increased to $l_{segm} - l'_{bonded}$. The same length values could also have been achieved without straining if the fiber width were smaller and segment length larger in the first place, $w'_{fiber} < w_{fiber}$ and $l'_{segm} > l_{segm}$ (c).

for inactive fiber segments that can be straightened and loaded when paper strain increases.

The elastic modulus of paper usually changes little when paper is strained to the plastic regime (cf. Chapter 2). The modulus may decrease by less than 10% or increase even less. At the microscopic level these small changes arise from at least two opposite effects. In highly activated cases (activation parameter α_{act} is high) and in shear-lag dominated cases (inter-fiber bonding is low), the contribution that crossing fibers make to segment stiffness decreases when bonds open gradually. Also, the critical length l_{crit} at fiber ends may increase. An opposite effect arises when initially unloaded segments start carrying load. This is seen as an increase in the elastic modulus of paper, especially when there is a lot of "slack" in fiber segments (i.e., small α_{act}).

The partial opening of bonds can start already during drying. For example, the light scattering coefficient of paper usually increases slightly when the drying shrinkage is decreased or the drying strain of paper is increased. This must be coupled with fiber segment activation because a bond should not begin to open so long as the fiber is slack. Simple model calculation (Salminen et al., 1996) shows that the transverse shrinkage of fibers leads to a parabolic distribution of inter-fiber stress within bonds, with maxima at the bond corners (points A and B, Fig. 11.17). If the peak value there becomes too large, then the bond area begins to open. The opening then continues as long as the transverse shrinkage continues, while the peak stress at the bond edge stays constant.

We conclude that in the end of drying, the local stress at the bond corners (A and B in Fig. 11.17) should be equal (or at least similar) in all active fiber segments. This threshold value of the local stress value is presumably related to the sub-micrometer morphology of fiber surfaces (see Chapter 2 in Niskanen, 2008) and the eventual presence of a gel of colloidal and polymeric material. The local stresses at bond corners are balanced by axial compressive stresses inside the bonded fiber segment and any macroscopic drying stresses that are applied on the paper in the end of drying. The stresses at bond corners give a microscopic representation for the drying stress and for the phenomenological internal stress that is sometimes used to characterize paper rheology (cf. Chapter 9). Internal stress is determined, in simple terms, as the minimum stress that must be applied for stress relaxation to occur in paper.

If the surface chemistry and morphology of fibers is constant, then the configurations in Figures 11.17b and 11.17c should be equivalent if the bonded lengths l'_{bonded} and total segment lengths l'_{segm} are equal. Consider then the definition of the activation parameter α_{act} in Equation 11.11. When bond opening is allowed for, fiber width in the equation must be replaced with l_{bonded}. It follows that the bonded and free lengths of all active segments must be connected through $l_{free} = \alpha_{act} l_{bonded}$, or in other words

$$l_{bonded} - l_{segm} = \frac{w_{fiber}}{w_{fiber} + l_{act}} = \frac{1}{1+\alpha_{act}}. \qquad (11.15)$$

In inactive fiber segments l_{bonded} is equal to w_{fiber}. Equation 11.15 can be used to calculate an estimate for the bonding degree RBA of the network. Integration over the exponential distribution of free segment lengths gives

$$\text{RBA} = \frac{1}{1+\alpha_{act}} + \exp\left(-\alpha_{act} \cdot \frac{\rho}{\rho_{fiber} - \rho}\right) \cdot \left[\frac{\rho}{\rho_{fiber}} - \frac{1}{1+\alpha_{act}}\right]. \qquad (11.16)$$

At low densities, this gives the same result as Equation 11.8, RBA $\to \rho/\rho_{fiber}$, while at high densities RBA becomes constant at $1/(1+\alpha_{act})$. This saturation of RBA means that while the number of bonds per fiber increases with increasing density, the average area of these bonds (proportional to l_{bonded}) decreases, and these two effects exactly cancel each other at high densities. In reality, the situation may be different because many microscopic effects may play a role. Nevertheless, the saturation of RBA explains why z-directional strength properties of paper (such as the z-directional tensile strength) do not always develop in linear proportion to increased paper density.

Equations 11.15 and 11.16 connect fiber segment activation with the reduction of the average area of an inter-fiber bond. This has implications on the stress–strain curve and creep curves of paper, even if we cannot calculate the effect exactly. The reduction in l_{bonded} implies that the tensile breaking strain of paper decreases with increased activation as well. This can be concluded from the fact that plastic straining of paper drives open inter-fiber bonds so that finally paper fails. Small values of l_{bonded} or RBA mean that less plastic strain is needed for this to happen.

In parallel with decreasing breaking strain, the elastic modulus of paper increases as fiber segment activation increases. This means that the tensile strength of paper may – in principle – increase or decrease with increasing fiber segment activation, depending on how fast the modulus increases compared to the decrease in breaking strain. In reality, tensile strength usually increases with increasing activation. In compressive loading, failure is caused by buckling of free fiber segments. The buckling threshold decreases with increasing segment length. Thus, increased activation can *reduce* the in-plane compressive failure strain of paper or paperboard.

11.5 Fracture process in the fiber network

11.5.1 Microscopic observations

The microscopic fracture process of paper was introduced in Chapter 5. When a crack propagates through paper, microscopic *damage* arises from inter-fiber bond rupture and the related fiber pull-out, or, alternatively, fiber breakage. These are the energy consuming processes in the fracture of paper. When a lot of fracture energy is consumed at the microscopic level to open bonds or break fibers, paper has generally high fracture toughness.

The damage corresponding to the opening of bonds is evident in the light scattering coefficient of paper. A silicone impregnation method can be used to make visible the spatial distribution of damage (Korteoja et al., 1996). The silicone reduces the reflectivity of paper so much that broken inter-fiber bonds can be detected from the light reflected by the newly revealed fiber surfaces. The silicone used forms a low modulus rubber that does not alter the mechanical properties of paper, except when the fiber volume fraction is very low. Damage can also be detected through acoustic emission, as discussed in Chapter 5.

Figure 11.18 shows a crack propagating through a silicone impregnated sheet. This example contains 90% of mechanical pulp and 10% chemical pulp fibers. The apparent fracture process zone (FPZ) is otherwise fairly uniform but extends far out along some long and presumably stiff fibers. Comparison with Figure 5.11 demonstrates that in pure chemical fiber sheets the crack tip is more homogeneous and wider than in sheets containing mechanical pulp. Note that the microscopically seen

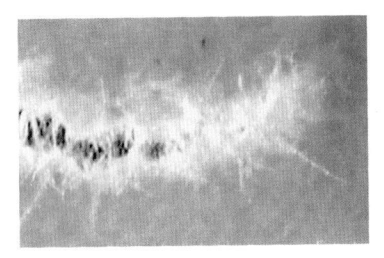

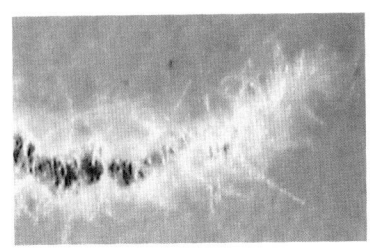

Fig. 11.18 Two successive images of fracture process in a silicone impregnated laboratory sheet of a mechanical pulp plus 10% chemical pulp fibers (Niskanen et al., 2001). The crack propagates from left to right. The images are taken against a black background that shows through the crack opening. The white color arises from broken inter-fiber bonds and fiber fractures that reflect light. The white fracture process zone is roughly 1 mm wide. Reproduced with permission from The Pulp and Paper Fundamental Research Society (www.ppfrs.org).

fracture process zone may be different from the phenomenological fracture process zone that is used in continuum fracture mechanics models.

The fracture process zone (FPZ) in Figure 11.18 shows where inter-fiber bonds have ruptured. A method has been developed to measure the mean width of the apparent FPZ following an in-plane tear test (Kettunen et al., 2000a). The result is called damage width w_{dama}. Damage width adds up the width of FPZ on both sides of the crack line. It is also equivalent to the diameter of the white circle in Figure 5.11.

In general, the damage width increases if fiber length increases, just as one might expect (Kettunen et al., 2000c). The ratio w_{dama}/l_{fiber} can vary significantly depending on, above all, fiber strength and fiber segment activation. If fiber strength is low, then many fibers break during the fracture process. This shows up indirectly as a reduction in damage width (Fig. 11.19). In addition to the increasingly narrow fracture process zone, the length of protruding fiber ends decreases as well. In the fracture process, these fiber ends have been "pulled out" from the paper structure on the opposite side of the fracture line. As fiber strength is reduced, fewer and shorter fiber ends survive extraction from the paper structure. It is then reasonable that damage width decreases because it represents the region from which fibers have been pulled out.

The effect of increased fiber segment activation as a result of the mechanical refining of a chemical pulp is shown in Figure 11.20. At a low level of refining, w_{dama} is large compared to l_{fiber}, and large variability in w_{dama} is observed along the fracture line. One can see in Figure 11.20a that many fibers extend out from the fracture line. However, quite a few fibers may fail even in this case, at least judging from the relatively sharp appearance of the fracture line.

11.5.2 Micromechanical description of the fracture process

Chapter 5 presented several models that can be used to evaluate how much the strength of a paper product or test piece is reduced by a defect of a certain size. Best accuracy is obtained with relatively complex models whose parameters are

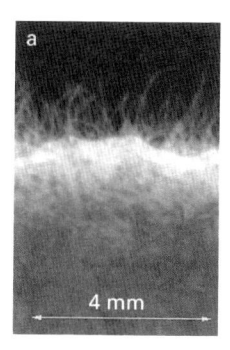

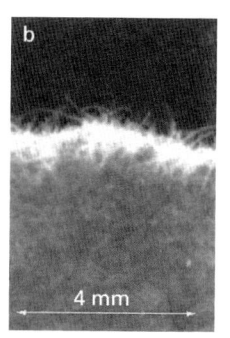

Fig. 11.19 Fracture process zones in laboratory sheets, where fiber strength was reduced by an acid treatment (Kettunen et al., 2000b). The reduction in fiber strength was characterized by a "zero-span tensile strength" values 145 Nm/g (a, untreated paper), and 97 Nm/g (b). These strength values are measured for a nominal zero open span length, so that the result should primarily depend on the tensile strength of fibers and not of the paper. Reproduced with permission from Pulp and Paper Technical Association of Canada (PAPTAC).

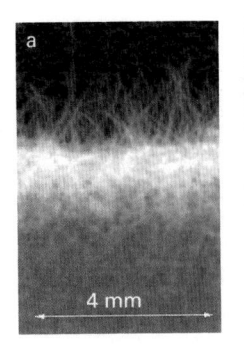

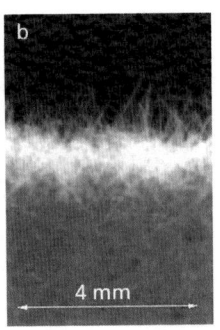

Fig. 11.20 Fracture process zones in laboratory sheets, where density was increased by fiber refining (Kettunen et al., 2000b). The paper densities were 563 kg/m^3 (a), and 681 kg/m^3 (b). Reproduced with permission from Pulp and Paper Technical Association of Canada (PAPTAC).

determined from experiments. In this chapter, we limit our discussion on simple approximations in order to be able to present evidence about the micromechanical effects of fiber properties and paper structure. The energy consumption during the fracture process is such a qualitative measure. Quite a lot of experimental data have been published for the total fracture energy, J-integral, and the area under the cohesive stress curve. Although marred by experimental complications, these measures of the energy consumption give, in general, similar values. Large fracture energy consumption is an indication of the toughness of paper.

Comparing different papers, measured values of the total fracture energy consumption generally follow damage width (Fig. 11.21). However, at low levels of activation, fracture energy is low even though damage width is high. In these cases, inter-fiber bonding is weak, at least when measured as the ZD tensile energy absorption (called "Scott bond" in the paper technology literature). Indeed, if the fracture energy values in Figure 11.21 were normalized with Scott bond values, then data on fracture energy vs. damage width often collapse on a single line (Niskanen et al., 2001). These observations suggest that the macroscopic fracture energy consumption per unit area of the apparent FPZ is accounted for by the energy consumed in inter-fiber bond failures (Tanaka et al., 2000). In contrast, fiber failures during the fracture process seem to primarily reduce the size of the FPZ but otherwise contribute little to the macroscopic fracture energy consumption.

Another easily interpreted macroscopic fracture property is the cohesive stress σ_{yy} that decays when a crack opening grows (Chapter 5.3.4). At the microscopic level in the fiber network, the decay of σ_{yy} arises from the breaking of bonds and fibers that reduces the elastic modulus of paper. Some examples of the σ_{yy}-curve were shown in Figure 2.11, but the shape of the curve has not been studied in detail. There are indications (Niskanen et al., 2001) that large values of the damage width w_{dama} imply that the cohesive stress decays slowly with the crack opening displacement. In other words, the cohesive stress–crack opening curve seems to have a long tail when the apparent FPZ is large.

The size of the fracture process zone gives a length scale that can be connected with the size of eventual defects in paper. Expressed in simple terms, one expects that

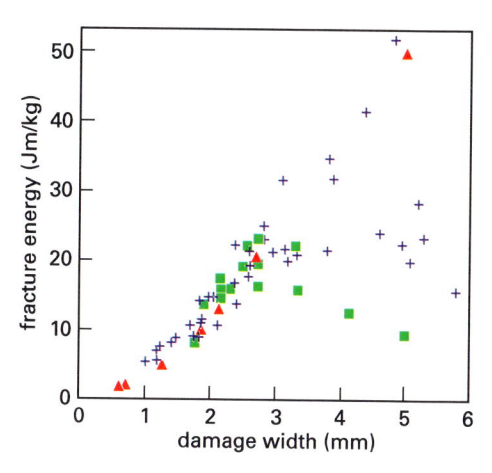

Fig. 11.21 Fracture energy against damage width for different pulp fiber mixtures (crosses), chemical pulp refining levels (squares), and fiber strength values (triangles). The in-plane fracture energy during stable crack propagation was measured using an in-plane tear method (Kettunen et al., 2000b).

a paper with a large FPZ can be expected to tolerate larger defects without failure than a paper with a small FPZ – provided that the fracture energy consumption per unit FPZ area is constant. Approaching the problem from the fiber network, it would be easy to accept that, keeping all other things constant, a paper made of long fibers would tolerate larger defects than a paper made of short fibers. As pointed out previously, long fibers give a generally large damage width w_{dama} (Kettunen, et al., 2000c), a measure for the apparent FPZ.

We conclude that a simplified characterization of the fracture properties of paper consists of two microscopic factors: the energy (per unit area of FPZ) consumed in inter-fiber bond failures and the size of the FPZ. Large values of the inter-fiber bonding energy and the damage width both contribute to the fracture toughness or defect tolerance of a paper product. Accurate quantitative evaluations of the failure load or critical defect size in any particular case can be made using models discussed in Chapter 5.

It is well-known what one can do in the papermaking process to improve the inter-fiber bonding energy. These include using particular bonding additives, such as starch; maximizing the positive effects of fiber surface fibrillation or fines material; as well as increasing the density of paper. The effects can be compared using some measure of ZD strength or fracture energy (Niskanen et al., 2005). However, the net effect on the fracture toughness is more complicated because the size of the fracture process zone can decrease when inter-fiber bonding is increased. In particular this happens if one increases fiber refining. The result is a higher inter-fiber bonding but smaller FPZ (see Fig. 11.20).

We conclude our discussion of the fracture process with a demonstration of a microscopic mechanism that gives a plausible explanation for the size reduction of FPZ. We focus on the damage width w_{dama} as a measure of the FPZ.

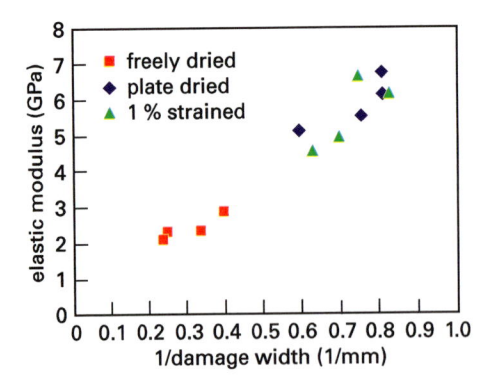

Fig. 11.22 Elastic modulus against the inverse damage width, l_{fiber}/w_{dama} (data from Sirviö et al., 2008) and comparing free drying shrinkage, drying against a steel plate, and 1% straining before constrained drying for different chemical pulp fibers.

If only the length of fibers is changed, damage width changes in linear proportion to the mean fiber length (Kettunen et al., 2000c). The ratio w_{dama}/l_{fiber} is usually larger than one, unless fibers are exceptionally weak. Often w_{dama}/l_{fiber} is close to two, but with unrefined chemical pulp w_{dama}/l_{fiber} can be as large as three so that the fracture process extends much further than just the span of fibers that cross the fracture line. On the other hand, with well-refined chemical pulps, w_{dama}/l_{fiber} can be close to one. These differences are presumably caused by the activation of fiber segments. For example, the elastic modulus of paper is typically small when damage width is clearly larger than fiber length (Fig. 11.22). In particular, the effect of drying shrinkage in the figure can only come from the activation of fiber segments.

The effect of fiber segment activation on the size of the fracture process zone can be qualitatively understood from the preceding sections. Suppose that a large fraction of fiber segments are inactive. Stress in the fracture process zone is then carried by a sparse "backbone" formed by the active fiber segments. The distance x from the crack trajectory to the point where the first bond failure occurs should be equal to the distance where the first active fiber segment is encountered, or (cf. Eq. 11.12)

$$x = -\frac{\langle l_{segm} \rangle}{\ln(1-p_{act})} \propto \frac{w_{fiber}\, \rho_{fiber}\,(\rho_{fiber}-\rho)}{\alpha_{act}\, \rho^2}. \tag{11.17}$$

Damage width must then be larger than this distance x. The equation demonstrates qualitatively how fiber segment activation influences the size of the fracture process zone.

It has been shown (Kettunen et al., 1999) that simple numerical simulation can reproduce experimentally observed distributions of bond failures against the distance from the fracture line and measured values of damage width. This may sound surprising, considering the stochastic nature of fracture processes. However, the damage width is an average material property of paper, just as the cohesive stress σ_{yy}. In the analysis one determines the average value of w_{dama} over long fracture lines and, therefore, obtains a very stable value. Thus it would be straightforward to study the effects of fiber segment activation in the same manner as in Kettunen et al. (1999).

11.6 Hygroexpansion

The hygroexpansivity of paper comes from the swelling or contraction of the fiber wall material when moisture content changes. In real uses of paper and board, problems arise from changes in the moisture content or intake of liquid water. Therefore, the dynamics of moisture-induced hygroexpansion and water-induced hydroexpansion are usually more important than the equilibrium hygroexpansion (see Chapter 9). With time-dependent changes, the local moisture content inside the fiber network is nonuniform. Fiber segments with higher-than-average moisture content expand more than average and experience compressive stresses (except when the paper is under external tension). The dynamic dimensional changes of paper are therefore dependent on both the creep behavior and the hygroexpansion or hydroexpansion of the fiber segments.

Our discussion in this section is limited to the equilibrium hygroexpansivity that would be observed with very slow changes of moisture content. Once again, numerical simulations would be needed to describe time-dependent dimensional changes of paper. As explained in Chapter 9, the hygroexpansivity of fibers is anisotropic. Over the range of relative humidity (RH) 0%–100% at room temperature, fibers expand roughly 1% in the longitudinal direction and 20% in the lateral direction. When one compares these numbers with the corresponding expansion of paper, it is clear that the transverse hygroexpansion of fibers makes a strong contribution to the hygroexpansion of paper. Uesaka (1994) described the mechanism that transfers the transverse fiber expansion to the macroscopic expansion fiber network. His expression for the hygroexpansion coefficient of paper (see Chapter 9) is

$$\beta = \beta_L^{fiber} + f_{21}\beta_T^{fiber}. \tag{11.18}$$

Here β_L^{fiber} and β_T^{fiber} are the longitudinal and transverse hygroexpansion coefficient of fibers, and f_{21} is a stress factor describing the transfer of stress from the network to the transverse direction in a single fiber. Typical values for fibers are $\beta_L^{fiber} = 0.03$ and $\beta_T^{fiber} = 0.6$.

In Equation 11.18, all effects of network structure and drying shrinkage are contained in the stress transfer factor f_{21} of inter-fiber bonds. If a typical inter-fiber bond has a simple "lap-joint" shape (Fig. 11.17b), stress transfer is very ineffective ($f_{21} \ll 1$), and the hygroexpansivity of paper is controlled by the longitudinal expansion of fibers. On the other hand, if the bonds have the "wrapped-around" shape (Fig. 11.17a), then a substantial fraction of the transverse hygroexpansion of fibers transfers to the macroscopic dimensions of paper (Uesaka and Qi, 1994). In his original work, Uesaka assumed that the hygroexpansivity is controlled by the nature of a typical inter-fiber bond, which in turn is controlled by the drying shrinkage of paper. In reality, the distribution of the various bonding geometries determines the mean value of the stress transfer factor f_{21}, and, thereby, the hygroexpansivity of paper.

The effect of bonding geometry connects the hygroexpansivity of paper with the activation of fiber segments. An example of this is shown in Figure 11.23. With free drying shrinkage, activation is low and most bonds are probably of the "wrapped-around" type (Fig. 11.17a). As the drying shrinkage is reduced, fiber activation increases and the prevalent bond type changes into the "lap-joint" (Fig. 11.17b).

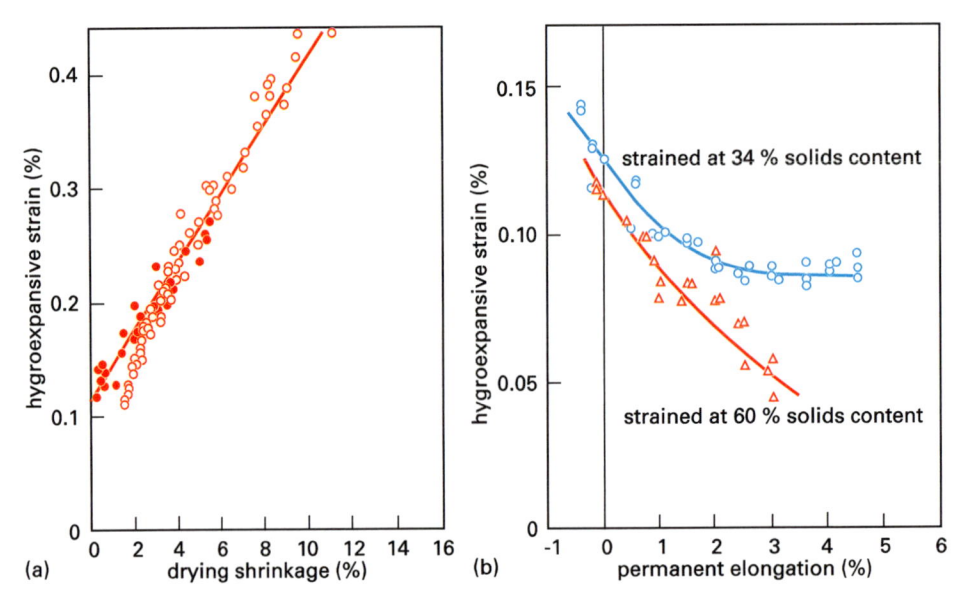

Fig. 11.23 Hygroexpansion of laboratory sheets between RH = 20% and 60% against drying shrinkage or the stretch during drying, redrawn after Nordman (1958). Reproduced with permission from TAPPI.

Corresponding reduction is observed in the hygroexpansivity of paper. With strong straining during drying, hygroexpansivity can be further reduced. The connection to fiber segment activation comes from Equation 11.11 that shows how the activation parameter α_{act} increases as the allowed paper shrinkage δ is decreased.

In principle, the hygroexpansion of paper should increase with increasing density because then there are more bonds per fiber where the transverse expansion of fibers can contribute to the paper expansion. However, this happens only if the geometry of the bonds – or the activation of fiber segments – does not change. However, as discussed in Section 11.4, inter-fiber bonds can gradually open during drying in cases dominated by fiber segment activation. If paper density is increased, this leads to a situation where the bonding degree RBA no longer increases with density (cf. Eq. 11.16). Experimental results on the effect of density on the hygroexpansion of paper in Figure 11.24 demonstrate how the hygroexpansion of paper increases with density when fiber segments are inactive (free drying shrinkage) but is constant or decreases slightly when fiber segments are active (restrained drying means shrinkage was prevented). The decrease of hygroexpansivity with density in activated sheets may well be consistent with our activation model, but that cannot be known in the absence of a detailed analysis of the stress transfer factor f_{21}.

The hygroexpansivity of paper, and more generally the moisture sensitivity, poses fundamental limitations to the use of paper as a material. One may benefit from the MD web tension on a paper machine that makes up a drying restraint and hence reduces the MD hygroexpansion caused by small changes of moisture (Fig. 11.24). However, this does not prevent the intrinsic dimensional instability of the papermaking fibers from reappearing if moisture content increases too much, even if this

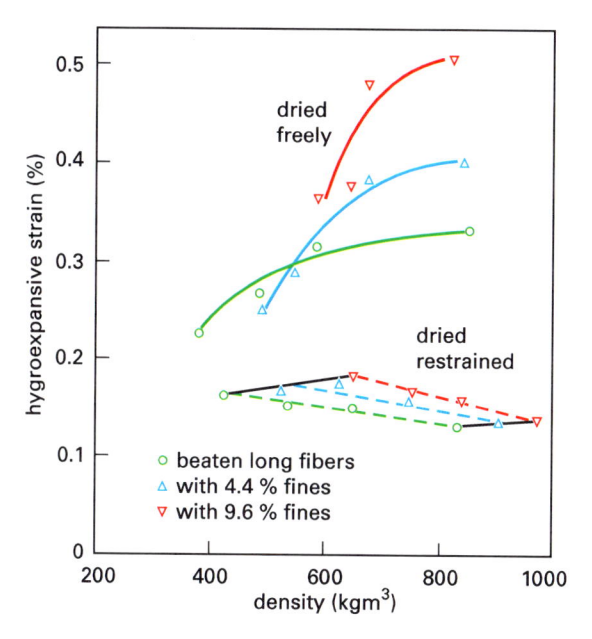

Fig. 11.24 Hygroexpansion of laboratory sheets between RH = 30% and RH = 60% against density (varied by wet pressing) in the case of restrained drying (almost zero shrinkage) and free drying shrinkage, and different levels fines addition, redrawn after Salmén et al. (1987). Fines are expected to improve inter-fiber bonding and therefore increase the stress transfer factor f_{21}. Reproduced with permission from Svenska Pappers- och Cellulosaingeniörsföreningen (SPCI).

happens for a short while or in a limited section of the paper product (Chapter 9). Likewise, there is little one can do to prevent creep compliance from increasing with the moisture content (Chapter 7).

Radical improvements in the moisture tolerance of paper materials would require changes in the raw material composition of paper. Such extensions to the property space of paper materials are discussed in the next chapter. These options must be weighed against the value that recyclability gives to products made of paper material. Recyclability back into the water-based manufacturing process requires that paper can be dispersed in water. It is not easy to see how one could achieve this without the preceding softening that happens when the moisture content of paper is increased.

11.7 Final remarks

In this chapter we have described the mechanisms that control some of the mechanical properties of paper. The discussion has been highly qualitative, mostly suggesting how one could approach the problem of modeling the mechanical properties of paper. The models we have presented are not intended to be quantitatively precise or complete; they merely demonstrate the microscopic mechanisms that underlie the macroscopic paper properties.

It should be clear to the reader that the mechanical properties of paper depend primarily on two mechanisms that act during the papermaking process:

- Network densification in pressing that controls inter-fiber bonding, fiber segment lengths, and geometric deformations such as fiber twist; and
- Fiber segment activation that controls the elastic modulus of fiber segments, as well as their stress–strain behavior and hygroexpansivity, which are strongly influenced by the bonding with other fibers.

We have shown how these effects can be described in physical terms and also given a qualitative demonstration how all the properties we considered depend on the same physical description of fiber segments and inter-fiber bonds. The same description of the fiber network should also apply to properties we have not discussed here, such as the behavior of paper under in-plane compression or out-of-plane delamination.

It should also have become clear that the relevant mechanical properties of fibers and inter-fiber bonds hardly can be determined by doing experiments on single fibers or inter-fiber bonds. In such experiments, it would be very difficult to reproduce the conditions that prevail in the fiber network during the papermaking process. Rather, we believe that the only reasonable approach is to measure paper properties and then infer the fiber and bond properties from the macroscopic results. Here one needs the kind of physical description of the fiber network that was qualitatively sketched above.

While the mechanical properties of fibers should not be considered disconnected from the papermaking process, it should go to determine the geometric dimensions of fibers from a large number of measurements performed on single fibers. There is a problem with many of the measurement methods in that they detect fibers only in the wet state where they are swollen by some amount. It might be more efficient to determine fiber dimensions microscopically from paper sheets.

Numerical simulations must be used if one wishes to go beyond the qualitative discussion presented here. This is also necessary when different fiber types are mixed or when fines or fillers are considered. Fines and fillers influence the fiber network by altering the densification process. This is relatively easy to quantify (Sirviö et al., 2003).

References

Alexander, S. D., and Marton, R. (1968). Effect of beating and wet pressing on fiber and sheet properties. II. Sheet properties. Tappi 51(6), 283–288.

Baum, G. A., Pers, K., Shepard, D. R., and Ave'Lallemant, T. R. (1984). Wet straining of paper. Tappi 67(5), 100–104.

Drolet, F., and Uesaka, T. (2005). A stochastic structure model for predicting sheet consolidation and print uniformity. In: Advances in Paper Science and Technology, Vol. 2, S. J. I'Anson, ed. (Bury, UK: Pulp and Paper Fundamental Research Society), pp. 1139–1154.

Giertz, H. W. (1964). Contribution to the theory of tensile strength. In: Proceedings of the EUCEPA/European TAPPI Conference on Beating, Venice, Italy, pp. 39–47.

Giertz, H. W., and Røedland, H. (1979). Elongation of segments – bonds in the secondary regime of the load/elongation curve. Paper presented at the International Paper Physics Conference, CPPA, Montreal.

He, J., Batchelor, W. J., and Johnston, R. E. (2003). The behavior of fibers in wet pressing. Tappi 2(12), 27–31.
Hollmark, H., Andersson, H., and Perkins, R. W. (1978). Mechanical properties of low density sheets. Tappi 61(9), 69–72.
Htun, M., and de Ruvo, A. (1978). Correlation between the drying stress and the internal stress in paper. Tappi 61(6), 75–78.
Jentzen, C. A. (1964). The effects of stress applied during drying on some of the properties of individual fibres. Tappi 47(7), 412–418.
Kallmes, O. J., and Corte, H. (1960a). The structure of paper. Part I. The statistical geometry of an ideal two dimensional fiber network. Tappi 43(9), 737–752.
Kallmes, O. J., and Corte, H. (1960b). Errata: The structure of paper. Part I. The statistical geometry of an ideal two dimensional fiber network. Tappi 44(6), 448.
Kettunen, H., and Niskanen, K. (1999). On the relationship of fracture energy to fiber debonding. In: Proceedings of the TAPPI International Paper Physics Conference, San Diego, California, pp. 125–143.
Kettunen, H., and Niskanen, K. (2000a). Microscopic damage in paper, part 1: Method of Analysis. J. Pulp Pap. Sci. 26(1), 35–40.
Kettunen, H., and Niskanen, K. (2000b). On the in-plane tear test. Tappi 83(4), 1–8.
Kettunen, H., Yu, Y., and Niskanen, K. (2000c). Microscopic damage in paper, part 2: Effect of fiber properties. J. Pulp Pap. Sci. 26(7), 260–265.
Korteoja, M. J., Lukkarinen, A., Kaski, K., et al. (1996). Local strain fields in paper. Tappi 79(4), 211–216.
Luner, P., Kärnä, A. E. U., and Donofrio, C. P. (1961). Studies in interfibre bonding of paper: The use of optical bonded area with high yield pulps. Tappi 44(6), 409–441.
Niskanen, K. J. (1990). Model study of paper elasticity and drying restraints. In: Proceedings of the Materials Interactions Relevant to the Pulp, Paper and Wood Industries, Materials Research Society Symposium, San Francisco, California, pp. 203–212.
Niskanen, K., Nilsen, N., Hellén, E., et al. (1997). KCL-PAKKA: Simulation of the 3D structure of paper. In: The Fundamentals of Papermaking Materials, Vol. 2, C. F. Baker, ed. (Leatherhead, UK: Pira International), pp. 1273–1292.
Niskanen, K., Kettunen, H., and Yu, Y. (2001). Damage width: A measure of the size of fracture process zone. In: Proceedings of The Science of Papermaking, 12th Fundamental Research Symposium, Oxford, UK, pp. 1467–1482.
Niskanen, K., and Rajatora, H. (2002). Statistical geometry of paper cross-sections. J. Pulp Pap. Sci. 28(7), 228–233.
Niskanen, K., Sirviö, J., and Wathen, R. (2005). Tensile strength of paper revisited. In: Proceedings of the Advances in Paper Science and Technology: 13th Fundamental Research Symposium, Cambridge, pp. 563–589.
Niskanen, K. (2008). Paper Physics, 2nd Edition (Helsinki, Finland: Paperi ja Puu Oy).
Nordman, L. S. (1958). Laboratory investigations into the dimensional stability of paper. Tappi 41(1), 23.
Page, D. H., and Schulgasser, K. (1989). Evidence for a laminate model for paper. In: Mechanics of Cellulosic and Polymeric Materials, R. W. Perkins, ed. (New York, NY, USA: ASME), pp. 35–39.
Page, D. H., Seth, R. S., and De Grâce, J. H. (1979). Elastic-modulus of paper. 1. The controlling mechanisms. Tappi 62(9), 99–102.
Salmén, L., Boman, R., Fellers, C., and Htun, M. (1987). The implications of fiber and sheet structure for the hygroexpansivity of paper. Nordic Pulp Pap. Res. J. 2(4), 127–131.
Salminen, L. I., Räisänen, V. I., Alava, M. J., et al. (1996). Drying-induced stress state of interfiber bonds. J. Pulp Pap. Sci. 22(10), 402–407.
Sanborn, I. B. (1962). A study of irreversible, stress-induced changes in the macrostructure of paper. Tappi 45(6), 465–474.

Seth, R. S. (1990). Fiber quality factors in papermaking – I The importance of fiber length and strength; II The importance of fiber coarseness. In: Material Interactions Relevant to the Pulp, Paper and Wood Industries, D. F. Caulfield, J. D. Passaretti, and S. F. Sobczynski, eds. (Pittsburgh, PA, USA: Materials Research Soc.), pp. 125–162.

Sirviö, J., Nurminen, I., and Niskanen, K. (2003). A method for assessing the consolidating effect of fines on the structure of paper. In: Proceedings of the International Paper Physics Conference, Victoria, BC, Canada, pp. 187–192.

Sirviö, J., Pohler, T., and Niskanen, K. (2008). A simple physical model for tensile stiffness of paper based on fiber activation. In: Proceedings of the Progress in Paper Physics Seminar, Espoo, Finland.

Tanaka, A., Kettunen, H., Niskanen, K., and Keitaanniemi, K. (2000). Comparison of energy dissipation in the out-of-plane and in-plane fracture of paper. J. Pulp Pap. Sci. 26(11), 385–390.

Uesaka, T. (1994). General formula for hygroexpansion of paper. J. Mater. Sci. 29(9), 2373–2377.

Uesaka, T., and Qi, D. (1994). Hygroexpansivity of paper-effects of fibre-to-fibre bonding. J. Pulp Pap. Sci. 20(6), J175–J179.

Wuu, F., Mark, R. E., and Perkins, R. W. (1991). Mechanical properties of "cut-out" fibers in recycling. TAPPI International Paper Physics Conference Proceedings, Vol. 2, Atlanta, GA (Tappi Press), pp. 663–680.

12 Wood biocomposites – extending the property range of paper products

Lars Berglund

12.1 Introduction

The forest products industry relies heavily on paper, paperboard, and sawn timber, but also on a category of established products sometimes classified as "traditional wood composites": glulam, plywood, particleboard, fiber board, and so forth. It is helpful to consider paper and paperboard products in the same context as wood composites because this puts a stronger emphasis on the engineering materials nature of load-bearing materials. Traditional wood composites are often used in the building industry, where many structures are subjected to significant loads, which can be both static and dynamic. By using the term *biocomposites* in the heading of this chapter we want to underline that the discussion will extend far beyond the traditional uses of wood composites.

A widening of the perspective for wood fiber-based materials by inclusion of composite materials is of great interest because it may help to find new applications in large material volume areas such as the building, automotive, and packaging industries. Also, a context of composite *materials* rather than forest *products* is helpful because the science and engineering of materials puts a strong focus on the micro-scale structural organization of material constituents. Focus is on the relationships between processing and microstructure, and between microstructure and properties. Material components such as fibers and polymers are subjected to processing and combined into a material with a certain microstructure. During processing, the material is also given geometrical shape. Common examples in the context of structural mechanics include plates, beams, and cylinders. A material of a given shape can then serve simple or complex functions, such as transmitting loads, heat, and the ability to survive repeated folding or storing energy at minimum weight.

The term *composite material* does not have a unified definition accepted over all different categories of composites, but the following criteria have been presented (after Hull, 1981):

1. A composite material consists of two or more physically distinct and separable material components (constituents). Usually, the properties of different constituents are substantially different.
2. In order to optimize the properties, the composite can be prepared by mixing the constituents so that the structure, to some extent, can be controlled.
3. The properties are superior, and possibly unique, compared with properties of individual constituents.

Common constituents that can offer substantial potential for mechanical reinforcement include fibers, platelets, particles, and ribbons (i.e., fibers that have a rectangular cross-section and substantially larger width than thickness). Air is also a constituent so that foams and porous networks (paper, fiberboard) can be classified as composites. Table 12.1 presents some examples of current material categories that can be classified as wood composites. We have made the classification according to the micro-structural characteristics and the scales of constituent size or constituent type. Some application examples are also presented.

The orientation distribution of the reinforcement component as well as its size is important for the mechanical properties of the composite. Larger constituents tend to result in materials with larger defect size and, therefore, lower strength. Oriented reinforcement provides higher strength in the orientation direction, and this is one of the main advantages of composite materials. It makes it possible to tailor the anisotropy (orientation dependency) of the material properties.

The typical densities and mechanical properties of different wood composites are listed in Table 12.2. Density is important to consider in material comparisons because mechanical properties show strong dependency on density. Because wood composites tend to be porous, the correct parameter in a micromechanics context is the relative density or volume fraction, as will be discussed later.

Comparing the different materials in Table 12.2, spruce wood has good mechanical properties but is limited by the restricted geometric shape. Complex machining operations are required. The anisotropy and high porosity also results in locally weak regions in machined structures of complex shape. Laminated structures such as LVL beams and plywood sheets often show high strength due to the thin lamellae. In addition, the orientation distribution of the lamellae can be controlled. Again, there is little freedom with respect to shape. The comparison between high-density fiberboard and a linerboard is of some interest. Kraft linerboard is a paperboard made of chemical pulp fibers (Chapter 2.5.1) and used as the surface ply in corrugated boards (Chapter 3.2.1). It has higher strength than the fiberboard, despite lower density. The wood fiber/polypropylene composite (an example of wood plastics) is also interesting. This material category is very successful in North America for decking applications, replacing impregnated wooden boards. It can also be injection molded into complex geometrical shapes. The elastic modulus is quite high, and the strength is respectable compared with many other materials. The main structural advantage of wood plastics is low porosity. This is interesting because it indicates the potential of new types of composites based on wood fibers.

The wood fiber itself has attractive characteristics including high aspect ratio (the length-to-diameter ratio), high axial strength and elastic modulus in the fiber wall, as well as favorable fiber network forming characteristics (Chapter 11.3). Networks made of strong wood fibers or chemically tailored fibers can be used in new fiber architectures of designed orientation distributions and combined with new polymer matrices, foams, or other porous materials to form new types of wood composites. Interesting functions include thermal insulation and mechanical performance. Wood fiber composites could also provide new opportunities with respect to molding of intricate geometrical shapes. The "nanopaper" material in Table 12.2 represents some of the advantages that can be obtained with cellulosic

Tab. 12.1 Material categories that may be classified as wood composites.

Wood composite category	Specific wood composite material	Description	Example of applications
Polymer-modified wood	Impreg	Wood is impregnated by monomers, which are polymerized.	Flooring
Laminated wood	Plywood	Veneer layers are laminated and bonded with a certain veneer orientation distribution.	Building industry, furniture
	Laminated veneer lumber (LVL)	Veneer layers are stacked to form laminated beams.	High-strength beams for building industry
	Glulam	Board layers are stacked to form beams.	Beams for building industry
Strands or particles with adhesive	Particle board	Large wood particles are coated by adhesive and hot-pressed to porous particleboard.	Furniture, building industry
	Oriented strand board (OSB)	Anisotropic strands are coated by adhesive and compressed to oriented high-density boards.	Competes with plywood at lower cost
Porous wood fiber networks	High density fiberboard (HDF) (850–1100 kg/m^3)	Mechanical or Masonite pulps are hot-pressed and bonded by lignin or adhesive.	Flooring, siding, wall panels, furniture
	Medium density fiberboard (MDF) (600–800 kg/m^3)	Mechanical pulp is combined with an adhesive and hot-pressed.	Furniture, cupboards, doors flooring
	Paper (kraft paper 600–800 kg/m^3)	Wood pulp is filtrated and dried into network.	Printing, packaging
	Paperboard	Typically thicker than about 0.25 mm (ISO definition: >224g/m^2).	Packaging
Impregnated wood fiber networks	Paper laminates	Paper is impregnated by resin and polymerized.	Electric insulation boards, flooring
Short, discrete wood fibers in polymer matrix	Wood plastics	Saw dust or wood pulp is mixed with thermoplastic or resin and is extruded, injection molded, or foamed.	Decking, building industry, furniture, automotive

Tab. 12.2 Typical densities and mechanical properties of different wood composite materials.

Wood composite	Density (kg/m³)	Elastic modulus (GPa)	Bending strength (MPa)	Tensile strength (MPa)
Spruce board	400	11	9.3	n.a
Spruce glulam	400	12.4	16.6	n.a
Spruce LVL	≈550	14	51	42
Spruce plywood	≈550	6.9–13	21–48	6.9–13
Oriented strand board (OSB)	≈550	4.8–8.3	21–28	6.9–10.3
High-density fiberboard (HDF)	850–1100	2.8–5.5	31	15
Medium-density fiberboard (MDF)	600–800	2	20	n.a
Particleboard	550–750	2–4 (from bending)	15–25	n.a
Kraft linerboard	600–800	2.8–4.1	n.a	25 (4% failure strain)
40wt% wood fiber/ PP and MA-PP	1030	4.2	n.a	52 (3.2% failure strain)
Nanopaper	1300	13	n.a	200–300 (10% failure strain)

nano-fibers. Their dimensions are three orders of magnitude smaller than regular wood fibers. The elastic modulus is 13 GPa, and the strength in tension exceeds 200 MPa due to the fine structure of the material. This will be discussed in Section 12.4.

The development of new wood composites should preferably be motivated by new applications for wood-based materials. Property comparisons with other categories of materials will then be helpful and can be pedagogically made by the use of so-called property charts, such as Figure 12.1, that have been introduced by Michael Ashby and coworkers (Ashby, 1999). An interesting feature of these charts is that the material efficiency of specific structures can be compared for different classes of materials. For instance, the efficiency of tensile members ("stiff ties") is given by the ratio E/ρ of elastic modulus and density. A higher value for this index gives a lower-weight tie for the same stiffness. The efficiency of a beam loaded in bending is measured by the index $E^{1/2}/\rho$. The corresponding index for flat plates loaded in bending is $E^{1/3}/\rho$. Three stiffness guidelines corresponding to these indices are plotted in

Figure 12.1 together with the elastic modulus and density data for different material categories, all on log–log scales to facilitate comparison.

Different values for the stiffness indices appear as a set of parallel lines in Figure 12.1 so that one can compare material categories. E/ρ for wood parallel to grain is almost as high as for steel. The corresponding values of $E^{1/2}/\rho$ and $E^{1/3}/\rho$ are much higher than for steel, which means that the efficiency in bending of beams and plates is much higher for wood than for steel. Wood yields lighter-weight structures for a given stiffness. It is also interesting to note that in bending wood is stiffer than many man-made synthetic fiber composites. Related property charts can be worked out for other properties and characteristics in order to identify if there is "free" property space in the charts, where novel wood composites may be designed to

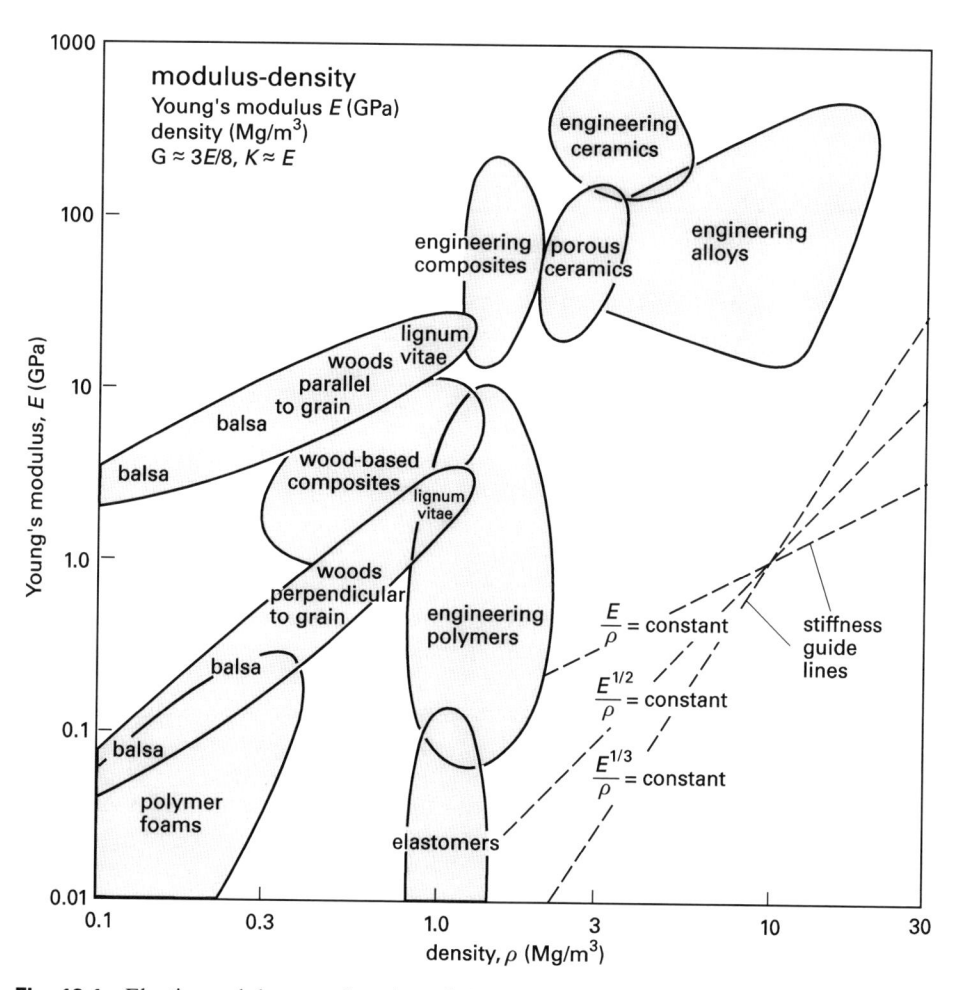

Fig. 12.1 Elastic modulus as a function of density for different material categories, plotted on log–log scale. Note the materials efficiency guidelines explained in the text (Reprinted from Wegst and Ashby (2004) with permission from Taylor & Francis).

give new combinations of elastic modulus and density. Some particularly interesting new wood composites are those that extend the current range of properties, as exemplified by the nanopaper material in Table 12.2.

12.2 Material components: fibers and polymers

12.2.1 Plant fiber structure

Even in an engineering context, it is helpful to consider that the plants, from which we derive wood and other fibers, are biological organisms. Fibers have specific functions in the plant organism, and these functions, together with growth aspects, explain the structure of plants. Plant fibers in trees and grasses (i.e., annual plants such as flax) are single cells. The fibrous shape indicates that they have a mechanical function in the plant. Fiber geometry provides anisotropy, and the load-carrying ability is high in the direction of fiber axis. In addition, the fiber cell wall itself is anisotropic due to the organization of its components. The cell wall is much stiffer in its length direction than in the transverse directions.

Figure 12.2 presents a micrograph of a plant fiber or cell. Softwoods (coniferous trees such as pine and spruce) are most relevant in the context of composites because they consist of long fibers called *tracheids*. Their length is typically 2–4 mm and diameter 20–40 μm, and thus the aspect ratio (length/diameter) is around 100. The lumen is the empty space at the center of the cell. The fiber cell wall consists of a thin primary wall layer and a thick secondary wall layer. The latter divides further into thin S_1 and S_3 layers and a thick S_2 layer that occupies about 85% of the fiber cell wall thickness. An important structural feature is that the cell wall is composed by cellulose microfibrils with a diameter of 5–15 nm, depending on the plant. Cellulose microfibrils reinforce the cell wall and are oriented at a certain angle to the fiber direction (microfibril angle, MFA). The smaller the microfibril angle, the stiffer and stronger is the fiber. The S_2 layer typically has MFA = 10–30°, but at chest height of the trunk the outer part in a mature coniferous tree contains tracheids (fibers) of very small MFA. This is the region of the tree that is subjected to high stresses when the tree trunk is bent by heavy winds.

The cell wall has an organization of a laminated composite material with cellulose microfibrils in a matrix of highly hydrated lignin-hemicellulose complex. Softwoods typically have around 42% cellulose, 27% hemicelluloses, and 28% lignin out of the dry matter and a few percent extractives (fatty acids and phenolics). The water content in a native wood fiber is around 30%. Most likely, the hydrated lignin-hemicellulose matrix is strongly associated with the cellulosic microfibril by physical adsorption of hemicelluloses. Lignin and hemicelluloses are also linked and form a polymer network. The mechanical behavior of a wet wood cell wall is poorly understood. It shows interesting features, including an impressive combination of strength, stiffness, and toughness despite its hydrated state.

In a mechanical property sense, cellulose microfibrils need to be stiff and strong because their main function in plant cell walls is to provide tensile performance. Obviously, wood tracheids also carry compressive stresses, but the tensile material

12.2 Material components: fibers and polymers

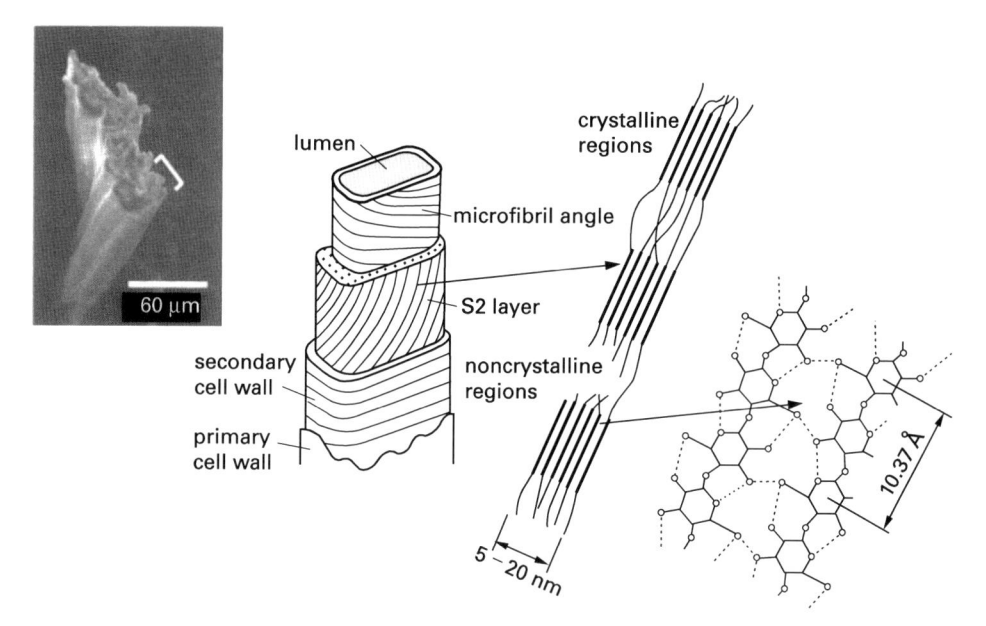

Fig. 12.2 Micrograph of a plant fiber cell (left). The cell wall structure consists of the primary cell wall; the secondary cell wall with the S1, S2, and S3 layers of different organization and microfibril angle of the cellulose(center); and the cellulose microfibril with ordered and less ordered regions (right). Courtesy of Prof. T Nishino, Kobe University.

function is still the most important one for cellulose in plant organisms. The estimated axial elastic modulus of the cellulose crystal is 134 GPa. It arises from the extended chain conformation of cellulose molecules, giving high density of strong inter-molecular covalent bonds, and strong intra-molecular hydrogen bonds that stiffen the molecule. The hydrated lignin-hemicellulose network, which constitutes the cell wall matrix, is amorphous in nature. In spite of some degree of hemicellulose orientation, the elastic modulus of the amorphous network is unlikely to exceed 1 GPa.

Geometric features of some plant fibers, including the microfibril angle (MFA) and cellulose degree of polymerization (DP) (a measure of the molar mass or length of cellulose molecules) are presented in Table 12.3. The high DP of cellulose is helpful because it means that a lot of the strong covalent bonds have to be broken in order to break microfibrils of the cell wall. However, during the cooking of wood fibers into a chemical pulp, some degradation of cellulose DP takes place, and this will decrease the strength of cellulose. It is interesting to note that hemp, jute, flax, and ramie have very low microfibril angles. Although it is tempting to relate this to plant stem function, we have to keep in mind that these are plants used for thousands of years as fiber sources for textiles. As a consequence, the breeding of these grasses has been strongly focused on the production of stronger fibers, probably selecting for small MFA and high cellulose content. It is also interesting to note the very large aspect ratio of grass fibers. In a practical context, it is difficult to process fibers that are 15 cm long (ramie). Usually the fibers used in textiles are made by spinning so that the plant fiber cells are intertwined into a continuous thread of larger diameter. Cotton is not a tensile material but has a seed hair function. Wood is very interesting due to its comparably low cost and the considerable infrastructure for harvesting and processing that is present in many regions of the world. The aspect ratio (length-to-diameter) is typically more than sufficient for biocomposite requirements, although continuous wood fiber products (such as yarns) are not readily available.

The mechanical properties of plant fibers are interesting for estimates of the property potential of biocomposites. In Table 12.4, some data are provided. It is interesting to note that the dry elastic modulus E of plant fibers compares well with glass fiber (E_{glass} = 70 GPa). As a consequence, specific stiffness (elastic modulus divided with density) for plant fiber biocomposites compares well with glass fiber composites. The elastic modulus of the wood cell wall is around 30 GPa, which is about one order of magnitude higher than in glassy polymers. The elastic modulus of different fibers is controlled by the cellulose content and microfibril angle (higher cellulose content and lower MFA increase the elastic modulus). All plant fibers have significantly reduced elastic modulus in the completely wet state. However, the tensile strength is not significantly reduced because moisture increases the plasticity and toughness of the cell wall so that the fibers become less sensitive to defects. The breaking strain ε^{max} of fibers is not very high because the cellulose molecules in microfibrils are in extended chain conformation and aligned with the fiber axis and cannot stretch very much without failing. Wood fibers with high MFA (occurring in so-called compression wood) are an interesting exception together with coir fibers because at high MFA the microfibrils can slide and reorient relative to one another, providing high ductility to the cell wall.

The wood products industry supplies many different types of wood fibers for different applications. It is interesting to consider the suitability of different fibers for new types of wood fiber biocomposites. Table 12.5 lists the most common types of wood fibers, with comments on their characteristics. Due to the low cost, saw dust and other particles from saw mills and machining of wood are in widespread use in particleboards and melt-processed thermoplastic wood composites. However, because the typical aspect ratio of these particles is low (≤10), the reinforcement

Tab. 12.3 Geometrical parameters, microfibril angle (MFA), and cellulose degree of polymerization (DP) in some plant fibers (after Wainwright et al., 1982).

Fiber	Width (μm)	Length (cm)	Aspect ratio	MFA	DP
Hemp	16–50	1.2–2.5	500–700	2.3°	9300
Jute	20–25	0.15–0.5	75–200	7.9°	n.a
Flax	11–20	1–3	900–1500	6°	7000–8000
Ramie	20–75	12–15	2000–6000	6°	10800
Cotton	15–30	2.5–6	1700–2000	28°–44°	15300
Wood	20–30	0.1–0.3	33–150	10°–40°	10000

Tab. 12.4 Elastic modulus, tensile strength, and breaking strain of some plant fibers in the wet and dry state (after Wainwright et al., 1982).

Fiber	Elastic modulus E (GPa) Wet	Dry	Elastic modulus ratio, E_{dry}/E_{wet}	Tensile strength σ^{max} (MPa) Wet	Dry	Tensile strength ratio, $\sigma^{max}_{dry}/\sigma^{max}_{wet}$	Breaking strain (%) Wet	Dry
Hemp	35	70	2	n.a	920	n.a	n.a	1.7%
Jute	n.a	60	n.a	n.a	860	n.a	n.a	3.0%
Flax	27	80–110	3–3.7	880	840	0.96	2.2%	1.8%
Ramie	19	80	4.2	1080	900	0.89	2.4%	2.3%
Wood	n.a.	35	2	82	120	1.5		

Tab. 12.5 Chemical composition and characteristics of wood reinforcement particles.

Wood fiber type	Chemical composition	Characteristics
Saw dust, wood flour	Similar to wood	Short aspect ratio, large size since the particles often are tracheid bundles
Mechanical wood pulp	Similar to wood (high yield)	Individualized fibers 20–30 µm in width and 1mm (hardwood) to 3 mm (softwood) in initial length, mechanically cut or damaged, little fiber collapse
Chemo-thermomechanical pulp (CTMP)	Low extractives content, otherwise similar to wood (yield 90%–93%)	Less mechanical damage compared with TMP
Bleached sulphate pulp (kraft)	Low lignin conc., typically 15% hemicellulose, 85% cellulose	Individualized, mechanically intact, collapsed after refining
Sulphite pulp	Low lignin conc., 4%–15% hemicelluloses, 85%–96% cellulose	Individualized, mechanically intact, collapsed after refining, lower cellulose molar mass than kraft pulp
Nanofibrillated cellulose (NFC)	Typically 4%–15% hemi-cellulose, depending on pulp source	5–15 nm in width and several micrometers in length

potential of the stiff wood fiber is not utilized. In contrast, mechanical pulps such as TMP and CTMP (Chapter 2.5.1) have much higher aspect ratio and have better reinforcement efficiency in biocomposites. The chemical composition in TMP and CTMP is fairly similar to wood. This may cause problems with odor (thermal degradation of hemicelluloses and lignin) or discoloration (extractives) problems in biocomposites that are processed at high temperature (e.g., melt-processing or compression molding). Bleached kraft pulps are more stable. They can also have high molar mass of cellulose, which gives good fiber strength. The hemicellulose content is usually significant since it may be hydrolyzed in high-temperature processing. Sulfite pulps can have very low hemicellulose content, but at the same time, the DP of cellulose is lower than in kraft pulps.

Nano-fibrillated cellulose (NFC) is a new component for biocomposites. It is obtained by mechanical disintegration of wood pulp fibers. Usually, the source is a chemical sulphite pulp or bleached kraft pulp. A chemical or enzymatic pretreatment step is needed to reduce the energy consumption so that the manufacturing cost becomes reasonable. The dimensions of NFC are three orders of magnitude lower than for plant fibers (5–30 nm in dimater and 5–10 µm in length). As a consequence, biocomposites of very fine structure can be prepared. NFC-based composites have much greater strength and ductility than plant fiber-based biocomposites because defect size is governed by the particle size of the reinforcement phase.

12.2.2 Polymer matrices and binders

The polymer matrix improves the properties of composites. Compared with a fiber network without additional polymers, inter-fiber bonding and interaction can be better controlled. Also, the density range can be extended. The term *binder* refers to when the wood composite is a porous fiber network and the polymer binder is an adhesive, which primarily improves the inter-fiber bonding but does not completely fill the network.

The different categories of polymer matrices are listed in Table 12.6. The most common thermoplastics for melt-processed biocomposites are polypropylene (PP) and polyethylene (PE), which are widely used in large-volume products, such as the decking market in North America. There is a substantial market for recycled PE and PP, making them readily available. Their disadvantage is low strength, brittleness, odor, and surface appearance problems. Polylactic acid (or polylactide, PLA) is an interesting thermoplastic with biomass origin, but it is quite brittle unless plasticized and is also quite expensive.

Water-soluble thermoplastics are used in paper manufacturing and converting (such as the manufacture of corrugated boards). They include cellulose derivatives and starch. Starches are obtained from potato, corn, and other food sources, and this is a disadvantage. Amylopectin-rich starch is quite brittle without plasticizer, but the main problem is still the high water solubility. Although advantageous in processing, this limits the durability to the biocomposite material in humid conditions. Cellulose derivatives are interesting because they are made from sustainable resources. For example, hydroxyethyl cellulose (HEC) is a very tough polymer. However, as with starches, biocomposites based on cellulose derivatives need either protection against water or additional chemical modification to reduce water solubility.

Thermoset resins are chemically cross-linked network polymers, which cannot be dissolved or melted. They are prepared by chemical reactions that initiate either by the mixing of two components or by heating. They are widely used for wood composites in the form of phenol-formaldehyde (PF), melamine-formaldehyde (MF), and urea-formaldehyde (UF) resins or adhesives. PF has good hygro-thermal

Tab. 12.6 Categories of polymers used as polymer matrix or binder in biocomposites.

Polymer category	Polymer	Characteristics
1. Thermoplastics (melt-processed)	Polypropylene, polyethylene (PP, HDPE, LDPE)	Petroleum-based (not ethanol-PE), wide industrial use, ductile, low cellulose compatibility
	Polylactic acid, or polylactide (PLA)	Corn-based (starch), increasing industrial use, somewhat brittle
2. Water-soluble thermoplastics	Starch	Cheap, brittle without plasticizer, available as melt-processable grades and blends with other thermoplastics
	Cellulose derivatives	Examples; carboxymethyl cellulose (CMC), hydroxyethyl cellulose (HEC), good cellulose compatibility, interesting binder with cellulosic fibers
3. Thermoset resins	Formaldehyde cured resins	Currently used as binders in wood composites; phenol-formaldehyde, PF, melamine formaldehyde, MF, urea formaldehyde, UF.
	Epoxies, unsaturated polyesters	Dominating resins in current glass fiber composites, low viscosity, good mechanical properties, insensitive to moisture, petroleum-based although new grades are partly of bio-origin
	Furan resins	Bioresin based on sugar cane, good mechanical properties, complex curing chemistry, dark color, small current market
4. Wood cell wall matrix	Lignin-hemicellulose complex	Hydrated lignin-hemicellulose network. A unique feature is its highly hydrated state and strong association with cellulose microfibrils

stability, whereas MF is not as good. MF can be optically transparent. UF is of low cost but sensitive to hydrolysis when water is present. There is strong interest in replacing formaldehyde resins with more environmentally friendly alternatives, but this is a challenge because of associated cost increases which are difficult to accept in commercial applications.

Epoxies and unsaturated polyesters are well established as polymer matrices in the composites industry. Recently, several "green" resins have been developed that typically have an increased proportion of molecules from renewable resources. For instance, polyesters can be made partly from soybean oil that is chemically modified (epoxidized or acrylated) so that conventional curing chemistry can be used. Epoxies have high hygro-mechanical durability, but they are much more expensive than polyester resins.

Recently, there has been an increasing interest in furan resins. They usually derive from sugar cane and are prepared from sugars. However, complex chemistry is needed, and the cured thermoset resin is often brittle, has pores, and is black in color.

Still, there are strong development efforts underway to improve the characteristics of furan resins.

The final row in Table 12.6 describes the characteristics of the hydrated lignin-hemicellulose matrix complex in the wood cell wall. Although this complex is unlikely to be directly useful as a polymer matrix, we can learn from its function in the cell wall. The first interesting characteristic is that it is strongly associated with the cellulose microfibrils. Possibly, this intimate association or bonding is why the wood cell wall shows such high stiffness and strength, in spite of the hydration of the matrix at the 30% water content of the cell wall. In conventional micro-scale biocomposite materials, the fiber–matrix interface is likely to de-bond when the material adsorbs moisture and expands. In contrast, the wood cell wall is molecularly designed to perform in the wet state. The second interesting characteristic is the significant toughness of the wood cell wall. Local strain in the cell wall can exceed 20%, and the cell wall still has considerable load-carrying capacity. Most likely, this is related to the fine structure of the reinforcing microfibrils in combination with some mechanism whereby the microfibril–matrix interaction holds despite the high strain. This has not yet been achieved with micro-scale fiber composites, where the fiber–matrix interface will crack and de-bond at high strain.

12.3 Micromechanics of fiber composites

In this chapter, the potential of wood biocomposite materials is in focus. In order to appreciate this potential, it is important to understand the basic stiffening mechanisms of fiber reinforcement. Here, the micromechanics of fiber reinforcement of a polymer matrix are helpful. Let us consider the case of continuous cylindrical fibers embedded in a matrix (Fig. 12.3). An important effect of stiff and strong fibers added to a soft polymer matrix is an improvement of the mechanical properties that depend on the relative proportions of the matrix and fiber phase. It is the constituent volume fraction, not weight fraction, that determines the reinforcement effect of a certain amount of fibers. This is easily understood by considering the hypothetical comparison of lead "fibers" and glass fibers in a polymer matrix. Even if the two fiber types would have the same elastic moduli and geometric dimensions, a given weight fraction of lead fibers in the composite would give much lower elastic modulus than the same weight fraction of glass fibers because the number of lead fibers in the composite would be much smaller.

12.3.1 Weight fraction and volume fraction

Let us consider a composite material consisting of a volume fraction V_{fc} of fibers and V_{mc} of matrix material. The corresponding weight fractions are W_{fc} and W_{mc}, so by construction

$$V_{fc} + V_{mc} = 1 \text{ and } W_{fc} + W_{mc} = 1. \tag{12.1}$$

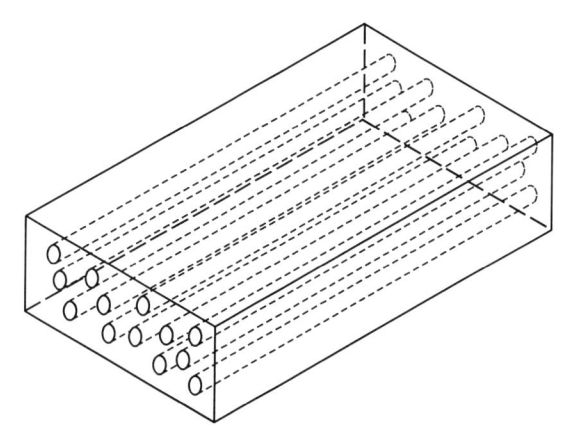

Fig. 12.3 The architecture of a unidirectional fiber composite. Cylindrical fibers are continuous, oriented parallel to each other in one direction, and surrounded by a matrix phase.

Weight fractions are easily estimated by weighing the constituents. However, theoretical analysis of composite properties (Agarwal and Broutman, 1990) is based on the knowledge of volume fractions because they control physical properties. A very important relationship expresses how the volume fractions of fiber and matrix are calculated from the weight fractions and the densities (ρ_{fiber} and ρ_{matrix}) of the constituents, and the density of the composite in the absence of voids ($\rho_{void\text{-}free}$):

$$V_{fc} = (\rho_{void\text{-}free}/\rho_{fiber}) \cdot W_{fc} \text{ and } V_{mc} = (\rho_{void\text{-}free}/\rho_{matrix}) \cdot W_{mc}. \tag{12.2}$$

The density of a void-free composite is

$$\rho_{void\text{-}free} = \frac{1}{(W_{fc}/\rho_{fc}) + (W_{mc}/\rho_{mc})}. \tag{12.3}$$

The volume fraction of voids Φ in a real composite is also important because it has a strong influence on the mechanical properties. It requires the experimental determination of the composite density ρ_{comp}:

$$\Phi = (\rho_{void\text{-}free} - \rho_{comp})/\rho_{void\text{-}free}. \tag{12.4}$$

12.3.2 Elastic properties in unidirectional composites

Fiber composites usually have high elastic modulus combined with low density. The reinforcing mechanism of stiff fibers in a polymer matrix can be understood by a micromechanics analysis of the stress and strain distributions at the scale of fibers. Composites are complex materials from the point of view of stress analysis. For this reason, a simple composite structure is a good starting point that provides an insight into reinforcement mechanisms.

We start from the unidirectional composite in Figure 12.3, where the longitudinal and transverse directions are defined as parallel and perpendicular to the fiber direction. The fibers are perfectly bonded to the polymer matrix so that in the longitudinal direction the fiber strain and matrix strain are identical with the composite strain, and the elastic modulus E_L of the composite in the longitudinal direction is

$$E_L = E_{fiber,L} \cdot V_{fc} + E_{matrix} \cdot V_{mc}, \qquad (12.5)$$

where the $E_{fiber,L}$ is the longitudinal (or axial) elastic modulus of fibers, and E_{matrix} is the elastic modulus of the matrix that can be assumed isotropic. For stiff fiber composites, the axial elastic modulus of fibers is typically much higher than the elastic modulus of the matrix, and therefore, the stress carried by the fiber phase is higher. This is the main reinforcement principle for fiber composite materials. For instance, the longitudinal cell wall modulus of wood fibers is typically 30 GPa, whereas glassy polymers have a elastic modulus of about 3 GPa.

In a void-free composite $V_{mc} = 1-V_{fc}$, and we can write

$$E_L = E_{fiber,L} \cdot V_{fc} + E_{matrix} \cdot (1-V_{fc}). \qquad (12.6)$$

This equation is usually called the rule of mixtures. Its agreement with experimental data is very good when the constituents are elastic. The reason for the good agreement is that the real structure is very similar to the structure assumed (unidirectional fibers in the loading direction) in the plane, and the fiber–matrix bonding is good at low strains.

A model for transverse elastic modulus E_T can be derived assuming constant stress in the matrix and the fibers. The final expression is

$$E_T = \frac{E_{fiber,T} E_{matrix}}{E_{fiber,T}(1-V_{fc}) + E_{matrix} V_{fc}}, \qquad (12.7)$$

where the correct fiber elastic modulus $E_{fiber,T}$ is in the transverse direction. The agreement between predictions based on Equation 12.7 and experimental data is poor, especially at high volume fractions of stiff fibers. The constant stress assumption is not very good in that case because the stress state in the matrix is highly nonuniform and often very different from the fiber stress. Instead, the Halpin-Tsai model is better suited as a predictive tool (Hull, 1981; Agarwal and Broutman, 1990). For cylindrical fibers the model is

$$\frac{E_T}{E_{matrix}} = \frac{1+2\eta_T \cdot V_{fc}}{1-\eta_T \cdot V_{fc}} \text{ with } \eta_T = \frac{(E_{fiber,T}/E_{matrix})-1}{(E_{fiber,T}/E_{matrix})+2}. \qquad (12.8)$$

This is an approximation based on finite element modeling results. The predictions agree quite well with experimental data, and the model is theoretically sound.

12.3.3 Elastic properties in short fiber composites

In the context of fiber composites, wood fibers are termed *short fibers* because they have a finite length. As may be expected, short fibers have a lower reinforcement ability than continuous fibers. When the composite material is subjected to a mechanical load, the externally applied stress is transferred from the matrix to the stiff fibers. The situation is similar to the fiber network of paper discussed in the preceding chapter.

In order to understand the nature of the stress distribution in a short-fiber composite, consider a single fiber oriented parallel to a uniaxial external load. The fiber has a finite length l_{fiber} and is embedded in a matrix. Figure 12.4a illustrates the strain distribution along the fiber embedded by showing deformation lines of the matrix. Close to the fiber ends, the fiber strain is lower and the matrix strain larger than elsewhere because the fiber is stiffer than the matrix. The tensile stress in the fiber (Fig. 12.4b) is highest in the middle of the fiber and decreases at the ends, where in turn the interfacial (fiber/matrix) shear stress is high. The length fraction at a fiber end where tensile stress is reduced is $l_{crit}/2$.

Figure 12.4 illustrates the importance of the aspect ratio of fibers l_{fiber}/d_{fiber}, where l_{fiber} is fiber length and d_{fiber} fiber diameter. High aspect ratio means that only small proportions of the fiber ends carry lower stress. For example, in saw dust composites the reinforcing particles can have an aspect ratio as low as 10 or lower, which gives very low reinforcement efficiency.

Prediction of the elastic modulus of real short-fiber composites is a complex problem because the elastic modulus depends on the distributions of fiber length and fiber orientation (compare with Chapter 11.3.2). A simple case can help us to understand the important parameters. Consider the elastic modulus of a short-fiber composite where all the fibers are aligned. The Halpin-Tsai equation has been extended to the longitudinal elastic modulus of short-fiber composites

$$\frac{E_L}{E_{matrix}} = \frac{1 + 2 \cdot \frac{l_{fiber}}{d_{fiber}} \cdot \eta_L \cdot V_{fc}}{1 - \eta_L \cdot V_{fc}} \quad \text{with} \quad \eta_L = \frac{(E_{fiber,L}/E_{matrix}) - 1}{(E_{fiber,L}/E_{matrix}) + 2 \cdot \frac{l_{fiber}}{d_{fiber}}}. \tag{12.9}$$

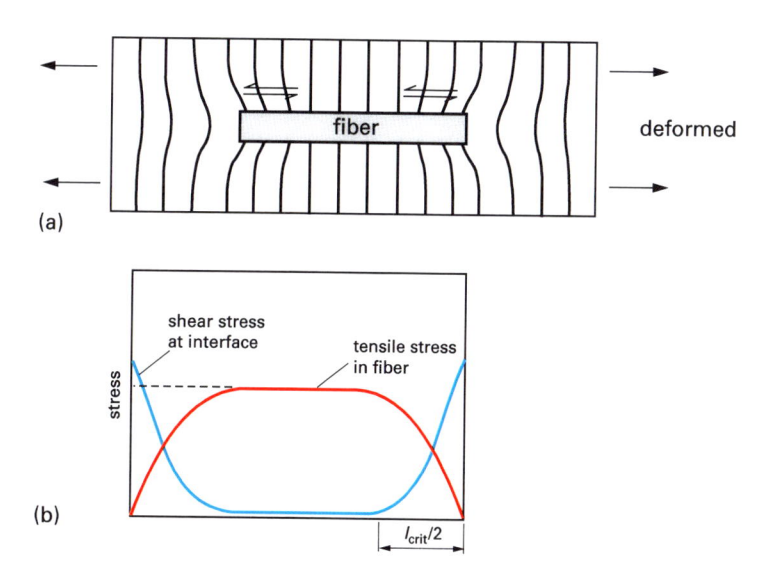

Fig. 12.4 Displacement field created by a stiff fiber of finite length that is perfectly bonded in a continuous matrix when tensile stress is applied to the fiber-matrix assembly in the direction parallel to the fiber axis (a), and the corresponding distribution of tensile stress in the fiber and shear stress at the fiber-matrix interface. Half the critical length $l_{crit}/2$ denote the section where the tensile stress is lower than in the center.

This expression is at best empirical. Note the similarity with Equation 12.8. For E_T the fiber aspect ratio l/d is simply 1 because a cylindrical fiber loaded in the transverse direction is considered. The orientation distribution of the fibers will obviously also influence the moduli (see Hull, 1981). For the practically important case of isotropic fiber orientation distribution in the plane, a simple empirical expression can be used for the elastic modulus of the short-fiber composite:

$$\overline{E} = \frac{3}{8}E_L + \frac{5}{8}E_T, \tag{12.10}$$

where E_L and E_T are obtained from the Halpin-Tsai expressions (Eqs. 12.8 and 12.9).

12.3.4 Interfacial strength in short fiber composites

The stress distribution along a short fiber influences the composite strength. For a fiber with short aspect ratio, the stress in the fiber never reaches the fiber strength, and when the composite fails, intact fibers are pulled out of the matrix. Some minimum fiber length is therefore needed in order to utilize the fiber strength. This gives a definition for the critical fiber length l_{crit} illustrated in Figure 12.4. It is the shortest length of fibers in a given composite that allows the fibers to be loaded maximally so that they start failing. Consider a short cylindrical fiber embedded in a block of the matrix material. As the length of the fiber is increased, the critical fiber length l_{crit} is reached where the fiber stress at composite failure is only marginally smaller than the fiber strength σ_{fiber}^{max}. The force balance between the tensile force in the fiber and the mean shear force at the fiber–matrix interface can be analyzed in a straightforward manner, resulting in the following expression for the critical aspect ratio

$$\frac{l_{crit}}{d_{fiber}} = \frac{\sigma_{fiber}^{max}}{2 \cdot \tau^{max}}, \tag{12.11}$$

where τ^{max} is the shear strength of the fiber–matrix interface or the yield strength of the matrix, whichever fails first. If we assume that the wood fiber cell wall strength is 300 MPa and the interfacial shear strength is 10 MPa, we end up with an aspect ratio of 15 in order to ensure fiber fracture. For a 30 μm diameter wood fiber this is well below 1 mm, the typical length of hardwood fibers.

We can conclude that if the interfacial strength or matrix strength is low, the minimum aspect ratio (Eq. 12.11) is high. Low fiber strength has the opposite effect. If the strength of the short-fiber composite is considered, higher interfacial shear strength increases the composite strength, as will become apparent from the experimental data in the next section. Note also that the discussion here only considered the conditions for barely reaching a stress in the fiber that corresponds to the fiber strength. In order to have the load-carrying efficiency of short fibers approaching continuous fibers, a much higher aspect ratio is required than that in Equation 12.11. More advanced treatments also consider the statistical distribution in fiber strength.

12.4 Composites data: wood fiber/thermoplastic

Thermoplastic composites are of great interest to industry. One important reason is that thermoplastic biocomposites can be melt-processed, for instance, by continuous extrusion of profiles such as sheets or rods. In addition, injection molding of three-dimensional components can be carried out at great production speed for use in automotive applications, including interior components. The most well-known example of thermoplastic biocomposites in the forest products sector is wood flour composites with polyethylene (PE) or polypropylene (PP) as thermoplastic matrix. In North America, PE-based materials have been widely used in the large decking market, where the "wood plastics" provide an alternative to impregnated wood. IKEA has launched a few injection-molded furniture products based on a wood flour/PP composite.

Table 12.7 shows the tensile strength and elastic modulus for different compositions of the wood/PP composites. Using 40 weight-percent of wood flour reinforcement, the strength is virtually unchanged, whereas elastic modulus is increased from 1.5 to 3.9 GPa. Interestingly, the strength increases considerably if a little maleic-anhydride modified polypropylene (MA-PP) is added. The MA-PP serves as a compatibilizer that improves the interfacial strength τ^{max} and also helps the dispersion of wood flour particles, which often improves the strength of the composite. In terms of Equation 12.11, the increase in the interfacial shear strength increases load transfer efficiency of wood flour particles by making the critical aspect ratio smaller. Similar effects are seen with wood fibers. Without the MA-PP addition, wood fibers influence only the elastic modulus but not the strength of the composite. When the MA-PA is added, the tensile strength increases also compared to the unreinforced PP. With 60 weight-percent of wood fibers and 3% of MA-PP, the elastic modulus is close to 7 GPa and the material is quite brittle.

Polyamide-6 (PA6) is an interesting alternative to PP, as demonstrated in Table 12.8. The composition of PA6 with 33 weight-percent of sulfite pulp reinforcement can have an elastic modulus of 5.7 GPa and a tensile strength of 87 GPa. This is quite encouraging because the values are close to those for PA6 reinforced with glass fibers but with a lower density. One reason for better performance of PA6/wood composites compared with PP/wood composites is that PA6 itself has better properties. The interaction between PA6 and cellulose is more favorable, and the interfacial shear strength is likely to be higher. The sulfite pulp with high cellulose content (95%) was chosen because of the high processing temperature needed with PA6, which would lead to thermal degradation effects if the fibers contained hemicelluloses.

Other polymer matrices such as PLA and starch can also be used for thermoplastic composites. PLA tends to be brittle if plasticizers are not used. On the other hand, plasticizers reduce the thermal stability and strength of the composite. Thermoplastic starch is used in some packaging applications. Amylopectin-rich starch from potato can be used and mixed with glycerol plasticizer or another thermoplastic. The use of thermoplastic starches combined with plant fibers is hampered by the high solubility of starch in water.

Tab. 12.7 Elastic modulus and tensile strength of PP/wood flour and PP/wood fiber composites for different mixing ratios, with and without added MA-PP compatibilizer (Stark and Rowlands, 2003).

Composite	Elastic modulus (GPa)	Tensile strength (MPa)
PP	1.5	28.5
40 WPP wood flour	3.9	25
40 WPP wood flour +3% MA-PP	4.1	32
40 WPP wood fiber	4.2	28
40 WPP wood fiber +3% MA-PP	4.2	52
50 WPP wood fiber	5.8	28
50 WPP wood fiber +3% MA-PP	6.7	53

Tab. 12.8 Melt-processed thermoplastic composites where chemical sulfite pulp fibers ($\approx$95% cellulose content) made of different wood materials are combined with PP or PA6. A glass fiber/PP composite is given as a reference (Jacobson et al., 2002).

Composite	Elastic modulus (GPa)	Tensile strength (MPa)
PP	1.39	27.6
PP/33WF	3.38	33.1
PA6	2.75	60.2
PA6, 33%WF, HW	5.71	86.5
PA6, 33%WF, SW	5.35	81.9
PA6, 33%GF	8.01	111

12.5 Composites data: wood fiber/thermoset

Wood fiber composites with thermoset matrix are already used industrially in moldable compounds with polyphenol-formaldehyde (PF) or melamine-formaldehyde (MF) resin as the matrix. The first black telephones were made of wood fiber/PF compounds, and white wood fiber/MF compounds are used in electrical insulation components (wall sockets, plugs). The compound is premixed and cured under pressure and elevated temperature in the injection molding or compression molding process. Another fabrication route is to impregnate paper sheets with a water solution of the resin, dry the impregnated paper, and then laminate several layers for curing under pressure. Such laminates are used in electronic circuit boards but also in flooring and kitchen surfaces. The main problem with these wood fiber composites is that the use of formaldehyde requires careful safety precautions. Formaldehyde-free wood fiber composites are therefore an important vision for future wood thermoset composites.

When cured, the PF and MF resins form a cross-linked structure that is stiff and brittle, resulting in stiff and brittle wood composites. Phenol, melamine, and formaldehyde are soluble in water and in the wood fiber cell wall and swell the cell wall more than water does. Therefore, the volume fraction of fibers in formaldehyde matrices can be quite high, which serves to improve the elastic modulus and strength of the composite. The fiber–matrix interface is also very strong because the resin polymer enters the fiber wall. When the composite is loaded, failure occurs in the brittle matrix.

Figure 12.5 shows a micrograph of a composite structure where paper has been impregnated with a unsaturated polyester (UP) and cured. The polymer has entered in the lumen of fibers. The volume fraction of fibers is relatively low because UP cannot swell the wood cell wall. One can see that the microstructure of a real composite is quite different from what was assumed previously in the elastic modulus model calculations.

The elastic moduli of various wood fiber composites are compared in Figure 12.6. One can make several observations. First, paper impregnated with thermoset resins shows very high elastic modulus, 16–25 GPa, when the fiber content is high. These high-modulus materials have strongly oriented paper and phenol-formaldehyde matrix and are studied for potential use in aircraft structures. The unidirectional UF/wood fiber composite in Figure 12.6 is the same as in Figure 12.5. The longitudinal elastic modulus is as high as 8 GPa although the weight fraction of fibers is only 25%. The modulus values for wood flour are generally low due to the low aspect ratio. In summary, Figure 12.6 illustrates that low aspect ratio fibers make inefficient reinforcement and that paper with preferred orientation in the testing direction gives highest modulus. For a more detailed analysis one would need quantitative information on fiber length, orientation, and volume fraction (rather than weight fraction), as well as the elastic modulus of the matrix (see Section 12.3).

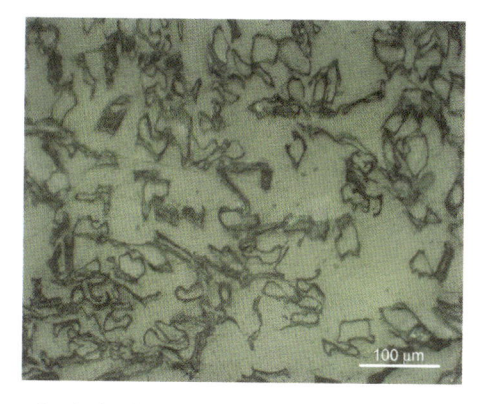

Fig. 12.5 Optical micrograph of a cross-section of a composite of wood fibers embedded in an unsaturated polyester matrix. Courtesy of LO Nordin, Lulea Technical University.

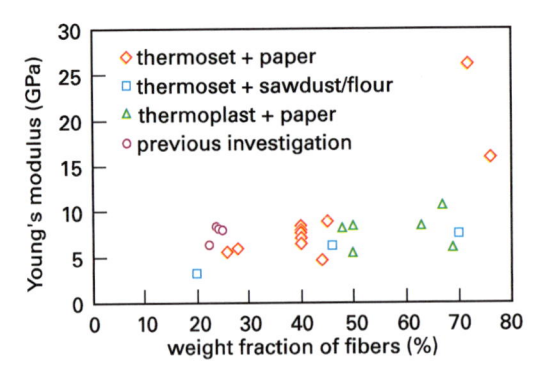

Fig. 12.6 Elastic modulus against fiber weight fraction in a variety of wood composites.

12.6 Nano-fibrillated cellulose materials

Plant fibers, including wood fibers, are widely used in man-made materials. The mechanical properties of paper and board materials, as discussed in the present text, are quite good. Comparisons with other polymer-based materials at similar porosities are usually favorable for plant fiber materials. This is due to the geometric shape of the fibers and the stiff and strong cellulose nano-fibers in the cell wall. In fact, the good strength and elastic modulus of wood fibers comes primarily from the cellulosic reinforcement of the cell wall. For this reason, the disintegration of wood into cellulosic nano-fibers provides us with a stronger and stiffer fibrous component and increases freedom in materials design because we are no longer limited by the plant fiber geometry.

New methods have been developed for disintegration of nano-fibrillated cellulose (NFC) from wood. The energy requirement is lower than earlier thanks to a chemical or enzymatic pretreatment prior to mechanical homogenization (Henriksson et al., 2007, 2008). The homogenization itself can be carried out in homogenizers commonly used for fruit juices and vegetable soups in food industry. Wood NFC is particularly interesting because of its commercial potential. The elastic modulus of the cellulose crystal is 130 GPa or higher. The elastic modulus of NFC is lower because the molecular disorder increases during mechanical disintegration. The strength of individual nano-fibers is difficult to estimate, but the fine dimensions suggest that it exceeds the highest strength known for plant fibers, which is around 1 GPa. Reductions in the cellulose molar mass may reduce this value. The breaking strain of individual nano-fibers in pure tension is low, probably below 1%, because of the extended molecular chain conformation that gives little opportunity for plastic deformation. The dimensions of NFC nano-fibers are typically 5–20 nm in diameter and over 2 μm in length.

12.6.1 Cellulosic "nanopaper"

Nano-fibrillated cellulose (NFC) can be used in a filtration process similar to ordinary papermaking. Although significant practical problems (difficult water removal, slow filtration) need to be solved before such a process becomes industrially feasible, the interesting properties of nanopaper structures push such a development. Figure 12.7 shows the surface of nanopaper. The extent of the pore volume fraction depends on the preparation conditions. The small size of nano-fibers is apparent; they have a diameter that is more than 3 orders of magnitude smaller than ordinary wood fibers (5–20 nm instead of 20–30 µm). The tensile properties are reported in Table 12.9. One may note the high tensile strength (≈220 MPa) and elastic modulus.

Comparison of the breaking strain (≈7%; even much higher values have been reached recently) with conventional paper is interesting. The most likely deformation mechanism that enables large strain is that individual nano-fibers can slip and rotate with respect to each other without damage creation. This is possible because of porosity in the material combined with the favorable interaction of nano-fibers at 50% relative humidity, giving rise to friction that is suitable for strain hardening. The small size of nano-fibers is helpful because individual nano-fibers can fail without creating large micro-cracks. Besides, the inherent strength of nano-fibers is much higher than that of the wood fibers used in conventional paper and board products.

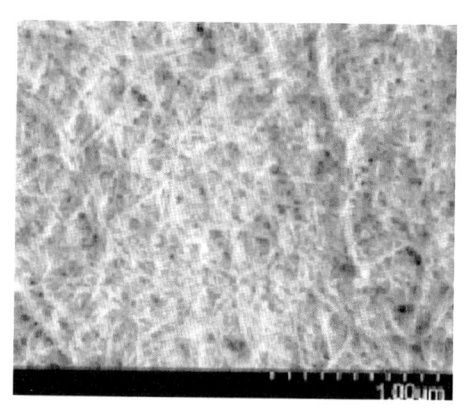

Fig. 12.7: Field-emission scanning electron micrograph of a nanopaper made of nano-fibrillated cellulose (NFC). Courtesy of Dr. H. Sehaqui, Wallenberg Wood Science Center, KTH Royal Institute of Technology, Stockholm.

Tab. 12.9 Mechanical properties of cellulose nanopapers dried from different liquids to reach different porosities. The nano-fibrillar cellulose is made from pretreated wood fibers of high purity (Henriksson et al., 2008).

Liquid	Porosity (%)	Elastic modulus (GPa)	Strain hardening coeff. n	Yield stress, (MPa)	Tensile strength (MPa)	Breaking strain (%)	Work to fracture (MJ/m^3)
Water	19	15	1.9	91	205	6.9	9.8
Methanol	28	11	1.4	76	114	5.4	5.3
Ethanol	38	9.8	1.1	58	106	4.7	3.6
Acetone	40	7.4	0.83	48	95	6.2	4.2

The strain-hardening coefficient n in Table 12.9 is the ratio of the secondary slope of the stress–strain curve (in the plastic region) to the elastic modulus (initial slope of the curve). The tensile energy absorption is the area under the stress–strain curve, measuring the total deformation energy per unit volume that is consumed when straining the material to failure. This property is remarkably high.

12.6.2 Nano-composites

The group of Professor Hiroyuki Yano at Kyoto University in Kyoto, Japan, pioneered the use of nanopaper structures in combination with the water-soluble polyphenol-formaldehyde (PF) resin. The elastic modulus of the nano-composite is linearly proportional to the volume fraction of NFC. The same has been observed for other polymer matrices. Linear extrapolation then yields an estimate for the longitudinal elastic modulus of NFC in the limit of 100% NFC content. A typical result is in the range of 30–60 GPa for the longitudinal elastic modulus of NFC. This is a significantly lower value than the 130 GPa in the axial direction of the cellulose crystal, and the observation needs further study. The elastic modulus of wood fiber/PF composites is similar to NFC/PF composites. This is expected because elastic theories are independent of the size of the reinforcement as long as the fiber volume fraction, matrix porosity, and fiber orientation distribution remain unchanged.

The tensile strength of biocomposites with isotropic in-plane fiber orientation distribution cannot be predicted from simple micromechanical considerations. Empirical phenomenological models are available, but they do not help to understand the failure processes. Nevertheless, the experimental observations in Table 12.10 are interesting. If one compares the NFC/PF composite with a wood fiber/PF composite, the NFC composites have significantly higher strength and breaking strain. The probable reason is that defects are much smaller in the NFC composite.

The high breaking strain of the nanopaper made of NFC can be maintained in the composite if a ductile polymer is used; see the data for glycerol plasticized starch and HEC as matrices in Table 12.10. In contrast, brittle matrices (PF and MF) lead to a low breaking strain and lower tensile strength because the strain-hardening region after the yielding point is very limited. Note that all the NFC composites in Table 12.10 are based on a nanopaper that has been impregnated with the polymer matrix. In the case of NFC/HEC, the material is also porous, which seems to facilitate ductility.

NFC has also been used in novel materials, such as NFC-reinforced polymer foams, NFC foams, and aerogels, and inorganic hybrids, such as clay nanopaper, and magnetic nanopaper (Berglund and Peijs, 2010). They represent interesting new material opportunities based on forest products, with intended uses outside the market for existing forest products markets.

12.7 Conclusions

Wood composites traditionally include established forest products such as particleboard, fiberboard, and plywood. The interest for forest products has been limited in

Tab. 12.10 Mechanical properties of cellulosic nanopaper and nano-composites based on nanopaper.

Matrix	NFC content (% by volume)	Elastic modulus (GPa)	Tensile strength (MPa)	Breaking strain (%)	Reference
–	80 (nanopaper)	13 GPa	220	10	Henriksson et al., 2008
PF (phenol-formaldehyde)	80 (NFC)	15 GPa	200	5	Nakagaito and Yano, 2008
PF (PF with pulp fiber)	80 (pulp fiber)	≈15 GPa	130	2.5	Nakagaito and Yano, 2008
MF (melamine-formaldehyde)	80	15	140	3	Henriksson and Berglund, 2007
Starch/glycerol 50/50	60	6	80	8	Svagan, Samir, and Berglund, 2007
HEC (hydroxyethylcellulose)	60	10	180	20	Sehaqui et al., 2011

the community of materials science and engineering. One reason may have been that the creative possibilities with fibers, particles, and veneer are somewhat limited. This is even more correct if the current cost-levels and production technologies are set to constrain creativity. The use of the term *material* instead of forest products is a good start for change because it implies that the products have a microstructure that can be altered to extend the property range.

The full potential of wood fibers is not utilized in many of the present forest products. The combination of separate long wood fibers with polymers is of interest in this context. The example of PA6 combined with high-purity wood fibers for melt-processed biocomposites is one such example. Papermaking processes have the advantage that they preserve the fiber length and make it possible to combine high wood fiber content with a polymer matrix. The existing materials include wood fiber/thermoset laminates, but in the future the flora of materials may increase to include thermoplastic matrices.

The micromechanical mechanisms of biocomposites are the same as for synthetic fiber composites. The mechanical properties of short fiber composites depend on the fiber volume fraction, matrix porosity, interfacial shear strength, fiber length distribution, and fiber orientation distribution. The difference is in the application potential because wood fibers have highly nonideal shape and variability of properties. For this reason, the application of theoretical predictions requires some caution. If the precision in the measurement of constituent properties is poor, improved modeling accuracy does not help.

The development of nano-fibrillated cellulose (NFC) can potentially change the existing paradigm. It is a new material component of a very small size. It has appealing properties such as high elastic modulus, high tensile strength, low thermal expansion, and potential for optical transparency, at the same time, it preserves or even improves the network bonding characteristics of wood pulp fibers. It can be used in nanopaper films, coatings, fibers, foams, aerogels, and organic/inorganic hybrids, as well as in tough moldable biocomposites (Berglund and Peijs, 2010). The improvements obtained in the strength and tensile energy absorption compared with conventional wood composites, paper, and board materials are very impressive. The remaining challenge is to develop processing technologies.

References

Agarwal, B. D., and Broutman, L. J. (1990). Analysis and Performance of Fiber Composites (New York, NY, USA: Wiley-Interscience).

Ashby, M. F. (1999). Materials Selection in Mechanical Design (Oxford, UK: Butterworth-Heinemann).

Berglund, L. A., and Peijs, T. (2010). Cellulose biocomposites – from bulk moldings to nanostructured systems. MRS Bulletin 35, 201–207.

Henriksson, M., and Berglund, L. A. (2007). Structure and properties of cellulose nanocomposite films containing melamine formaldehyde. J. Appl. Pol. Sci. 106, 2817.

Henriksson, M., Berglund, L. A., Isaksson, P., Lindstrom, T., and Nishino, T. (2008). Cellulose nano-paper structures of high toughness. Biomacromolecules 9, 1579.

Henriksson, M., Henriksson, G., Berglund, L. A., and Lindstrom, T. (2007). An environmentally friendly method for enzyme-assisted preparation of microfibrillated cellulose (MFC) nanofibers. Euro. Pol. J. 43, 3434.

Hull, D. (1981). An Introduction to Composite Materials (Cambridge, UK: Cambridge University Press).

Jacobson, R., Caulfield, D., Sears, K., and Underwood, J. (2002). Low temperature processing of ultra-pure cellulose fibers into nylon 6 and other thermoplastics. In Proceedings of the 6th International Conference on Woodfiber-Plastic Composites, Madison, WI (Forest Prod Society), pp. 127–133.

Nakagaito, A. N., and Yano, H. (2008). Toughness enhancement of cellulose nanocomposites by alkali treatment of the reinforcing cellulose nanofibers. Cellulose 15, 323–331.

Sehaqui, H., Zhou, Q., and Berglund, L. A. (2011). Soft Matter, 7, 7342–7350.

Stark, N. M., and Rowlands, R. E. (2003). Effects of wood fiber characteristics on mechanical properties of wood/polypropylene composites. Wood & Fib. Sci. 35, 167–174.

Svagan, A. J., Samir, M.A.S.A., and Berglund, L. A. (2007). Biomimetic polysaccharide nanocomposites of high cellulose content and high toughness. Biomacromolecules 8, 2556.

Wainwright, S. A., Biggs, W. D., Currey, J. D., and Gosline, J. M. (1982). Mechanical Design in Organisms (Princeton, NJ, USA: Princeton University Press).

Wegst, U.G.K., and Ashby, M. F. (2004). The mechanical efficiency of natural materials. Phil. Mag. 84, 2167–2181.

Index

activation parameter α_{act} 211, 219, 224, 226
adhesion force 95, 101
air drag 95, 106
anisotropy 7, 23, 62, 81, 165, 166, 174, 84, 169, 213, 215

basis weight 23, 105, 152
— of fiber 198
— values of 5, 198, 209
beating *see* refining of pulp
bending stiffness *see* stiffness
board
— corrugated board 29, 30, 33, 38, 124, 43, 129, 149, 186
— fiber board 231, 232
— paperboard 5, 198
— particle board 231
board grades
— carton board 34, 53, 55, 60, 73, 84
— container board 30
— fluting 31, 43, 87
— liner board 30, 18, 43, 87
Box Compression Test (BCT) 42
breaking strain ε^{max} 15, 13, 15, 62, 125, 184, 157, 219
— shear 17, 64
— tensile 13, 15, 62, 125, 234, 247, 251
buckling
— of board panel in compression 39, 124, 130
— of fiber segment in compression 220
— of paperboard in compression 39, 45, 85
— of paperboard in folding 53, 60
— of paper in tension 70, 87, 172, 173

cell wall *see* fiber, wall structure
cohesive stress 13, 18, 73, 79, 81
cohesive zone model 13, 80, 84, 222

compliance
— creep J 113, 115, 117, 118, 120, 124, 127, 132, 227
— curve J 122
— elastic, tensor S 7
composite material 231
coverage 202
crack, cracking 53, 59, 63, 73, 75, 84, 87, 140, 145, 220, 224
creasing
— of carton board 59, 87
— of corrugated board 33
— of paper web *see* wrinkling
creep 112
— accelerated 48, 127
— compliance *see* compliance, creep J
— compressive 124
— curve 113, 217
— failure 48, 143, 147, 148
— log-linear 122
— master curve 116, 121
— mechanosorptive 127
— primary 118
— rate 128, 217
— secondary 118, 124, 131, 142
— tensile 120
— tertiary 118
— time shift 121
critical length *see* fiber, critical length
cross-machine direction CD 24, 106, 172, 183
curl
— of fiber 157, 216
— irreversible 167
— of paper 24, 163

damage
— in delamination 84
— of fiber 201, 206
— of fiber network 57, 220
— parameter 60, 81, 148

Index

- width w_{dama} 221, 222
- zone   79, 80

defect   *see* crack, cracking
defect-sensitivity   80
deformation theory of plasticity   78
degree of polymerization DP of cellulose   237
delamination   53, 60, 63, 72, 86
density (mass density) ρ 23, 24, 35, 86, 124, 200, 204, 206, 207
- apparent density   205
- effective density   205
- of fiber   204
- of fiber wall   205
- values of   9, 10, 205, 212, 213, 215

die-cutting of corrugated board   33
draw   *see* speed v, difference, draw

Edge Crush Test (ECT) of corrugated board   38, 43
elastic modulus E 15, 25, 102, 124, 167, 207, 214, 222, 224
- against moisture content   11
- of cellulose crystal   237
- of composite   247, 249
- of fiber E_{fiber} 208, 213
- of nano composite   252
- of nano paper   251
- of printing plate   186
- values   9, 55
- of wood   238
- in z-direction (ZD)   10

fiber
- basis weight   198
- bonding   *see* inter-fiber bonding
- conformability, flexibility   20
- critical length (l_{crit})   208
- length (l_{fiber})   19, 221
- orientation   *see* orientation distribution of fibers
- segment length, free (l_{free})   201, 209
- segment length, total (l_{segm})   205
- shrinkage   210, 225
- strength   240, 246. See also strength, of fiber
- thickness (d_{fiber})   202
- wall structure   21, 200
- wall thickness   236
- width w_{fiber} 19

fibril of fiber wall   21, 198
fillers   21
fines   21, 198, 200
finite element model   44, 59, 78, 81, 107, 183, 186
fluting
- in printing paper   171
- waves of corrugated core   31, 45, 73

flutter of web   105, 107
folding
- of carton board   53, 55, 59, 60, 87
- of corrugated board   33
- permanent   57

formation, uniformity of mass distribution   23, 152
fracture
- energy   80, 222
- linear elastic fracture mechanics (LEFM)   75
- mechanics   67
- mode   67
- process zone (FPZ)   74, 79, 220, 222
- toughness   67
- toughness, J-integral   76

furnish of fibers   21, 124

global load sharing (GLS)   145

Halpin-Tsai model   244, 245
handsheet   *see* laboratory sheet
humidity   *see* relative humidity (RH)
hydro expansion   171, 225
hydrophilic   163
hygro expansion   23, 25, 93, 165, 127, 173, 185, 225
- coefficient β 8

hygroscopic strain   *see* hygro expansion

inter-fiber bonding   13, 57, 74, 200
- bond area   219
- failure   217, 220
- relative bonded area (RBA)   208, 219, 225
- shear modulus G bond   208

internal stress   168, 219
isochronous curve   115, 124, 127

Index

J-integral   *see* fracture, toughness

kraftpaper   *see* paper grades

laboratory sheet   5, 215
lifetime of box (t_{BL})   48, 128, 129, 131, 133, 142, 147
local load sharing (LLS)   145
lumen   200

machine direction (MD)   5, 72, 124, 183
master curve   *see* creep
MD/CD ratio   *see* anisotropy
mechanosorption   127
microfibril
— angle (MFA)   213
— dimensions   236
moisture
— application   166
— change   174, 185
— content (MC)   10, 15, 105, 119, 126, 127, 128, 133, 164, 168
— content hystresis   167
— difference   175
— diffusion   168, 170
— expansion   *see* hygro expansion
— history   167
— loss   172
— resistance   47
— sensitivity   131, 226
— variation   178

nano-fibrillated cellulose (NFC)   240, 250

orientation distribution of fibers   7, 23, 105, 141, 215
orthotropic   7

packaging
— corrugated container   33
— primary   29
— Regular Slotted Container (RSC)   32
— secondary   29
paper grades
— baking paper   12
— glassine   12
— kraftpaper   13, 14
— newsprint   5, 13, 14, 23, 87, 150, 153

— office paper   5, 23
— sack paper   12, 87
paper machine
— dry end   22, 215
— dryer section   22
— drying shrinkage   24, 210, 211, 218, 225
— drying tension   24, 105, 127, 168, 210, 211
— forming section   22, 54
— wet end   22, 93, 101, 105
— wet pressing   22
Poisson ratio v 9
pore volume fraction Φ 16, 189, 191, 202, 204
porosity   *see* pore volume fraction Φ
printing   33, 72, 93, 96, 139, 154, 188
— flexo   186
— offset   171, 175, 182
— web tension   70, 93, 141, 151, 154, 172
pulp   19
— chemical   13, 35, 74, 164, 198, 203, 210, 212
— chemical, kraft   20, 21
— chemical, sulfite   20, 240
— chemi-thermomechanical (CTMP)   20
— mechanical   13, 35, 151, 184, 198, 203
— mechanical groundwood (GW)   20
— mechanical pressure groundwood (PGW)   20
— recycled   35
— thermomechanical (TMP)   20, 21

refining of pulp   20, 206, 210, 221
relative humidity (RH)   9, 47, 127, 142, 164, 165
relaxation
— curve   113
— modulus G 113, 115
— of stress   97, 112, 118, 122, 129, 171, 219
— time shape   121, 129

segment   *see* fiber
shear modulus G
— of inter-fiber bond   208
— values   55

shives 151
Short Span Compression Test (SCT) of carton board 38, 43
speed v
— difference, draw 93, 96
— difference, jet-to-wire 23
— of paper machine 22, 101
— of printing press 96, 173
— of sound waves c 96, 99
— of web 107
stiffness
— bending S_b 24, 45, 57, 60, 187
— of corrugated core 45
— of fiber 208, 211
— of ink layer 176
— of interface 85
— normal 83
— shear see shear modulus G
— specific 238
— tensile see elastic modulus E
— tensor C 7, 81
strain
— components 11, 77, 117
strain rate 15, 96, 123, 147
strain-to-failure see breaking strain ε^{max}
strength
— of box 43, 46, 131, 133. See also Box Compression Text (BCT)
— of cellulose 237
— of composite 232, 247
— compressive 16, 43, 86. See also Short Span Compression Test (SCT) of carton board
— of fiber 221, 238, 242
— of interface 85
— multi-axial see Tsai-Wu failure criterion
— of paper web 70
— safety factor 131, 149
— stacking 29, 36, 73, 133
— of structure 67, 87
— tensile see breaking strain ε^{max}, tensile
— testing 108

— variability 133, 141, 143, 150, 157
— in z-direction 17, 83, 219
stress intensity factor K 75

temperature Θ 38, 119, 126, 142, 168, 176, 177
thermal
— degradation 239
— expansion 165
— expansion coefficient α 8
— insulation 232
— stability 240
thickness d 5, 23, 24, 31, 35, 54, 55, 202, 204
— apparent thickness 205
— direction See z-direction (ZD) 5
— effective thickness 205
— of fiber wall 200
— piling thickness 205
time-dependency 11, 95, 101, 111, 114, 124, 126, 147, 225
Tsai-Wu failure criterion 18, 45, 85

velocity see speed v

washboarding 187
weakest-link scaling (WLS) 144
web break 69, 93, 98, 102, 105, 139, 150, 154, 184
Weibull
— distribution 144
— modulus k 144
— Weibull plot 144
wrinkling 93, 141
wrinkling of paper 107, 172

yield
— criterion 46
— of pulping 20
— small-scale yielding 75
— strain 183
— stress 187

z-direction (ZD) 5, 16, 24, 72, 73, 83, 182